Springer Geology

The book series Springer Geology comprises a broad portfolio of scientific books, aiming at researchers, students, and everyone interested in geology. The series includes peer-reviewed monographs, edited volumes, textbooks, and conference proceedings. It covers the entire research area of geology including, but not limited to, economic geology, mineral resources, historical geology, quantitative geology, structural geology, geomorphology, paleontology, and sedimentology.

Tatiana Chaplina
Editor

Processes in GeoMedia—Volume VIII

Springer

Editor
Tatiana Chaplina
Alicante, Spain

ISSN 2197-9545 ISSN 2197-9553 (electronic)
Springer Geology
ISBN 978-981-97-6626-0 ISBN 978-981-97-6627-7 (eBook)
https://doi.org/10.1007/978-981-97-6627-7

This Springer imprint is published by the registered company Springer Nature Singapore Pte Ltd.
The registered company address is: 152 Beach Road, #21-01/04 Gateway East, Singapore 189721, Singapore

Contents

Physical Modeling of the Formation and Chemical Composition of Cryogenic Structures in Marine Sediments Within the Shelf Zone

V. E. Gagarin, A. V. Koshurnikov, and D. M. Frolov

Abstract The paper presents the findings of a physical model of the process of ground freezing under various boundary conditions, specifically in a subaquatic environment in a shallow zone. The study analyzed the characteristics of ice formation within the ground, the redistribution of salt, as well as the temperature, moisture content, and electrical resistivity distribution at different depths during the experiment. This allowed for the determination of a certain correlation between these parameters when the ground freezes in a subaquatic setting. To address this problem, a laboratory setup was developed and constructed, enabling experimental investigations to be conducted under well-controlled boundary conditions. Measuring the electrical resistivity of the ground during the experiments yielded values that, through modeling, can implicitly be linked to the physical and chemical characteristics of the ground on a given territory. This, in turn, could be confirmed through full-scale geophysical surveying of the area.

Keywords Physical modeling · Soil freezing · Shelf zone · Soil physical and chemical properties · Electrical resistivity · Laboratory setup

1 Introduction

Marine sediments are one of the most widespread genetic types of Quaternary sediments. Formed in subaqual conditions, they can be subaerial and freeze as a result of sea level fluctuations. Marine sediments form extensive plains adjacent to the Arctic Ocean. Despite the diversity of occurrence conditions, structure and material composition, marine sediments have a number of common properties, first of all, high salinity.

V. E. Gagarin · A. V. Koshurnikov · D. M. Frolov (✉)
Faculty of Geography, Lomonosov Moscow State University, st. Leninskiye Gory, 1, Russia 119331 Moscow
e-mail: frolov@geogr.msu.ru

T. Chaplina (ed.), *Processes in GeoMedia—Volume VIII*, Springer Geology,
https://doi.org/10.1007/978-981-97-6627-7_1

The cryogenic strata of marine sediments located in different zones are very different from each other in terms of their structure, mineral composition and thermal state.

The main difference of frozen ground as a new natural formation is not just the presence of ice coexisting with unfrozen liquid phase, but the formation of a specific cryogenic spatial structure that characterizes new associations and interrelationships of ground components, new porosity, new state of liquid phase, etc., which determines the integrity of this medium as a complex macrosystem. That is why the physical and other properties of frozen ground are not equal to the additive sum of the corresponding properties of its components, taking into account their volume content.

The complex, factor-dependent conditions of cryogenic strata formation and the difficult accessibility for their study explain the fact that a holistic concept of oceanic cryolithogenesis and the importance of physical modeling under laboratory conditions have not yet been established.

The present work is devoted to the experimental study of the epigenetic freezing of marine sediments located in the coastal part of the sea area.

In the framework of research according to (Baulin 1981; Baulin 1997; Danilov 1983; Dubikov 2002; Frolov 2005; Grigoriev 1981; Grigoriev 1982; Khimenkov 2003; Melnikov 1995; Markov 1961; Neshiba 1991; Romanovsky 1993; Saks 1952; Shpolyanskaya 1991; Shpolyanskaya 1980; Vasilchuk 2006; Vasilchuk 1985; Zhigarev 1997) we had the following goals and objectives in front of US:

Purpose: Studying the features of formation of cryogenic textures of marine sediments freezing in subaquatic conditions in the shelf zone of the Kara Sea.

Objectives:

1. Establishment of a physical modeling facility to record the progress of schliering, temperature regime and salt transfer.
2. Based on literature data, justify the methodology for conducting physical modeling.
3. Conducting ground freezing experiments under different boundary conditions.
4. From the data obtained, identify the relationship between freezing process, salt transport, moisture changes and electrical conductive properties.
5. Provide recommendations on the use of physical modeling data in field studies on the Arctic shelf.

2 Experimental Study of Features of Formation of Cryogenic Textures of Marine Sediments

2.1 Description of the Experimental Setup

To model the process of soil freezing and to observe the formation of cryogenic textures in soils, an installation was assembled, including: transparent prefabricated shells made of Plexiglas with a cross section of 7 × 7 cm, height of 20 cm, with electrodes in the side wall (Fig. 1); cryostat with an external cooling circuit (Fig. 2), cooling chamber; thermal braid (Fig. 3); geophysical equipment for measuring soil resistivity (Fig. 4), photo camera for recording the freezing process (Fig. 5).

An experimental methodology was developed, including preparation of soil samples and observation of their freezing under subaquatic marine conditions.

Freezing was carried out from above under boundary conditions of the 1st kind. For this purpose, the external circuit of the cryostat was connected to the freezing plate, on which the negative temperature was maintained during the experiment. During the experiment, the soil was in a transparent prefabricated cage, which, in order to exclude freezing from the sides, was placed in a refrigerating chamber with a constant temperature of 0 °C.

The thermal braid consisted of 10 sensors. The 3rd-10th sensors of the thermobraid were located at equal distance along the whole height of the cage, the 1st sensor was

Fig. 1 Electrodes in the side wall of the experimental cage

Fig. 2 Cryostat with external cooling circuit

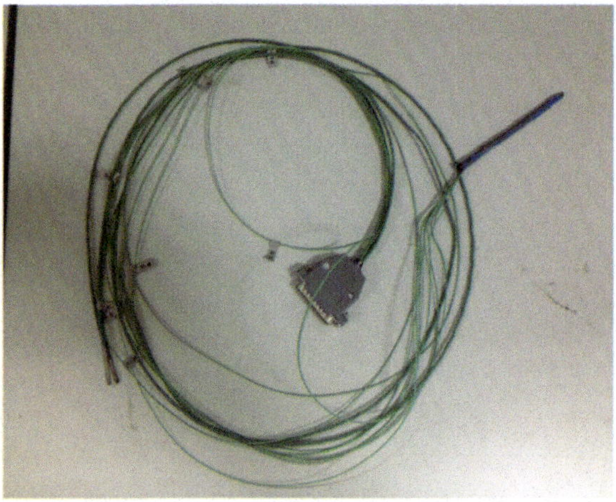

Fig. 3 Thermal braid of 10 sensors

located directly in the frost plate, sensor No. 2 was located at the bottom of the sample in the cold chamber.

To measure the specific electrical resistance of the soil during the experiment, electrodes were used, which were located in the side wall of the cage with the soil. Electrodes A (№. 6) and B (№. 17) were used as feeding electrodes. The receiving electrodes M and N were alternately changed from top to bottom: M1(№. 7) N1(№. 8); M2(№. 8) N2(№. 9); … M9(№. 15) N9(№. 16). The geometric coefficient K was

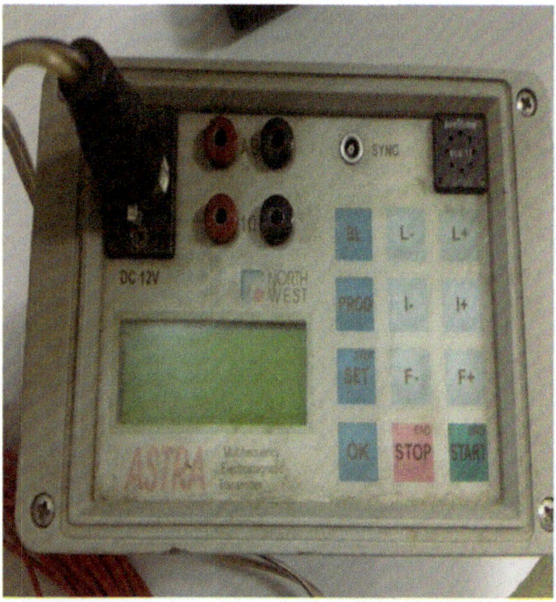

Fig. 4 Electrical exploration geophysical complex consisting of ASTRA generator and APL-02 receiver

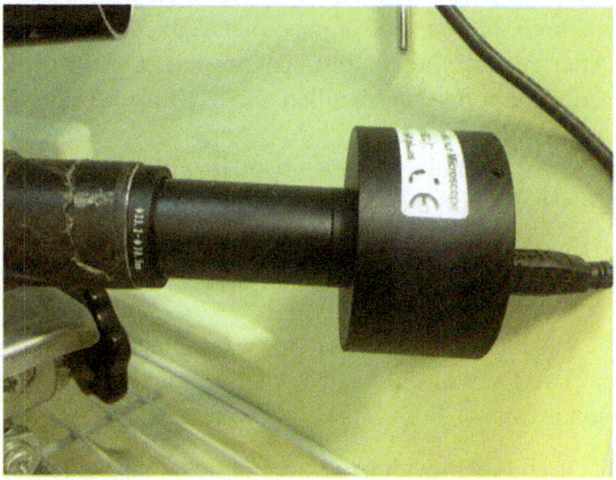

Fig. 5 Ocular camera for registration of freezing process

calculated for this scheme of the ground resistivity measurement setup, using a fluid with known resistivity (Table 1).

A DCM 500 ocular camera with a 300 s image rate was used for photographic recording of the freezing process.

Table 1 Calculation of geometric factor K for circuit AB6_17

AB 6_17				
MN	U, мV	I, мA	R, O_{M*M}	K
7_8	724.6	200	0.1636	0.045156
8_9	154.5	200	0.1636	0.21178
9_10	143.5	200	0.1636	0.228014
10_11	130.4	200	0.1636	0.25092
11_12	140.3	200	0.1636	0.233215
12_13	139.8	200	0.1636	0.234049
13_14	176.6	200	0.1636	0.185277
14_15	259.3	200	0.1636	0.126186
15_16	487.8	200	0.1636	0.067077

2.2 *Methodology of Sample Preparation for Experimentation.*

Before starting the experiment, the tested soils were studied. The granulometric composition, particle density (pycnometric method), initial salinity (water extraction method), presence of organic material (muffle furnace calcination) were analyzed. After determining the physical properties of the soil, the sample was moistened to the yield point with distilled water. Moisture samples were taken after the sample had stood for one hour.

Then the shells were evenly filled with soil and its density values were determined. The shells with the soil were placed in a refrigerated chamber and kept at 0 °C for a day.

Before freezing, the soil sample was covered with a layer of water of salinity 33 ‰, which was pre-cooled to 0 °C.

Distilled water was used for water preparation where sodium chloride was added.

Two types of soils were used to conduct the soil freezing experiment under coastal conditions.

1. Light loam with admixture of peat (experiment 1 and 2).
2. Light loam (experiment 3).

The physical characteristics of the soil used in the experiments are summarized (Tables 2 and 3).

Table 2 Physical characteristics of the soil used in the experiment 3

№.	№. production	Sampling depth, м	Name of soil	Total moisture, (%)	Moisture of min. layers, %	Density of soil particles, g/см3	Salinity by water extract, %	Organic matter content, %	Plasticity limits (%)		Plasticity number	Fluidity rate
									Yield point	Rolling boundary		
13	G.29.6.0	3.5	Light loam	13.2	13.27	2.70	0.06	1.8	30	12	8.4	0.19

Table 3 Physical characteristics of the soil used in experiments 1 and 2

№.	№. production	Sampling depth, м	Name of soil	Total moisture, (%)	Moisture of min. layers, %	Density of soil particles, g/cм3	Salinity by water extract, %	Organic matter content, %	Plasticity limits (%)		Plasticity number	Fluidity rate
									Yield point	Rolling boundary		
22	1313 I 13	4	Light loam	39	22.27	2.57	0.01	5.5	43	31	7.3	1.10

3 Modeling of Soil Freezing in Subaquatic Conditions

3.1 Experimental Studies

After appropriate preparation of the samples, their freezing began. As already mentioned, the laboratory setup prepared by us allowed us to make visual observations of the formation of cryogenic textures directly during the experiment itself. And also, using the method of electrical profiling, to observe the measurement of the specific electrical resistance of soils, which, as it is known (Zykov 2007) depends on both the ice content in the soil and their salinity. We modeled three variants of freezing conditions.

(1) Conditions of complete freezing of water and formation of ice-soil boundary:

  (A) a layer of water with salinity 33‰ – 2 cm. The temperature set on the plate is – 9 °C (Fig. 6 1A).
  (B) power of water with salinity 33‰ – 5 cm. The temperature set on the slab is – 9 °C (Fig. 6 1B).

(2) Conditions of incomplete water freezing and ice-water-soil boundary formation:

  (A) power of water with salinity 33‰ – 5 cm. The temperature set on the stove is – 6 °C. (Fig. 6 2A)

During the experiment, changes in water and soil temperature were recorded along the entire transect. The samples were frozen for 5 days. After that, the shells with soil were removed and moved to the freezing chamber, where the samples were characterized in detail.

Before sampling the soil sample for salinity and moisture, a visual description was made, structural features were determined, photographs were taken, areas of

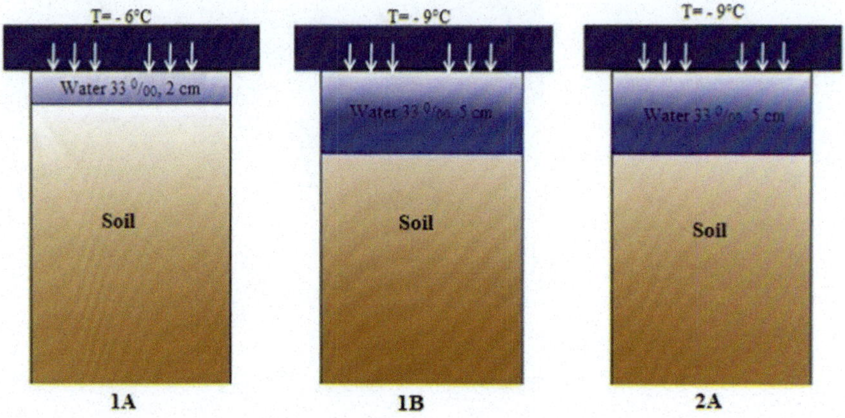

Fig. 6 Three types of experimental conditions

schliering, schlier size and orientation, void formation were identified, and the ice-soil contact zone was described.

After description of the sample in the casing, layer-by-layer extraction of soil was carried out in parallel into two boxes, each centimeter for testing for moisture and salinity. Moisture of the sample was determined by weight method. Soil salinity was determined by means of conductometer "cond 51" (Fig. 7) calibrated for chloride ion determination in the laboratory. Soil salinity was determined in laboratory conditions at room temperature + 23°.

Experiment №. 1

Initial moisture content of the soil sample is 43%

Temperature gradient: at the bottom 0 °C, at the top – 9 °C.

Freezing time 5 days.

Salt content in the upper water layer 33‰. Water thickness 2 cm.

Water was colored with ink for easy registration of migration processes.

The experiment was conducted in such a way that during the experiment the upper water layer froze completely and a contact zone was formed: ice-soil.

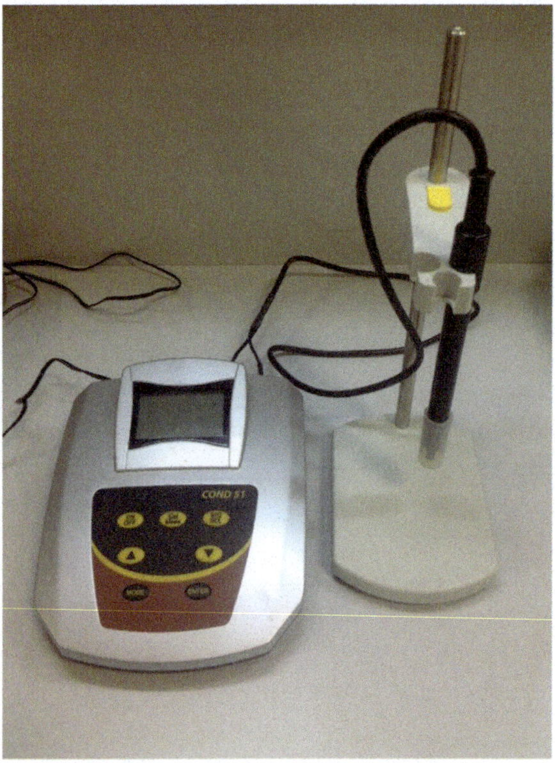

Fig. 7 Cond 51 Conductometer

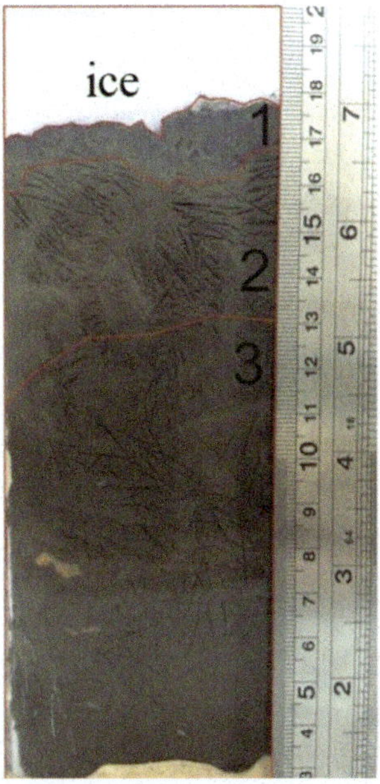

Fig. 8 Sample №. 1 after the experiment

1. The first zone-ice-soil contact (0–2 cm) is observed to be free of ice formation, the presence of a zone of cooled rocks, which indicates a high concentration of salts. The complete temperature distribution by depth during freezing can be observed in Fig. 8.
2. The second zone—intensive ice formation (2–5 cm) we observe reticulated cryogenic texture, mainly dominated by frequent vertically oriented schlirs up to 1 mm in thickness and 2 cm in length.
3. The third zone (5–15 cm) is a zone of reticulated cryogenic texture with sparser schlieres of different orientation (Fig. 9).

The distribution of NaCl in the soil after the experiment had the following character.

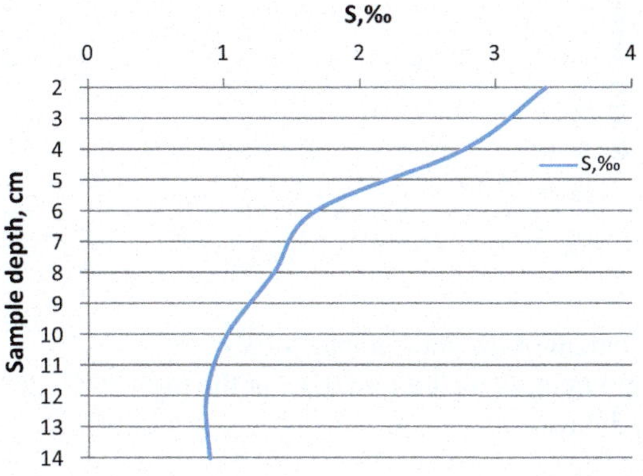

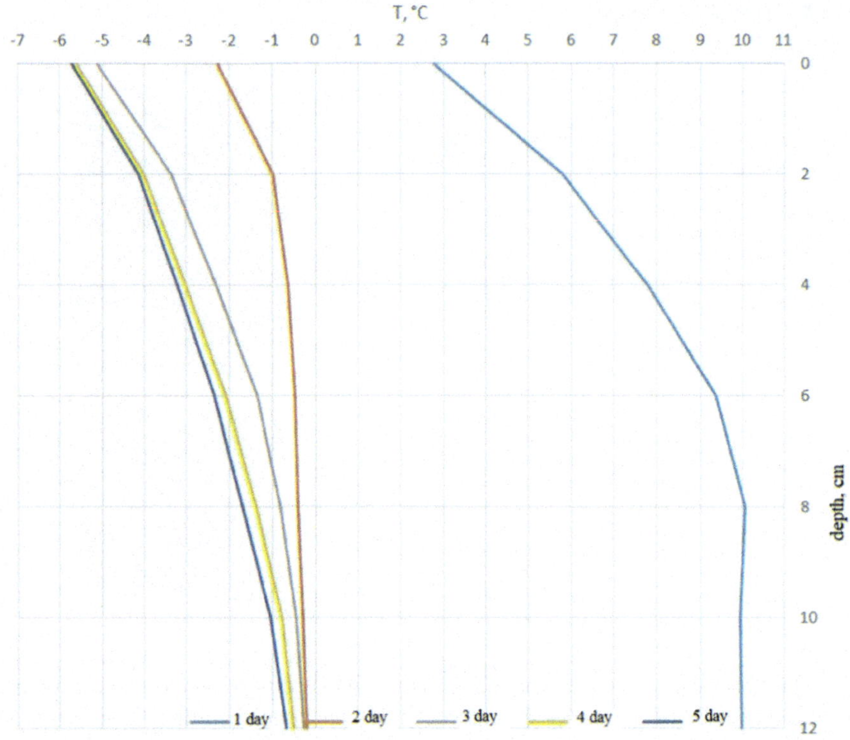

Fig. 9 Graph of temperature distribution by depth with time

The distribution of moisture content by depth after the experiment was as follows.

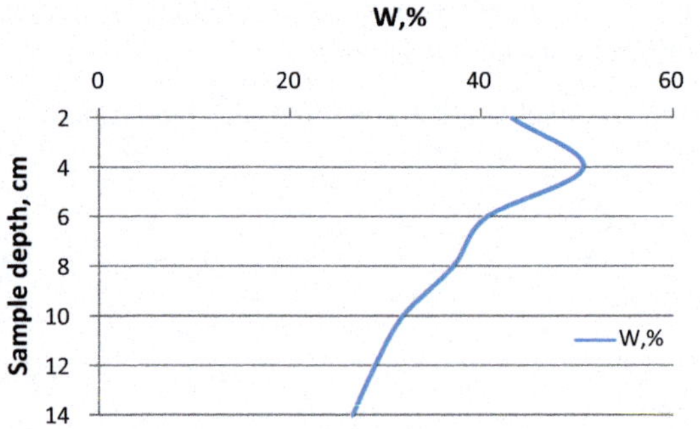

Experiment №. 2

Initial moisture content of the soil sample is 43%

Temperature gradient: at the bottom 0 °C, at the top – 9 °C.

Freezing time 5 days.

Fig. 10 Sample №. 2 after the experiment

Salt content in the upper water layer 33‰. Water thickness 5 cm.

The experiment was conducted in such a way that during the experiment the upper water layer froze completely and a contact zone was formed: ice-soil.

During the whole experiment, measurements of specific electrical resistance were carried out (Fig. 10).

1. The first zone-ice-soil contact (0–3.5 cm) shows absence of ice formation, presence of a zone of cooled rocks, which indicates high salt concentration. The complete temperature distribution along the depth during freezing can be observed in Fig. 11.
2. The second zone of intensive ice separation (3–5 cm) we observe a reticulated cryogenic texture, mainly dominated by frequent vertically oriented slurries of different thicknesses.
3. The third zone (5–15 cm) is a zone of reticulated cryogenic texture with sparser schlieres of different orientation up to 2–3 cm long, 1 mm wide.

The distribution of moisture content by depth after the experiment was as follows.

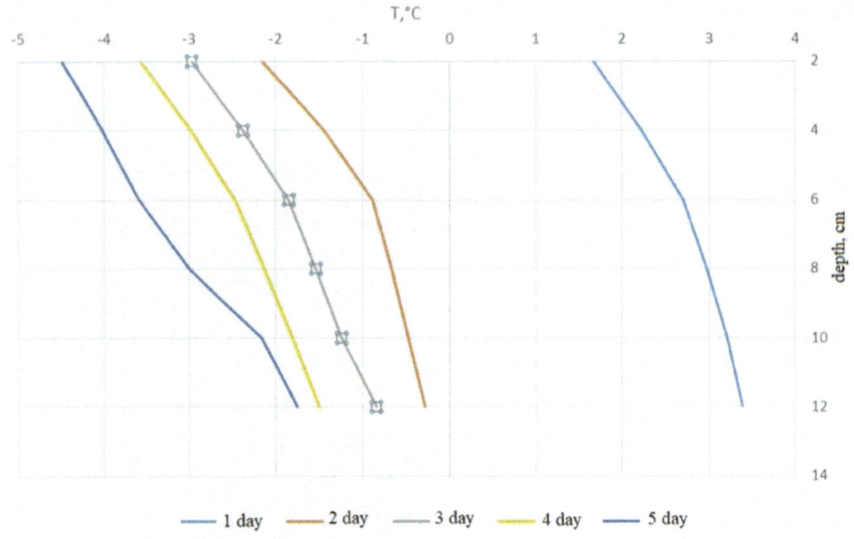

Fig. 11 Graph of temperature distribution by depth with time

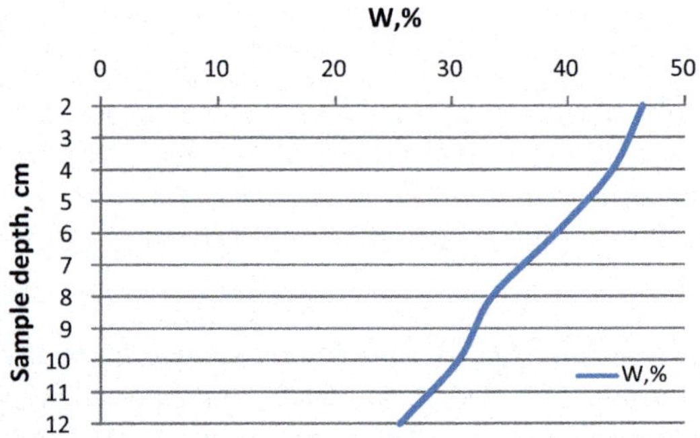

The distribution of NaCl in the sample after the experiment had the following character.

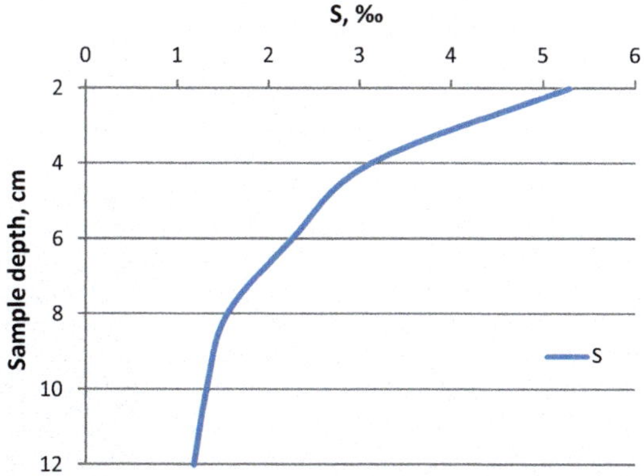

Experiment №. 3

Initial moisture content of the soil sample-30%

Temperature gradient: at the bottom 0 °C, at the top – 6 °C.

Freezing time 5 days.

Salt content in the upper water layer 33‰. Water thickness 5 cm.

Water was colored with ink for easy registration of migration processes.

The experiment was conducted in such a way that the upper water layer was frozen not completely and a contact zone was formed: ice–water–ground.

The salinity of the upper ice layer was 7.2‰

The salinity of the lower ice layer was 2.6‰.

Water salinity amounted to 16.3‰ (Fig. 12).

After the experiment is completed, the sample can be divided into two zones:

1. First zone—(0–0.8 cm) Zone of unfrozen water.
2. Second zone—(0.8–12 cm) Zone of no schliering. Soil in the cooled state (Figs. 13 and 14).

Humidity and salinity after the experiment were distributed as follows (Figs. 15 and 16).

Experiment №. 4

The experiment was conducted according to the scheme of experiment №2, but instead of salt water distilled water was used. In the course of the experiment only measurements of specific electrical resistance were made (Fig. 17.)

The point was to separate the effect of salt migration from the effect of schliering during freezing on resistivity.

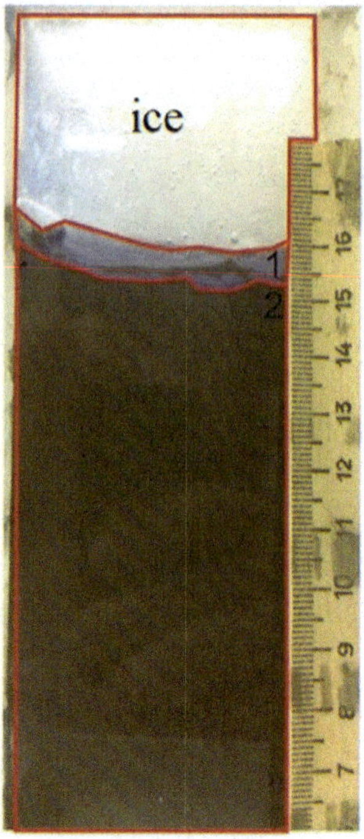

Fig. 12 Sample №. 3 after the experiment

4 Results of Studies of Formation of Cryogenic Structure and Composition of Freezing Sediments

Thus, within the framework of this work, physical modeling of the process of ground freezing under different boundary conditions was carried out for subaquatic soil in the shallow water zone, where the interaction of sea ice, seawater and bottom soil is most noticeable. Such freezing is possible either during transgression of the sea on melted soils with subsequent temperature decrease and their transition to the frozen state, or it can be the result of shoaling of the Arctic water body and the possibility of ground freezing directly through the ice. It should be noted that a closed system was modeled, which is close to the conditions of freezing of coastal basins.

The analysis of the peculiarities of ice formation in the ground, redistribution of salts, distribution of temperature, humidity and specific electrical resistance by depth during the experiment allowed us to determine some relationship of these parameters among themselves in the process of freezing of the subaqueous soil.

A common feature for all samples is the presence of a highly saline negative-temperature zone in the upper near-contact part, which is in the cooled state (zone 1 Figs. 8 and 10). In this zone, ice formation does not occur under these thermal conditions. Also, the indicator of this zone is a gradual decrease in the electrical resistance of the ground in the process of freezing (Fig. 14).

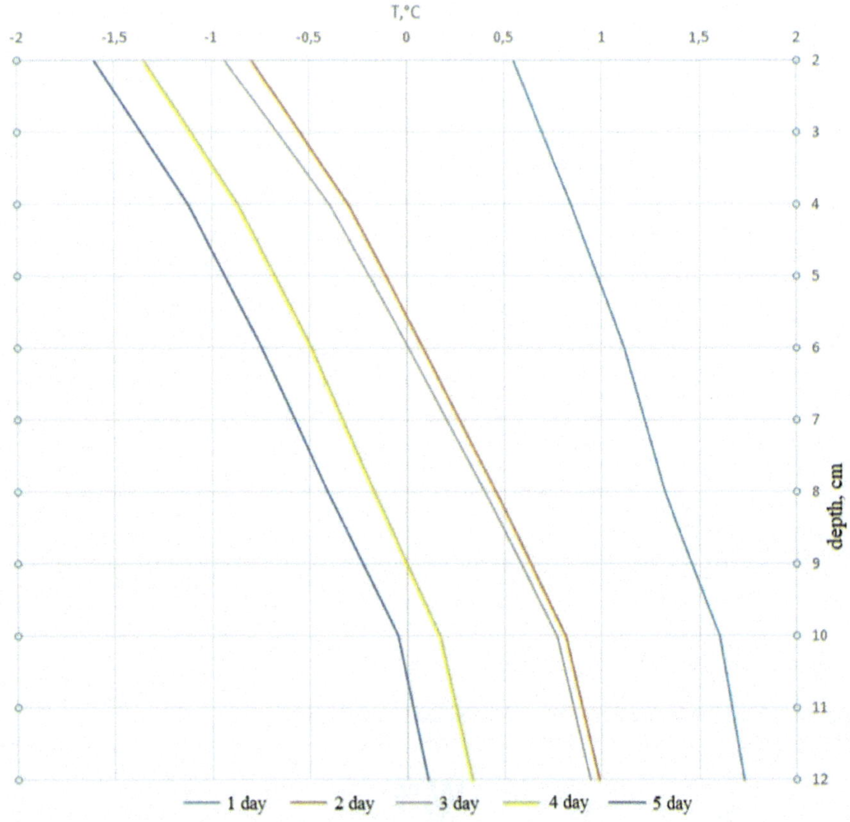

Fig. 13 Graph of temperature distribution by depth with time

Observations showed that in the ground, in the case of direct ice-soil contact (Experiment 2), the diffusion of salts into this region is 20–25% higher than in the case when there is a layer of water between ice and ground (Experiment 3).

In experiments 1 and 2, we can distinguish a zone of intensive ice formation (zone 2 Figs. 8 and 10), which is emphasized by the redistribution of moisture content (Fig. 15). The cryogenic structure of this area indicates that we are dealing with ice growth under conditions of supercooling and increased mineralization. This is evidenced by the abundance of thin deep penetrating subvertical slurs with mineralized soil layers between them.

Rarer ice slits and decreased humidity in the 3rd zone (Fig. 10) indicates that moisture migration to the intensively growing ice took place from this zone.

Experiment 3 showed that in the presence of ice-water-soil contact and certain boundary conditions, we obtain a soil that is completely in a cooled state.

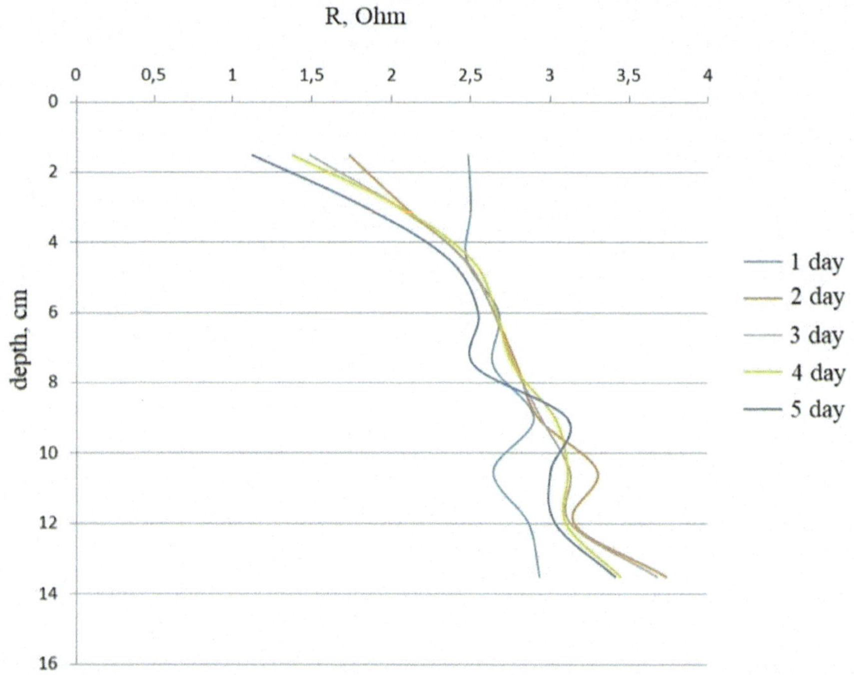

Fig. 14 Graph of electrical resistivity distribution by depth with time

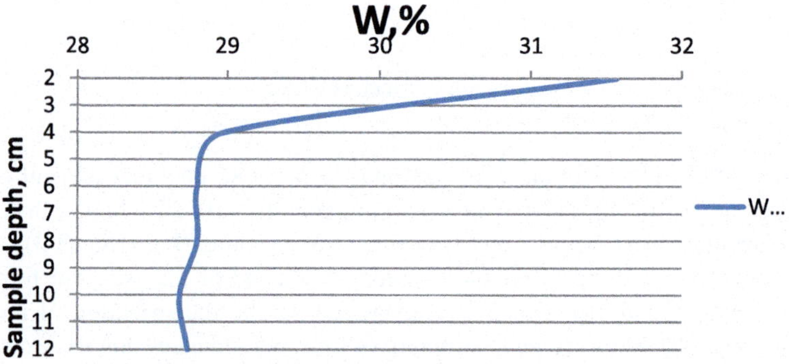

Fig. 15 Moisture distribution by depth after the experiment

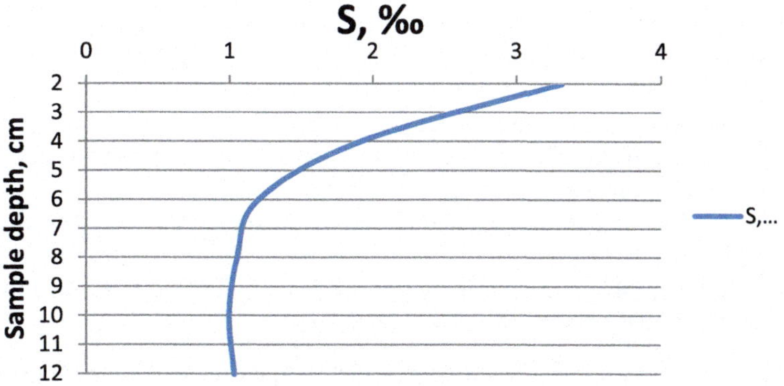

Fig. 16 NaCl distribution in the sample after the experiment

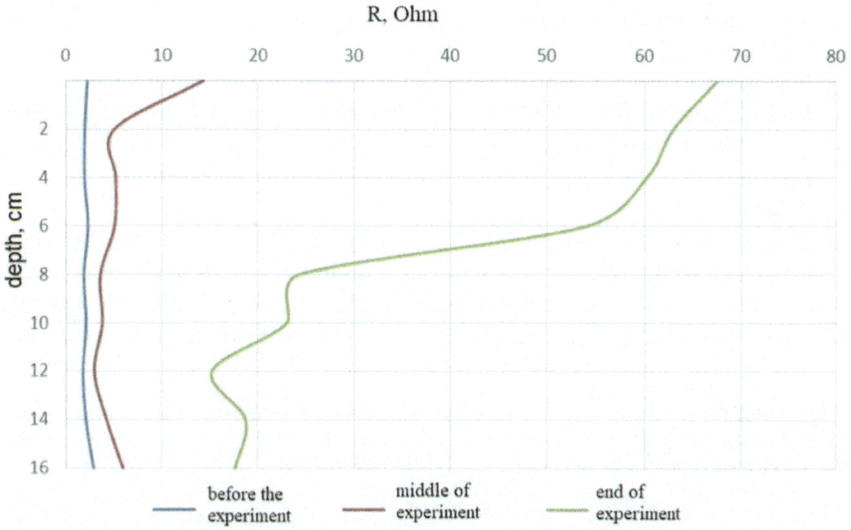

Fig. 17 Graph of UES distribution by depth with time

5 Conclusion

The analysis of literature data carried out within the framework of this work allows us to assert that we are dealing with a poorly studied area of permafrost distribution, namely, the shelf zone of the Arctic seas. This is despite the fact that the interest in the study of this object, in addition to its undoubted scientific value, is fueled by the need to solve the problems of developing the widespread hydrocarbon deposits here.

The lack of data is explained by the difficult conditions of research work and the complexity of the geocryogenic environment, which does not allow us to make an unambiguous judgment about the genesis and dynamics of the development of permafrost ground massifs in the conditions of the oceanic Arctic shelf.

Particular attention is drawn to the transitional coastal shelf zone, where the processes of soil freezing and thawing are directly related to global geological processes—sea transgressions and regressions and local exogenous and endogenous geocryologic phenomena.

Physical modeling of soil behavior in this particular zone is the focus of our research.

To solve this problem, a laboratory setup was designed and fabricated to allow experimental studies under well-defined boundary conditions.

1. The study of the cryogenic structure of the samples made it possible to characterize the dynamics of ice formation and growth in soils freezing under subaquatic conditions. The orientation of ice slits and their morphology in soil samples freezing in ice-soil contact conditions allowed us to judge about their growth under supercooling conditions. Such phenomena are not observed during ground freezing under ice-water-soil conditions.
2. In all soil samples, a surface layer 0.5–1 cm thick with increased mineralization (zone 1) and where ice formation does not occur is noted. This feature can be useful for analyzing the structure of soils in natural conditions and its presence may indicate subaquatic formation of permafrost in the conditions of closed systems.
3. Measurement of the specific electrical resistance of the soil during the experiments allowed to obtain values that as a result of modeling can be implicitly associated with the physical and chemical properties of the soil of the territory. This, in turn, can be confirmed by in-situ geophysical studies of the territory.

Acknowledgements The work was carried out in accordance with the state budgetary theme "Evolution of cryosphere under climate change and anthropogenic impact" (121051100164-0), "Danger and risk of natural processes and phenomena" (121051300175-4).

References

Baulin VV, Chekhovsky AL, Sukhodolsky SE (1981) Main stages of development of permafrost rocks of the north-east of the European part of the USSR and Western Siberia. In: History of development of permafrost rocks of Eurasia. Nauka, Moscow, pp 24–41

Baulin VV, Dubikov GI, Komarov IA et al (1997) Natural conditions of the Baidaratskaya Guba. Main results of research for construction of the underwater crossing of the Yamal-Center trunk pipeline system. GEOS, Moscow

Danilov ID, Zhigarev LA (1983) Cryogenic rocks of the arctic shelf. Nauka, Moscow

Dubikov GI (2002) Composition and cryogenic structure of permafrost strata of Western Siberia. GEOS

Frolov AD (2005) Electrical and elastic properties of frozen rocks and ice. Second supplement and revised edition. ONTI PSC RAS, Pushchino

Grigoriev NF (1981) Modern permafrost rocks on the Kara Sea shelf. In: Cryolithozone of the arctic shelf. IM YaF SB AS USSR, Yakutsk

Grigoriev NF (1982) Temperature characterization of bottom sediments in the coastal part of the Kara Sea (West coast of the Yamal Peninsula). Thermics of soils and rocks in cold regions

Khimenkov AN, Brushkov AV (2003) Oceanic cryolithogenesis. Nauka, Moscow

Melnikov VP, Spesivtsev VI (1995) Engineering-geological and geocryological conditions of the shelf of the Barents and Kara Seas. Nauka, Siberian Publishing Company of the Russian Academy of Sciences, Novosibirsk

Markov KK (1961) Paleogeography of the quaternary period of the USSR. Moscow State University Publishing House, Moscow

Neshiba S (1991) Oceanology. Mir

Romanovsky NN (1993) Fundamentals of lithosphere cryogenesis. Moscow State University Publishing House, Moscow

Saks VN (1952) Conditions of bottom sediments formation in the arctic seas of the USSR. Proc NIIGA 35:140

Shpolyanskaya NA (1991) Submarine cryolithogenesis in the arctic. In: Materials of glaciological studies. Chronicle, discussions. Vyp. 71, pp 65–70

Shpolyanskaya NA, Minkin MA, Feldman GM (1980) Dynamics of permafrost in Western Siberia in the Pleistocene. In: Natural conditions of Western Siberia. Izd-vo MSU, Moscow, Vyp. 7, pp 127–141

Vasilchuk YK, Krylov GV, Podborny EE (2006) Cryosphere of oil and gas condensate fields of the Yamal Peninsula, vol 3. Under general. Nedra, S.-Peterb. Department, Tyumen, TyumenNIIgirogaz LLC, St. Petersburg

Vasilchuk YK, Trofimov VT (1985) Features of the formation of permafrost strata in the north of Western Siberia in the Kargin and Sartan epochs of the late Pleistocene. In: Development of the Eurasian cryolithozone in the Upper Cenozoic

Zhigarev LA (1997) Oceanic cryolithozone

Zykov YD (2007) Geophysical methods of cryolithozone research. Izd. vo MSU

Neural Network Modeling as a Tool for Analyzing the Actual Development of Wood Resources in Forest Planning

V. N. Baklagin and P. V. Budnik

Abstract The work considers a fundamentally new approach to the analysis of the actual development of wood resources. This approach consists of using an artificial neural network as a tool for predicting the volume of harvesting of forest resources in the territory under consideration. The work provides a complete mathematical description of the artificial neural network for predicting the actual volume of timber harvesting, substantiates the methodology for developing the architecture of the artificial neural network, and selects forest indicators used as a natural basis for the input signals of the neural network model.

Keywords Artificial neural network · Forest resources · Prediction

1 Introduction

An urgent task of planning in the field of forest use is to determine efficiency indicators for the implementation of forest development activities, in particular the actual volume of wood harvesting or the share of development of the estimated logging area. For production forests, the maximum permissible volume of timber harvesting is established—the estimated cutting area. The estimated cutting area includes the supply of wood in mature and overmature stands. The procedure for calculating the estimated cutting area is approved by order of the Federal Forestry Agency. At the same time, the assessment of procurement volumes is carried out by experts. Such an assessment is subjective, and its accuracy depends on the experience and intuition of the expert. This approach does not contribute to increasing the efficiency and quality of forest planning.

V. N. Baklagin (✉)
Northern Water Problems Institute Karelian Research Centre Russian Academy of Sciences, Petrozavodsk, Republic of Karelia, Russia
e-mail: slava.baklagin@mail.ru

P. V. Budnik
Petrozavodsk State University, Petrozavodsk, Republic of Karelia, Russia

T. Chaplina (ed.), *Processes in GeoMedia—Volume VIII*, Springer Geology,
https://doi.org/10.1007/978-981-97-6627-7_2

There are a number of methods developed by scientists to obtain more accurate information about the potential volume of harvesting forest resources from the developed territory. For example, the work (Voronkov et al. 2009) attempts to calculate the economic accessibility of forest resources for individual regions of the Russian Federation, taking into account their natural and economic characteristics. To assess the possible volume of timber harvesting, statistical analysis methods were used, in particular the method of regression and correlation analysis. Such methods show good results in assessing economically possible timber harvesting in a specific area, but cannot take into account a number of factors, such as, for example, the density and length of roads, the number of wood processing enterprises, population size and others.

At the same time, in the last decade, with the growth of computer power, artificial neural networks have been actively used to solve problems of time series forecasting, pattern recognition, as well as to determine natural production indicators. They show good results in various fields of knowledge and technology. Based on a certain historical sample, artificial neural networks are capable of simulating various natural production phenomena and processes. In the field of forest resource use, artificial neural networks have become particularly widespread as a tool for predicting the risk of forest fires (Stankevich 2018; Pham et al. 2020; Pan et al. 2020; Naderpour et al. 2021; Eslami et al. 2021). In addition, neural network technologies are used to assess factors influencing the investment attractiveness of a territory (Rodionov and Samarin 2015).

It should also be noted that to date, a fairly large amount of information has been accumulated on the actual development of forest areas (central forest districts) in the regions of the Russian Federation, which includes data on the actual annual volumes of timber harvesting, as well as a number of natural, production and economic indicators. Sources of information on harvesting volumes for each central forestry are forest plans prepared on the basis of forest management documentation, information contained in the state forest register, territorial planning documents, and materials from the state forest inventory. Each subject of the Russian Federation has a forest plan. Forest management involves the delineation of forest lands into central forest districts. For each central forest district, forestry regulations are developed, according to which the use, protection and protection of forests is carried out. Forestry regulations contain data on available timber resources, natural production conditions, as well as forest infrastructure.

Information on the population was collected according to the Federal State Statistics Service by municipalities. Information on logging and wood processing enterprises was obtained from the Federal Tax Service.

Based on the available data, an attempt was made to create a methodological basis for predicting the actual volume of timber harvesting from the territory of forestry areas based on the use of artificial neural networks.

2 Theoretical Basis

In this work, to build a model for calculating the actual volume of timber harvesting, it is proposed to use artificial neural networks with direct signal propagation—multi-layer perceptrons without bias (Fig. 1). The architecture of such neural networks is represented by several layers: an input layer with the number of neurons (m), several hidden layers (w) with the number of neurons in each h layer (n_h), where$h = 1 \ldots w$, and an output layer with the number of neurons (p).

The functioning of an artificial neural network consists of transmitting signals from neurons of the previous layer to neurons of the next layer through synaptic connections that have certain weights, different for each pair of neurons (Fig. 1) (Simpn 1999):

1. The input neurons receive signals $a = [a_1, a_2, \ldots a_i, \ldots a_m]$. By means of established weights of the synaptic connections, signals from each input neuron are transformed, summed and fed forward into the neurons of the hidden layer as an input signal: $b_{inputj_1} = \sum_{i=1}^{m} a_i \bullet x_{i,j_1}$, where x_{i,j_1} is the weight of synaptic connection between i neuron of the input layer and j neuron of the first hidden layer, $i = 1 \ldots m, j_1 = 1 \ldots n_1$.
2. In each neuron of the hidden layer, signals are transformed through the activation function $b_{outputj_h} = f_a\left(b_{inputj_h}\right)$. Similarly, signals from the neurons of the hidden layer are converted, summed up and supplied to the neurons of the subsequent hidden layer: $b_{inputj_{h+1}} = \sum_{j_h=1}^{n_h} b_{outputj_h} \bullet y_{hj_h,j_{h+1}}$, where $y_{hj_h,j_{h+1}}$ is the weight of the synaptic connection between the j_h neuron of the hidden layer h and the j_{h+1} neuron of the hidden layer $h+1, j_h = 1..n_h, j_{h+1} = 1 \ldots n_{h+1}$.
3. Signals from the neurons of the last hidden layer are converted, summed and supplied to the neurons of the output layer $c_{inputk} = \sum_{k=1}^{p} b_{outputj_w} \bullet y_{zj_w,k}$, where $y_{zj_w,k}$—the weight of the synaptic connection between the j_w neuron of the last hidden layer w and the k neuron of the output layer c, $k = 1 \ldots p$.

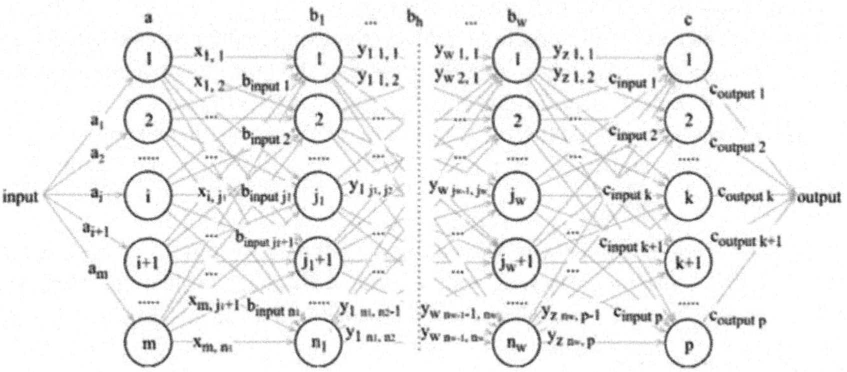

Fig. 1 General scheme of operation of an artificial neural network with direct propagation of signals

4. Within each neuron of the output layer, the signal is transformed with an activation function: $c_{outputk} = f_a(c_{inputk})$. Then, the result is interpreted.

Training of artificial neural networks in this work is assumed using the backpropagation method using a teacher (Simpn 1999). The backpropagation method assumes the use of some training samples consisting of training sets (input and corresponding output signals). For training purposes, training epochs are formed. They include training sets of the given sample, which are randomly mixed for each epoch. The process of training of perceptrons is carried out for several training epochs with a number of iterations for each. The number of iterations corresponds to the number of training sets included in the training epoch. Each iteration includes the following procedures:

1. A training set is randomly selected from the training epoch (each training set is used only once in each training epoch), which consists of an array of input signals $a = [a_1, a_2, \ldots a_i, \ldots a_m]$ and an array of target values (teacher) $t = [t_1, t_2, \ldots t_k, \ldots t_p]$.
2. For each neuron of the output layer, the values of the output signals $c_{outputk}$, and the local gradient are calculated using the formula: $\sigma_k = e_k \bullet f_a\prime(c_{inputk})$, where e_k is the error signal, calculated using the formula: $e_k = t_k - c_{outputk}$. In addition, the amount of adjustment of the weight of the synaptic connection between the neurons of the last hidden layer and the neurons of the output layer is calculated: $\delta y_{zj_w,k} = \alpha \bullet \sigma_k \bullet b_{outputj_w}$, where α is the learning rate of the artificial neural network.
3. For each neuron of the last hidden layer, the local gradient is calculated using the formula: $\sigma_{j_w} = f_a\prime(b_{inputj_w}) \bullet \sum_{k=1}^{p} \sigma_k \bullet y_{zj_w,k}$, as well as the amount of adjustment of the weight of the synaptic connection between the neurons of the last and previous hidden layers: $\delta y_{w-1j_{w-1},j_w} = \alpha \bullet \sigma_{j_w} \bullet b_{outputj_{w-1}}$.
4. For each neuron of the previous hidden layer, the local gradient is similarly calculated using the formula: $\sigma_{j_h} = f_a\prime(b_{inputj_h}) \bullet \sum_{j_{h+1}=1}^{n_{h+1}} \sigma_{j_{h+1}} \bullet y_{hj_h,j_{h+1}}$, and the amount of adjustment of the weight of the synaptic connection between the neurons of the last and previous hidden layers: 〚$\delta y_{hj_h,j_{h+1}} = \alpha \bullet \sigma_{j_{h+1}} \bullet b_{outputj_h}$.
5. For each neuron of the first hidden layer, the local gradient is calculated using the formula: $\sigma_{j_1} = f_a\prime(b_{inputj_1}) \bullet \sum_{j_1=1}^{n_1} \sigma_{j_1} \bullet x_{i,j_1}$, and the amount of adjustment of the weight of the synaptic connection: $\delta x_{i,j_1} = \alpha \bullet \sigma_{j_1} \bullet a_{outputi}$.
6. The weight of each synaptic connection of the perceptron is changed by adding the corresponding adjustment value: $x_{i,j_1}^{new} = x_{i,j_1}^{old} + \delta x_{i,j_1}$, $y_{hj_h,j_{h+1}}^{new} = y_{hj_h,j_{h+1}}^{old} + \delta y_{hj_h,j_{h+1}}$, $y_{zj_w,k}^{new} = y_{zj_w,k}^{old} + \delta y_{zj_w,k}$.

In each iteration, within the process of training of the perceptron, the mean square error (*MSE*) of the output signal with respect to the target values (teacher) was calculated. The arithmetic mean of the mean square errors $\overline{MSE}$ was calculated for each training epoch. The criterion for stopping of training was the restriction (Simpn et al. 2014): $\frac{|\overline{MSE}_{ep} - \overline{MSE}_{ep-1}|}{\overline{MSE}_{ep-1}} \bullet 100\% \leq 0.01\%$, where *ep* is the number of the current training epoch.

3 Justification for the Choice of Artificial Neural Network Input Signals

The main difficulty in selecting the input signals of an artificial neural network included in the model is the selection and formalization of natural, production and economic factors that have a direct impact on the actual volume of timber harvesting in the forestry. These factors were selected on the basis of previous studies, taking into account available sets of information, as well as on the basis of expert assessment based on an analysis of the activities of logging enterprises. In particular, a number of studies note that the efficiency of logging is determined by a number of factors (Shabaev et al. 2020; Gerasimov et al. 2013): the volume of available wood, production capacity and demand. The volume of available wood is determined by the size of the estimated cutting area, which is established for deciduous and coniferous farms for operational and protective forests. The estimated cutting area is assigned separately by type of felling: clear and selective. The volume of the calculated cutting area cannot be exceeded. Transportation constitutes a significant portion of the operating costs of timber harvesting (Melchiori et al. 2022; Mokhirev and Medvedev 2020). Therefore, some of the wood may not be economically available for development. Demand is determined by woodworking and wood processing enterprises, especially large ones, such as pulp and paper mills.

It is known that the performance of machines and mechanisms. influenced by natural production conditions. The influence of the average wood supply per hectare on the productivity of forestry machines, in particular the harvester (Kormanek and Baj 2018; Kärhä et al. 2004) and forwarder (Kellogg and Bettinger 1994; Proto et al. 2018a; Louis et al. 2022), is well known. In addition, the efficiency of forestry machines, especially transport ones, depends on soil conditions (George et al. 2019; Proto et al. 2018b). On waterlogged soils, the productivity of forestry machines decreases. Species composition can also significantly influence the productivity of machinery and equipment (Olivera et al. 2016; Hiesl and Benjamin 2013; Labelle and Soucy 2016), and also determines the economic value of a forest area.

The type of cutting significantly influences the productivity of forestry machines (Spinelli and Magagnotti 2013; Lageson 1997; Eliasson et al. 1999). In clear felling, the productivity of the harvester and forwarder is higher than in selective felling.

In this study, as an indirect assessment characterizing natural production conditions, it is proposed to use the distribution of forest lands by classes of natural fire hazard of forests. In the Russian Federation, five classes are distinguished, each of which is determined by characteristic types of forest, for example, the second class includes pine-lingonberry forests growing on drier soils, and the fifth class includes long-term spruce forests, in which the soil is always waterlogged and the forest reserve per hectare is relatively smaller, the trunks are tapering and difficult to clear of branches. Therefore, in the fifth class, the natural production conditions for harvesting will be predominantly worse than in the second.

Thus, it was decided to use the following indicators as a natural basis for generating the input signals of the artificial neural network a in the developed model: data on the forestry area $S_{forestarea}$, forest land area S_{forest}, estimated cutting areas for all types of felling V, average stock per hectare q, shares of areas belonging to different forest areas S_{fr}, road density for each type of road surface ρ_{roads}, areas by fire hazard classes $Sfire$, population size living in the territory of the forestry P, number of logging enterprises operating in the territory of the central forestry FF, number of wood processing enterprises (consumers) operating on the territory of this and neighboring forestries CF:

$$a = [nsf, nv, nq, nsfr, n\rho, nfire, npop, nff, cff], \tag{1}$$

where nsf is an array of input signals characterizing the ratio of the area of forest lands of the forestry to the total area of the forestry:

$$nsf = \left[nSf_0, nSf_1, \ldots, nSf_i, \ldots, nSf_n\right], \tag{2}$$

$$nSf_i = \begin{cases} 1, \frac{S_{forest}}{S_{forestarea}} \geq 0.1 \bullet i \\ 0, \frac{S_{forest}}{S_{forestarea}} < 0.1 \bullet i \end{cases} \tag{3}$$

where n is the number of intervals into which the value of the indicator is divided, each of which corresponds to a neural network signal;

where nv is an array of input signals characterizing the volume of the estimated cutting area for all types:

$$nv = \begin{bmatrix} \left[nv_{0,0}, nv_{0,1}, \ldots, nv_{0,i}, \ldots, nv_{0,n}\right], \ldots, \\ \left[nv_{z,0}, nv_{z,1}, \ldots, nv_{z,i}, \ldots, nv_{z,n}\right] \end{bmatrix} \tag{4}$$

$$nv_{j,i} = \begin{cases} 1, \frac{V_j}{\sum_{j=1}^{z} V_j} \geq 0.1 \bullet i \\ 0, \frac{V_j}{\sum_{j=1}^{z} V_j} < 0.1 \bullet i \end{cases}, \tag{5}$$

where V_j is the volume of the estimated cutting area for the j type of felling; z—number of felling types;

where nq is an array of input signals characterizing the average stock per hectare of forestry:

$$nq = \left[nq_0, nq_1, \ldots, nq_i, \ldots, nq_n\right], \tag{6}$$

$$nq_i = \begin{cases} 1, q/q_{\max} \geq 0.1 \bullet i \\ 0, q/q_{\max} < 0.1 \bullet i \end{cases} \tag{7}$$

where $q_{\max}$ is the established maximum value of the average reserve per hectare (for example, the recommended value for the regions of the European North of the Russian Federation is 400 m^3/ha);

where *nsfr* is an array of input signals characterizing the shares of areas related to different forest areas:

$$nsfr = \begin{bmatrix} [nsfr_{0,0}, nsfr_{0,1}, \ldots, nsfr_{0,i}, \ldots, nsfr_{0,n}], \ldots, \\ [nsfr_{r,0}, nsfr_{r,1}, \ldots, nsfr_{r,i}, \ldots, nsfr_{r,n}] \end{bmatrix} \tag{8}$$

$$nsfr_{j,i} = \begin{cases} 1, \frac{S_{fr\,j}}{\sum_{j=1}^{r} S_{fr\,j}} \geq 0.1 \bullet i \\ 0, \frac{S_{fr\,j}}{\sum_{j=1}^{r} S_{fr\,j}} < 0.1 \bullet i \end{cases}, \tag{9}$$

where r is the number of forest areas;

where $n\rho$ is an array of input signals characterizing the density of roads for each type of road surface:

$$n\rho = \begin{bmatrix} [n\rho_{0,0}, n\rho_{0,1}, \ldots, n\rho_{0,i}, \ldots, n\rho_{0,n}], \ldots, \\ [n\rho_{p,0}, n\rho_{p,1}, \ldots, n\rho_{p,i}, \ldots, n\rho_{p,n}] \end{bmatrix} \tag{10}$$

$$n\rho_{j,i} = \begin{cases} 1, \rho_{roads\,j}/\rho_{\max j} \geq 0.1 \bullet i \\ 0, \rho_{roads\,j}/\rho_{\max j} < 0.1 \bullet i \end{cases} \tag{11}$$

where p is the number of types of road surfaces; $\rho_{\max j}$—the accepted limit for the maximum density of roads in forestry (for example, the recommended value for the regions of the European North of the Russian Federation is 10 km/thousand ha);

$$nfire = \begin{bmatrix} [nfire_{0,0}, nfire_{0,1}, \ldots, nfire_{0,i}, \ldots, nfire_{0,n}], \ldots, \\ [nfire_{f,0}, nfire_{f,1}, \ldots, nfire_{f,i}, \ldots, nfire_{f,n}] \end{bmatrix} \tag{12}$$

$$nfire_{j,i} = \begin{cases} 1, \frac{Sfire_j}{\sum_{j=1}^{f} Sfire_j} \geq 0.1 \bullet i \\ 0, \frac{Sfire_j}{\sum_{j=1}^{f} Sfire_j} < 0.1 \bullet i \end{cases}, \tag{13}$$

where *npop* is an array of input signals characterizing the population size in the forestry area:

$$npop = [npop_0, npop_1, \ldots, npop_i, \ldots, npop_n], \tag{14}$$

$$nq_i = \begin{cases} 1, npop/npop_{\max} \geq 0.1 \bullet i \\ 0, npop/npop_{\max} < 0.1 \bullet i \end{cases} \tag{15}$$

where $npop_{\max}$ is the accepted limit of the maximum population size living on the territory of the forestry;

where nff is an array of input signals characterizing the number of logging enterprises operating in the territory of the central forestry:

$$nff = \left[nff_0, nff_{1,} \ldots, nff_i, \ldots, nff_n\right], \tag{16}$$

$$nff_i = \begin{cases} 1, FF/FF_{\max} \geq 0.1 \bullet i \\ 0, FF/FF_{\max} < 0.1 \bullet i \end{cases} \tag{17}$$

where $FF_{\max}$ is the accepted limit for the maximum number of logging enterprises operating in the territory of the central forestry;

where cff is an array of input signals characterizing the number of logging enterprises operating in the territory of the central forestry:

$$cff = \left[cff_0, cff_{1,} \ldots, cff_i, \ldots, cff_n\right], \tag{18}$$

$$cff_i = \begin{cases} 1, CF/CF_{\max} \geq 0.1 \bullet i \\ 0, CF/CF_{\max} < 0.1 \bullet i \end{cases} \tag{19}$$

where $CF_{\max}$ is the accepted limit for the maximum number of wood processing enterprises operating in the territory of the central forestry and adjacent to it.

4 Justification of Parameters and Architecture of Artificial Neural Network

The methodology for substantiating the parameters and architecture of an artificial neural network for predicting the actual volume of harvesting in the territory of the central forestry includes the following steps:

1. The number of neurons in the input layer of the neural network is determined by the number of input signals, according to Eq. (1). Each neural network has n neurons in the output layer corresponding to output signals c, characterizing the ratio of the volume of actual timber harvesting to the estimated cutting area in the central forestry under consideration.
2. The number of hidden layers, as well as the number of neurons in each hidden layer, is selected based on computational experiments in the process of training neural networks in several steps. At the first step, a neural network with the simplest architecture is trained. At each subsequent step, a neural network is trained that has one more neuron in the hidden layer than the neural network at the previous step. In this case, at each step 1000 iterations are performed, at each of which the value $\overline{MSE}$ for the last training epoch was calculated. If at the current step the mathematical expectation $\overline{MSE}$ is less than at the previous step, then the next step is performed, otherwise the procedure for selecting the optimal

neural network architecture ends, and the network trained at the previous step is selected.

3. The speed of training α is chosen by experiment: $\alpha = 0.001$.
4. The initial weights of synaptic connections of neurons were selected randomly from a uniform distribution with a mathematical expectation equal to 0, and the variance, which is determined by the formula (Simpn 1999): $D = \frac{1}{\sqrt{s}}$, where s is the number of synaptic connections of the neuron. Considering the variance formula for the uniform distribution: $D = \frac{(u-v)^2}{12}$, where u, v are the upper and lower limits of the interval respectively, and also taking into account the fact that the mathematical expectation is 0, the interval is as follows: $\left[-\sqrt{\frac{3}{\sqrt{s}}}; +\sqrt{\frac{3}{\sqrt{s}}}\right]$.

5 Conclusion

The forecast of the actual values of harvested volumes of forest timber resources is the most important assessment of the economic efficiency of the implementation of measures to implement the planned forest development. Such an assessment is carried out when preparing forest planning documents. However, in practice, its implementation is difficult due to the presence of hidden dependencies between forest planning indicators and the volumes of development of the estimated cutting area (harvesting volumes). This paper offers a detailed mathematical description of the concept and methodology for creating an artificial neural network, on the basis of which such a forecast can be obtained. At the same time, it is important to note that the proposed neural network mathematical model makes it possible to analyze the impact of planned activities on the economic efficiency of forest development of certain forest planning indicators and, on this basis, to adjust the forest plan.

Acknowledgements The research was supported by grant No. 23-21-00143 by the Russian Science Foundation, https://rscf.ru/en/project/23-21-00143/.

References

Eliasson L, Bengtsson J, Cedergren J, Lageson H (1999) Comparison of single-grip harvester productivity in clear and shelterwood cutting. J for Eng 10(1):43–48. https://doi.org/10.1080/08435243.1999.10702723

Eslami R, Azarnoush M, Kialashki A, Kazemzadeh F (2021) GIS-based forest fire susceptibility assessment by random forest, artificial neural network and logistic regression methods. J Trop for Sci 33(2):173–184

George AK, Kizha AR, Kenefic L (2019) Economic feasibility of timber harvesting in lowlands. In: Czupy I (ed) Exceeding vision: forest mechanisation of the future. University of Sopron Press, Sopron, Hungary, 379–393

Gerasimov Y, Sokolov AP, Fjeld D (2013) Improving cut-to-length operations management in russian logging companies using a new decision support system. Balt for 1:89–105

Hiesl P, Benjamin JG (2013) Applicability of international harvesting equipment productivity studies in Maine, USA: a literature review. Forests 4(4):898–921. https://doi.org/10.3390/f4040898
Kärhä K, Rönkkö E, Gumse S-I (2004) Productivity and cutting costs of thinning harvesters. Int J for Eng 15(2):43–56
Kellogg LD, Bettinger P (1994) Thinning productivity and cost for mechanized cut-to-length system in the Northwest Pacific coast region of the USA. J for Eng 5(2):43–52
Kormanek M, Baj D (2018) Analysis of operation performance in the process of machine wood harvesting with FAO far 6840 mini-harvester. Agric Eng 22(1):73–82
Labelle ER, Soucy M (2016) Effect of tree form on the productivity of a cut-to-length harvester in a hardwood dominated stand. Croat J for Eng 37(1):175–183
Lageson H (1997) Effects of thinning type on the harvester productivity and on the residual stand. J for Eng 8(2):7–14
Louis LT, Kizha AR, Daigneault A (2022) Factors affecting operational cost and productivity of ground-based timber harvesting machines: a meta-analysis. Curr Forestry Rep 8:38–54. https://doi.org/10.1007/s40725-021-00156-5
Melchiori L, Nasini G, Montagna JM, Corsano G (2022) A mathematical modeling for simultaneous routing and scheduling of logging trucks in the forest supply chain. For Policy Econ 136:102693. https://doi.org/10.1016/j.forpol.2022.102693
Mokhirev A, Medvedev S (2020) Assessment of road density in logging areas using geographical information systems. IOP Conf Ser Earth Environ Sci 507:012022. https://doi.org/10.1088/1755-1315/507/1/012022
Naderpour M, Rizeei HM, Ramezani F (2021) Forest fire risk prediction: a spatial deep neural network-based framework. Remote Sens 13(13):2513. https://doi.org/10.3390/rs13132513
Olivera AR, Visser M, Acuna MJ (2016) Automatic GNSS-enabled harvester data collection as a tool to evaluate factors affecting harvester productivity in a Eucalyptus spp. harvesting operation in Uruguay. Int J For Eng 27(1):15–28. https://doi.org/10.1080/14942119.2015.1099775
Pan H, Badawi D, Zhang X (2020) Additive neural network for forest fire detection. SIViP 14:675–682. https://doi.org/10.1007/s11760-019-01600-7
Pham BT, Jaafari A, Avand M, Al-Ansari N, Dinh DuT, Yen HPH, Phong TV, Nguyen DH, Le HV, Mafi-Gholami D (2020) Performance evaluation of machine learning methods for forest fire modeling and prediction. Symmetry 12(6):1022. https://doi.org/10.3390/sym12061022
Proto AR, Macrì G, Visser R (2018a) Factors affecting forwarder productivity. Eur J Forest Res 137:143–151. https://doi.org/10.1007/s10342-017-1088-6
Proto AR, Macrì G, Visser R, Harrill H, Russo D, Zimbalatti G (2018b) Factors affecting forwarder productivity. Eur J for Res 137:143–151. https://doi.org/10.1007/s10342-017-1088-6
Rodionov MG, Samarin AM (2015) Application of neural network technologies for the development of cluster structure of the territorial and administrative entities. Russ Automob Highway Industry J 2(42):150–158 (In Russian)
Shabaev A, Sokolov A, Urban A, Pyatin D (2020) Optimal planning of wood harvesting and timber supply in Russian conditions. Forests 11(6):662. https://doi.org/10.3390/f11060662
Simpn H (1999) Neural network: a comprehensive foundation, 2nd edn. Pearson Education, Inc.
Simpn H, Demuth HB, Beale MH, Jess OD, Hagan MT (2014) Neural network design, 2nd edn., 800
Spinelli R, Magagnotti N (2013) The effect of harvest tree distribution on harvesting productivity in selection cuts. Scand J for Res 28(7):701–709. https://doi.org/10.1080/02827581.2013.821517
Stankevich TS (2018) Application of convolutional neural networks to solve the problem of operational forecasting of the dynamics of forest fires. Bus Inform 4(46):17–27. https://doi.org/10.17323/1998-0663.2018.4.17.27(inRussian)
Voronkov PT, Voronkov AP, Belov AN, Dudina EA, Ilyukhina LA (2009) Modeling the economic availability of forest resources using regression analysis. Lesohozyajstvennaya Informaciya. 1–2:7–13 (in Russian)

Stability Analysis of Small Oscillations of a Weather Vane in a Quasi-Static Medium Flow

D. V. Belyakov

Abstract This paper presents a mathematical model for the motion of an aerodynamic weather vane. The structure is suspended on two springs, oriented into the wind direction, and undergoes oscillations under the influence of an incident quasi-stationary flow. The equations of motion for the considered model are derived, incorporating aerodynamic forces and moments determined solely through experimental tests in a wind tunnel. The considered system admits only numerical solution. Equilibrium equations are obtained, from which various stationary points are derived. Of particular interest is the central stationary point, where the body is at rest. First-order approximation equations are derived, and stability regions are determined using the Routh-Hurwitz criterion for various parameter values. A suite of programs implemented in the MATLAB computational environment is developed for numerical simulation, confirming the obtained results.

Keywords Weather vane · Stability · Small oscillations · Hurwitz criterion · Stability region

This article is dedicated to the investigation of stability loss phenomena in the mathematical model of oscillations of an aerodynamic pendulum in the shape of a weather vane (see Fig. 1). This problem is closely related to a physical phenomenon known as flutter and shares the same nature. Flutter is the occurrence of self-excited oscillations of wings and other elements of an aircraft under the influence of incident airflow. It arises with increasing flight speed and is analogous to the resonance effect. Similar problems were addressed in works (Belyakov 2015, 2016), where the plate was oriented orthogonally to the rod. The problem considered in this article can be solved using a simpler approach, employing methods of the first approximation theory (Bogolyubov and Mitropolsky 1955; Uryvskaya 2017).

D. V. Belyakov (✉)
Moscow Aviation Institute (National Research University), Moscow, Russia

T. Chaplina (ed.), *Processes in GeoMedia—Volume VIII*, Springer Geology,
https://doi.org/10.1007/978-981-97-6627-7_3

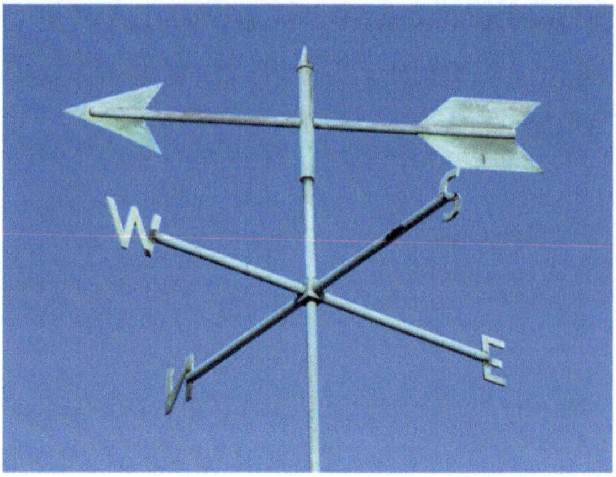

Fig. 1 Simple weather vane

1 Problem Statement

Let us consider the problem of a weather vane, which is a body represented by a thin plate installed along a rod of length r. The rod is fixed at the center of mass using two spring elements and undergoes self-excited oscillations in the medium flow (see Fig. 2).

We introduce a fixed coordinate system. We assume that in the resting position, the weather vane occupies an equilibrium position where it is oriented along the flow. We consider that the center of mass of the body can only move in the vertical plane, and the deforming forces of the attachment elements depend linearly on deviations and are reduced to a restoring force $F = -kx$ and a restoring moment $M = -c\theta$.

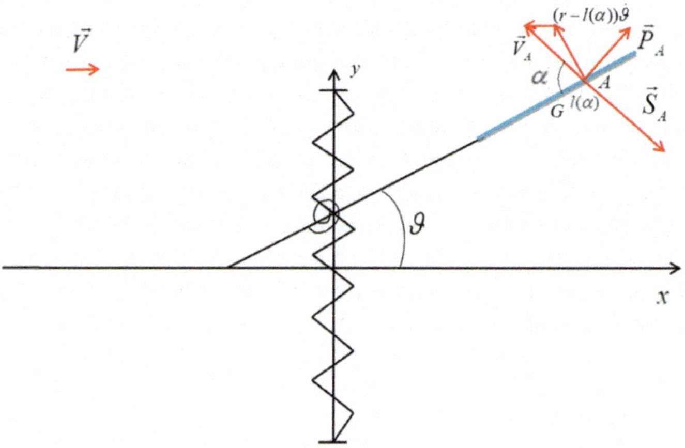

Fig. 2 Oscillations of a flat plate in the flow of a medium

Dynamic forces applied to the body are chosen in accordance with the "empirical theory of steady flow around a flat plate." In the studied scheme, it is assumed that the center of pressure of the plate, point A, can be considered as movable relative to the plate.

The shift of the center of pressure is described by the function $l(\alpha)$, which represents the distance between point A and the geometric center of pressure of the plate G. The dependence $l(\alpha)$, is determined from tests of rectangular plates with specified elongation in a wind tunnel and is based on experimental data published in Tabachnikov (1974). It is assumed that if the function $l(\alpha)$ is measured to the right of the center of mass, it is positive; if to the left of the center of mass, it is negative. Similar problems, with the same model of fluid-body interaction, have been considered in Pshikhopov et al. (2013), Lokshin et al. (1992). The aerodynamic forces acting on each plate are decomposed into components: the drag force $\overrightarrow{S_A}$, directed against the velocity $\overrightarrow{V_A}$ of point A relative to the flow, and the lift force $\overrightarrow{P_A}$, directed orthogonally to it. The magnitudes of the aerodynamic forces are given by:

$$|\overrightarrow{S_A}| = s(\alpha)V_A^2 = 0.5\rho\sigma c_x(\alpha)V_A^2$$

$$|\overrightarrow{P_A}| = p(\alpha)V_A^2 = 0.5\rho\sigma c_y(\alpha)V_A^2$$

where

- α is the angle of attack between the vector $\overrightarrow{V_A}$ and the plate,
- p, s are the aerodynamic angle of attack functions,
- c_x, c_y are dimensionless aerodynamic functions,
- ρ is the air density,
- σ is the area of one plate.

We will assume that the medium does not affect the rod.

Next, it is necessary to derive the equations of motion for the considered body. As generalized coordinates defining the position of the body, introduce the coordinate at the center of mass of the body and the angle θ of deviation of the plate from the horizontal.

In this case, the theorem on the motion of the center of mass projected onto the axis oY and the theorem on the change in the kinetic moment will have the form

$$\begin{aligned} m\ddot{y} &= -p(\alpha)V_A((r - l(\alpha))\dot{\theta}\cos\theta + V) \\ &\quad - s(\alpha)V_A\big((r - l(\alpha))\dot{\theta}\sin\theta - \dot{y}\big) - ky \\ J\dot{\theta} &= (\mathrm{r} - \mathrm{l}(\alpha))V_A^2 n(\alpha) - c\theta \end{aligned} \tag{1}$$

where the multiplier $\mathrm{n}(\alpha) = \mathrm{s}(\alpha)\sin\alpha + \mathrm{p}(\alpha)\cos\alpha$—will be the coefficient of the normal aerodynamic force.

The kinematic relations that relate V_A, α to $V, y, \theta, \dot{\theta}$ take the form:

$$V_{\text{A}}\sin\alpha = (r - l(\alpha))\dot{\theta} + V\sin\theta + \dot{y}\cos\theta$$
$$V_{\text{A}}\cos\alpha = V\cos\theta - y\sin\theta \quad (2)$$

Thus, a mathematical model of plate oscillations has been constructed, representing a closed system of Eqs. (1) ans (2).

2 Steady Points

Considering the translational motion from the kinematic relationships (2), taking into account that $\theta_* = const$, $y_* = const$, $\dot{\theta} = 0$ it follows that $tg\alpha_* = ctg\theta_*$, from which $\alpha_* + \theta_* = \frac{\pi}{2}$. Let's write down the equilibrium equations for the system (1).

$$\text{p}(\alpha_*)V - ky_* = 0$$

$$c\alpha_* + (r - l(\alpha_*))V^2 n(\alpha_*) = 0$$

Figure 3 depicts the solution of the second equation of the system in MATLAB for rectangular plates with an elongation of 8. With a sufficiently small stiffness coefficient of the spiral spring, only the following equilibrium positions exist:

$\theta_* = \pi k, k \in Z, y_* = 0$. As the angle of attack α increases, the number of equilibrium positions also increases, and when $\frac{\pi}{2} < \alpha_* < \pi$, another pair of equilibrium

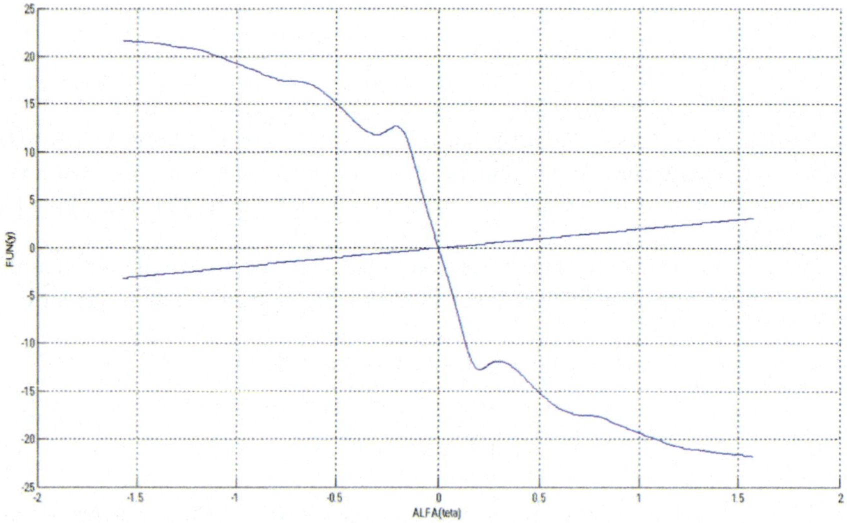

Fig. 3 Solution of equilibrium equations

positions symmetric about zero emerges. These positions can only be found numerically. It can be shown that when c = 0, we still have obvious families of equilibrium positions:

$$\theta_* = \pm\frac{\pi}{2} + \pi k, k \in Z, y_* = 0$$

3 Stability Regions

Let us consider the plate undergoing small oscillations around its equilibrium position. We will investigate the stability of the trivial equilibrium position $y = 0, \theta = 0$ to first order. To examine the stability of the trivial equilibrium position $y = 0, \theta = 0$, we will employ the Hurwitz criterion. We will conduct stability analysis for various parameter values m, J, k, c, V^2, r and perform a parametric stability analysis to first order.

We linearize equations of motion (5) and (6) at $y \to 0, \theta \to 0, \alpha \to 0$.

After a series of transformations, the equations of small plate oscillations take the form:

$$\begin{aligned} &m\ddot{y} + (p_0' + s_0)V\dot{y} + ky + p_0' lV\dot{\theta} - p_0' V^2\theta = 0 \\ &(p_0' + s_0)Vl\dot{y} + \dot{c}y + J\ddot{\theta} + ((p_0' + s_0) - c)Vl^2\theta \doteq 0 \end{aligned} \quad (3)$$

Let's write down the characteristic equation of system (3).

$$a_0\lambda^4 + a_1\lambda^3 + a_2\lambda^2 + a_3\lambda + a_4 = 0 \quad (4)$$

The stability conditions take the following form:

$$a_0 = mJ > 0 \quad \text{(Always satisfied)} \quad (5)$$

$$a_1 = (p_0' + s_0)V\left(m(r - l)^2 + J\right) > 0 \quad (6)$$

$$a_2 = (p_0' + s_0)mV^2(r - l) + mc + kJ > 0 \quad (7)$$

$$a_3 = (p_0' + s_0)V\left[V^2(r - l)(p_0' + s_0) + k(r - l)^2 - p_0' V^2\right] > 0 \quad (8)$$

$$a_4 = k\left[V^2(r - l)(p_0' + s_0) + c\right] > 0 \quad (9)$$

$$\Delta_3 = \begin{vmatrix} a_1 & a_3 & 0 \\ a_0 & a_2 & a_4 \\ 0 & a_1 & a_3 \end{vmatrix} > 0 \tag{10}$$

Let's plot the stability region on the parameter plane $(c,\ V^2)$ for various values of $r.l.$

- When condition $p_0' + s_0 < 0$ is satisfied, condition (6) is never fulfilled, and self-excited oscillations occur even at low flow velocities (conversely, if $p_0' + s_0 > 0$, then inequality (6) is always satisfied!).
- When conditions $r > l > 0$ and $p_0' + s_0 > 0$ are met, the parameter l is positive (indicating that the plate is aligned with the flow), and conditions (6)–(10) are automatically satisfied. The system is stable for any values of $(c,\ V^2)$.
- When conditions $r = l$ and $p_0' + s_0 > 0$ are met (indicating that the plate is aligned with the flow), conditions (6)–(9) are automatically satisfied, but condition (10) is not fulfilled. The system is unstable for any values of $(c,\ V^2)$.
- If $0 < r < l$ and $p_0' + s_0 > 0$ (indicating that the plate is aligned with the flow), to satisfy conditions (6)–(10), we have the system of inequalities (11)–(13) (see Fig. 4):

$$V^2 < \frac{mc + kJ}{m(l-r)\left(p_0' + s_0\right)} \quad \text{(depicted in red color)} \tag{11}$$

$$V^2 < \frac{c\left(p_0' + s_0\right)}{m\left(p_0' - (l-r)\right)\left(p_0' + s_0\right)^2} \quad \text{(depicted in blue color)} \tag{12}$$

$$V^2 < \frac{c}{(l-r)\left(p_0' + s_0\right)} \quad \text{(depicted in green color)} \tag{13}$$

Condition (11) is less restrictive than conditions (12) and (13), so it can be disregarded.

- When $r = 0$ and $p_0' + s_0 > 0$ (indicating that the plate is fixed at the center of mass), the system of inequalities (10)–(13) holds true if we are below the blue line in the green region (see Fig. 5).
- When $l < r < 0$ and $p_0' + s_0 > 0$ (indicating that the plate is oriented against the flow), the parameter becomes negative. The system of inequalities (10)–(13) is realized if we are below the steepest green line.
- As r decreases, the region will shrink.
- If $r = l < 0$, $p_0' + s_0 > 0$ (the plate is oriented against the flow), the system is always unstable! Condition (10) is not met!
- If $p_0' + s_0 > 0$ (the plate is oriented against the flow), stability conditions (10)–(13) are fulfilled if we are within the green region (Fig. 6).

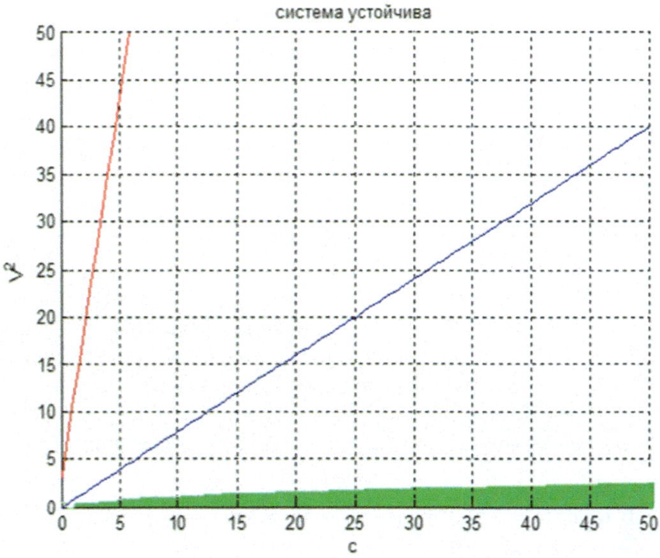

Fig. 4 Stability region (r = 0.5)

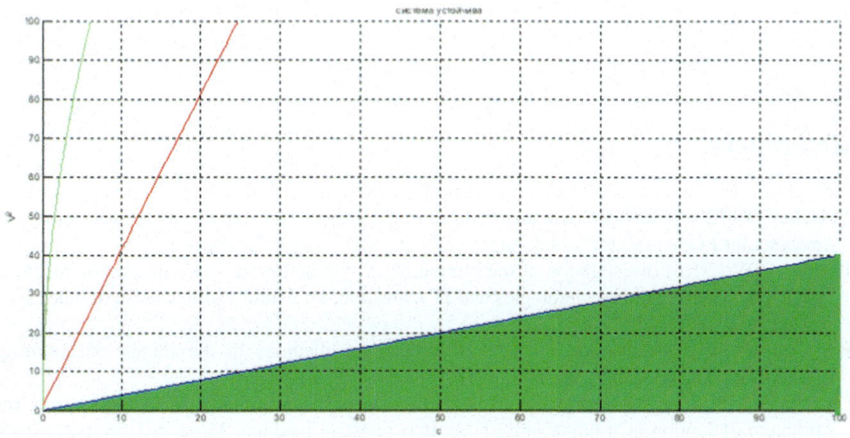

Fig. 5 Stability region (r = 0)

4 Conclusion

1. A mathematical model of the motion of an aerodynamic pendulum in the form of a weather vane has been created.
2. Equilibrium equations have been solved, and stationary points have been found.

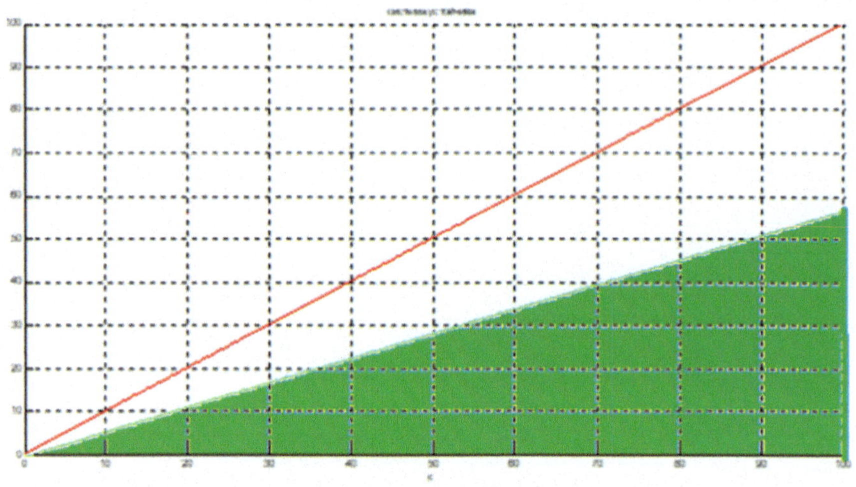

Fig. 6 Stability region (r = – 5) (the system is stable below the green line)

3. A parametric analysis of the dynamic system has been conducted, and stability regions have been plotted on the parameter plane (c, V^2) for various values of $r.l.$

References

Belyakov DV (2015) Development and features of the mathematical model of the Darrieus wind turbine. Int J Open Inf Technol 3. https://injoit.org/index.php/j1/article/view/226

Belyakov DV (2016) Investigation of the motion of axisymmetric body in a quasistatic medium. In: Modern information technologies and IT education, no 2. http://sitito.cs.msu.ru/index.php/SITITO/article/view/82

Bogolyubov NN, Mitropolsky YA (1955) Asymptotic methods in the theory of nonlinear oscillations. Gostekhizdat, Moscow, p 408

Lokshin BY, Privalov VA, Samsonov VA (1992) Model problem of flutter. In: Introduction to the problem of motion of a point and a body in a resistant medium. Moscow University Press, Moscow, p 38

Pshikhopov VK, Kulchenko AE, Chufistov VM (2013) Modeling of flight of a single-rotor helicopter under the control of positional-trajectory regulator. In: Engineer Herald of Don, no 2. https://ivdon.ru/ru/magazine/archive/n2y2013/1650

Tabachnikov VG (1974) Steady characteristics of wings at low speeds in the entire range of angles of attack. In: Proceedings of TsAGI. Nauka, Moscow, p 154

Uryvskaya TY (2017) Stability of linear systems with positive definite matrix. In:Engineer Herald of Don, no 4. https://ivdon.ru/ru/magazine/archive/n4y2017/4464

Change in the Organoprofile of Marsh Soils During the Evolution of the Sakhalin Island Seashore

V. V. Afanasiev, V. V. Demin, and Y. A. Zavgorodnyaya

Abstract The results of a study of soils of salt marshes located in the coastal zone of Salmon Bay in Aniva Bay, Sakhalin Island (46.74° N; 142.73° E) are presented. Five soil profiles were uncovered and samples were collected by horizon to analyse organic matter content and composition. The obtained rates of carbon accumulation in Aniva Bay marshes significantly (2–3 times) exceed most of the values of this value obtained for other coastal areas, which may be related to local climatic and geomorphological peculiarities of the study site. Given the size of the areas occupied by marsh soils, it is possible to make calculations of the total carbon sink in coastal landscapes on Sakhalin Island.

Keywords Estuary · Salt marsh · Intertidal mudflat · Coastal sedimentation · Carbon accumulation rate—CAR

1 Introduction

Salt marshes are tidal wetland ecosystems, dominating sheltered shores and estuarine margins in a wide range of climatic conditions (subarctic to tropical), but are most common in temperate latitudes. Coastal-marine wetlands are arguably the most effective biosequestration systems on the planet, and it is the bottom and peat-gley sedimentation of marshes and mudflats of marine estuarine-lagoon geosystems that will offer new opportunities to address GHG emission reductions (Najjar et al. 2018; Ouyang and Lee 2014; Spivak et al. 2019; Villa and Bernal 2018).

By 2100, average ocean levels could rise by 20–180 cm, according to various estimates. Salt marsh ecosystems can withstand sea-level rise: marsh vegetation

V. V. Afanasiev (✉)
SakhSU, Yuzhno-Sakhalinsk, Russia
e-mail: vvasand@mail.ru

Institute of Marine Geology and Geophysics FEB RAS, Yuzhno-Sakhalinsk, Russia

V. V. Demin · Y. A. Zavgorodnyaya
Lomonosov Moscow State University, Moscow, Russia

T. Chaplina (ed.), *Processes in GeoMedia—Volume VIII*, Springer Geology,
https://doi.org/10.1007/978-981-97-6627-7_4

captures sediment from the sea, which allows plants to grow better and in turn favours even faster accumulation of sediment and organic matter. This process is a major factor in allowing marshes ecosystems to keep pace with sea level rise. Sediment accretion rates in the marches' territories lie in the range of 0.07–4.2 cm/year (sediment accretion rate—SAR) and exceed the rate of global ocean rise in 45% of the studied territories. Simultaneously with sediment accumulation, carbon burial in marshes soils occurs. According to the latest data, the average carbon accumulation rate (CAR) is 244.7 gC-m^{-2}-$year^{-1}$, but for different coastal areas this value varies from 18 to 1713 gC-m^{-2}-$year^{-1}$. The carbon accumulation rate in marsh soils can be higher than in mangrove soils and 50 times higher than this parameter for forested continental ecosystems.

At present, data on coastal areas occupied by marshes, sediment accumulation rates and carbon accumulation rates remain estimates, and large areas of coasts are practically unstudied. The coastline of the Russian Federation is about 9.5% of the world coastline, slightly inferior to the length of the US coastline (11.5%), but while about 50% of the data on carbon accumulation rates relate to the US marshes, there is no information in the current literature on carbon accumulation rates in the soils of the Russian Federation marshes and, consequently, on the total annual carbon sink in coastal landscapes. Moreover, coastal marshes and mudflats are already included in the national inventories of the United States and many other countries, where they are sure of their cooling effect on the climate (Sapkota and White 2020; Lovelock et al. 2022).

2 Methods and Results

One of the main results of the reconnaissance stage of our work was the establishment of the fact of the highest sequestration capacity of salt marshes and drylands in the northern part of Aniva Bay. During 60 years, marsh peat bogs up to 80 cm thick were formed on an area of about 2 km^2. (Afanas'ev et al. 2023). Annual net carbon fluxes from 1 to 2 kg/m^2 or 3.7–7.4 kg/m^2 CO_2 were obtained from already dated sections of marsh peat bogs in the estuarine zone of rivers. An analytical review of the problem showed that the maximum rates of vertical growth of lowland (marsh) peatlands are observed in the first 20–30 years (Temmerman et al. 2004). Figure 1 shows a layer of turf brought to the surface of a marsh from a neighbouring site. Such "patches" are very common on the surface of muddy drylands and marches.

A large amount of sediment is retained by marsh vegetation during marine floods (Fig. 2). The unexplained fact is that the organic carbon content in these sediments is 2–3 times higher than in the surface sediments of adjacent muddy drylands.

Soils of salt marshes located in the coastal zone of Salmon Bay in Aniva Bay, Sakhalin Island (46.74° N; 142.73° E) were investigated. Five soil profiles were excavated 500 m from the mouth of the Susuya River along a transect running from the shoreline towards the forest area, from which samples were collected by horizon for analysis of organic matter content and composition.

Fig. 1 Turf layer deposited during the surge on the march surface

Fig. 2 Bottom sediment trapped by marshes vegetation during autumn storms

Soil *1*—marsh algal-detrital soil, located 1–2 m from the shore cliff in the zone of low marshes formation under sedge-sitnik plant community; soil *2*—marsh detrital soil, located at a distance of 70 m from the shore under cereal-sitnik-sedge meadow and also belongs to the ecosystems of low marshes; Soil *3*—marsh detrital peaty soil, formed 230 m from the shore under sedge-grass meadow in the zone of middle marshes development; soil *4*—marsh peaty soil under reed-grass meadow, middle marshes zone, 300 m from the shore; soil *5*—peaty soil, located 550 m from the bay shore on the edge of mixed forest (birch, pine, willow) with meadow-forest ground layer.

The lower layers of all studied soils were formed during the period of accumulation of marine sediments in the tidal zone. Compared to the overlying horizons, they have a lighter granulometric composition, contain a large number of organic inclusions represented by wood remains, as well as inclusions of pebbles and shells. This mineral mass with allochthonous organic material can be considered as soil-forming rock, on which peat bogs of different types are then formed due to the growth of dead plant mass. Marsh soils (*1–4*) are in the stage of synlithogenic soil formation, the formation of upper horizons occurs under conditions of regular precipitation input from the sea; peat soil *5* is located outside the tidal zone, its upper horizons are the result of postlithogenic processes.

Along the transect, the type of organic matter accumulation changes and the thickness of the upper organogenic horizons increases. For soil *1*, carbon accumulation in the upper thickness occurs due to conservation of plant residues, mainly roots, and in the uppermost horizon—as a result of algal mass die-off (Sorg% = 3–4%; C/N = 10). In the upper part of soil *2* profile formation of low thickness weakly corrugated horizons with high content of mineral mass is already noticeable (Sorg = 5–10%; C/N = 11). In the upper part of profiles of soils *3* and *4* there was well pronounced accumulation of reed peat with different degree of decomposition and content of mineral part (Sorg = 24–42%; C/N = 17). The upper horizons of soil *5* are layers of top peat formed as a result of forest phytocenosis development (Sorg = 35–45%; C/N = 19).

For the studied soils of marshes and peat soil of the forest ecosystem the average Carbon accumulation rate—CAR was calculated (Table 1). Determination of the carbon accumulation rate—SAR in the upper stratum was carried out on the reference layer with tsunamigenic deposits, the appearance of which was caused by earthquakes 15.11.2006 (with magnitude $M_w = 8.3$) and 13.01.2007 (with magnitude $M_w = 8.1$) in the area of the Central Kurils.

Table 1 Carbon accumulation rate—CAR

Soil	1	2	3	4	5
Height (EGM08), m	0.768	0.759	1.091	1.371	3.056
SAR, cm·year^{-1}	1.9	1.3	2.8	2.4	3.8
CAR, rC·m^{-2}· year^{-1}	576	449	929	919	2143

3 Conclusion

The obtained rates of carbon accumulation in Aniva Bay marshes significantly (2–3 times) exceed most of the values of this value obtained for other coastal areas, which may be due to local climatic and geomorphological peculiarities of the research object. Given the size of the areas occupied by marsh soils, it is possible to make calculations of the total carbon sink in coastal landscapes on Sakhalin Island.

The work was carried out within the framework of the State Task FEFF-2024-0004 “Creation of scientific bases for management of carbon uptake and accumulation processes in biomorpholithosystems of coastal-marine wetlands”.

References

Afanas’ev VV, Faustova AB (2022) The first results of the study of sequestration properties of coastal marine biomorpholithosystems (Sakhalin Region) Physica land mathematical modeling of earth and environment processes. Springer Proceedings in Earth and Environmental Sciences. https://doi.org/10.1007/978-3-031-25962-3_46

Lovelock CE, Adame MF, Bradley J et al (2022) An Australian blue carbon method to estimate climate change mitigation benefits of coastal wetland restoration. Restor Eco: e13739. https://doi.org/10.1111/rec.13739

Najjar RG, Herrmann M, Alexander R, Boyer EW, Burdige DJ, Butman D, Zimmerman RC (2018) Carbon budget of tidal wetlands, estuaries, and shelf waters of Eastern North America. Glob Biogeochem Cycles 32(3):389–416

Ouyang X, Lee SY (2014) Updated estimates of carbon accumulation rates in coastal marsh sediments. Biogeosciences 11(18):5057–5071

Sapkota Y, White JR (2020) Carbon offset market methodologies applicable for coastal wetland restoration and conservation in the United States: a review. Sci Total Environ 701(C):134497. https://doi.org/10.1016/j.scitotenv.2019.134497

Spivak AC, Sanderman J, Bowen JL, Canuel EA, Hopkinson CS (2019) Global-change controls on soil-carbon accumulation and loss in coastal vegetated ecosystems. Nat Geosci 12(9):685–692

Temmerman S, Govers G, Wartel S, Metre P (2004) Modelling estuarine variations in tidal marsh sedimentation: response to changing sea level and suspended sediment concentrations. Mar Geology 212(1–4):1–19

Villa JA, Bernal B (2018) Carbon sequestration in wetlands, from science to practice: an overview of the biogeochemical process, measurement methods, and policy framework. Ecol Eng 114:115–128

Structure and Productivity of Vegetation of Coastal Marshes in the Southern Part of Sakhalin Island (Salmon Bay Carboniferous Polygon)

I. O. Rozhkova-Timina, O. A. Korablev, and M. V. Pavlov

Abstract The chapter considers the productivity of plant communities of coastal marshes of the Sea of Okhotsk (Salmon Bay, upper Aniva Bay). The productivity of aboveground phytomass, aboveground mortmass (including litter and rags) and belowground phytomass was determined and this parameter was assessed in relation to the distance of the phytocoenosis from the sea. The study found no relationship between plant community productivity and its location relative to the sea line; nor between phytocoenosis productivity and its composition. A high unevenness of the herbage was noted.

Keywords Coastal marshes · Halophytic vegetation · Productivity · Sakhalin

1 Introduction

It is known that so-called marshes—salty and carbon-rich habitats inhabited by specific phytocenoses—are formed in estuaries and on muddy coastal drying areas of tidal seashores (Campbell et al. 2022). These phytocoenoses (also called marsh grasslands or coastal marsh vegetation) are regularly exposed to seawater through the tidal activity of the sea and storms (Zhang et al. 2022), as well as through groundwater (Sergienko et al. 2020).

In Russia, the vegetation of marsh meadows has been studied in sufficient detail on the coasts of the Arctic seas: the White Sea (Sergienko et al. 2020; Moseyev et al. 2021a), the Barents Sea (Lavrinenko 2020; Moseyev et al. 2021b; Lapina and Ivanova 2021), Bering Sea (Neshataeva 2014); there are data on marsh meadows of the Kara, East Siberian, Chukchi and Laptev Seas (Sergienko and Moseev 2015). Previously, we gave a classification of herbaceous vegetation of coastal marshes of the Sea of Okhotsk (Rozhkova-Timina 2023). At the same time, the literature is woefully

I. O. Rozhkova-Timina (✉) · O. A. Korablev · M. V. Pavlov
Sakhalin State University, Yuzhno-Sakhalinsk, Russia
e-mail: inna.timina@mail.ru

T. Chaplina (ed.), *Processes in GeoMedia—Volume VIII*, Springer Geology,
https://doi.org/10.1007/978-981-97-6627-7_5

short of data on their biological productivity, which is the most important functional characteristic of ecosystems and is used to reflect the stability of phytocoenoses.

In recent years, grasslands have been studied to assess the influence of climatic and hydrological factors on biodiversity (Zhang et al. 2021; Hossain et al. 2021) and net primary productivity of grasslands in connection with the study with the global carbon cycle (Zhang et al. 2019; Quan et al. 2020). In this case, the sea can be considered as a crucial regulator of the transport of dissolved carbon and related elements from land to ocean.

2 Materials and Methods

Surveys were conducted in July 2023 in the south of Sakhalin Island (coast of Salmon Bay at the head of Aniva Bay, vicinity of Solovyevka village). Following remote sensing of the area using large-scale satellite imagery, areas were selected to establish trial plots for long-term surveys.

On sample plots during the phase of maximum grass development, relevés and selection of cuttings were carried out to determine the value of above-ground and underground phytomass. Descriptions were carried out on an area of 100 m^2 and the total projective cover of the herbage, its height, floristic composition and abundance of plant species were recorded. Cuts were taken from 0.25 m^2 squares in fivefold repetition and then recalculated in g/m^2. The dead part of the aboveground phytomass (aboveground mortmass including litter and rags) and roots from a depth of 0–20 cm (without differentiation of live and dead roots) were counted separately (Biological Productivity…, 1988). The cuttings were dried to air-dry condition and weighed. Subsequently, the productivity of phytocenosis was determined in terms of cwt/ha.

Sampling was carried out at 14 points of marsh meadows at different distances from the sea line in the following community types (we used the classification we published earlier (Rozhkova-Timina 2023):

- *Triglochin—Carex* meadows. Occupies small areas on clay-silt drying marshes in close proximity to the sea and is subject to daily flooding by tidal waters.
- *Carex—Triglochin* meadows. Located slightly further (approximately 800 m) from the sea than the previous meadow type. As a result, they have a lower sedge content and a higher triostrene content. However, these communities are also developed on silty-clay sediments and are affected by daily tides.
- *Arctopoa* meadows. Developed on sandy substrates influenced by sisigian tides and storms.
- *Leymus* meadows. Developed on sand; some authors classify such communities as dune vegetation (Mucina et al. 2016).
- *Leymus—Stellaria* meadows. Closely related to intra-island psammophyte meadows and influenced by sisigian tides.
- *Phragmites—Rubus* meadows. Located in the zone of sisygian tides, although the species of these communities are not halophytes.

3 Results and Discussion.

The communities of this territory are small-species, the number of species in one description varies from 2 to 9. Characteristics of the phytocenoses under consideration are given in Table 1.

Table 1 Structural indices of the studied marsh grasslands

№.	Meadow	Grass height, cm	Total projective coverage (TPC), %	Dominant species	
				Composition	%
1	*Triglochin—Carex*	20	90	*Carex subspathacea*	80
				Triglochin maritimum	20
2	*Triglochin—Carex*	30	60	*Carex subspathacea*	90
				Triglochin maritimum	15
3	*Carex—Triglochin*	10	50	*Triglochin maritimum*	50
				Carex subspathacea	20
4	*Arctopoa*	120	50	*Arctopoa eminens*	80
				Leymus mollis	10
				Schoenoplectus tabernaemontani	10
				Potentilla anserina	10
5	s *Arctopoa*	150	95	*Arctopoa eminens*	30
				Potentilla anserina	30
				Stellaria longifolia	10
6	*Arctopoa*	н.д	н.д	*Arctopoa eminens*	н. д
				Potentilla anserina	
7	*Leymus*	120	60	*Leymus mollis*	50
				Carex cryptocarpa	30
				Rubia jesoensis	20
8	*Leymus*	150	100	*Phragmites australis*	40
				Leymus mollis	40
				Potentilla anserina	40
				Stellaria longifolia	15
9	*Arctopoa*	n.d.	n.d.	*Leymus mollis*	n. d
				Stellaria longifolia	

(continued)

Table 1 (continued)

№.	Meadow	Grass height, cm	Total projective coverage (TPC), %	Dominant species	
				Composition	%
10	*Phragmites—Rubus*	150	80	*Rubus arcticus*	50
				Phragmites australis	40
				Calamagrostis epigeios	20
				Sanguisorba stipulata	20
				Lathyrus pilosus	20
11	*Phragmites—Rubus*	160	100	*Thelypteris thelypteroides*	60
				Sanguisorba stipulata	50
				Phragmites australis	50
				Maianthemum bifolium	50
				Rubus arcticus	50
12	*Phragmites—Rubus*	n.d.	n.d.	*Phragmites australis*	n.d.
				Rubus arcticus	
13	*Leymus—Stellaria*	80	90	*Leymus mollis*	80
				Stellaria longifolia	60
				Artemisia littoricola	30
				Triglochin maritimum	20
14	*Leymus—Stellaria*	50	90	*Stellaria longifolia*	70
				Leymus mollis	50
				Lobelia sessilifolia	30
				Triglochin maritimum	10

The results of the study of marsh meadow phytocenoses yielded very diverse productivity data. Figure 1 presents a graph of productivity of aboveground (green) phytomass and shows the change in its productivity at each point. It is obvious from the figure that there is no relationship between phytocoenosis productivity and its composition. In addition, the high value of the standard deviation demonstrates the heterogeneity of grass density.

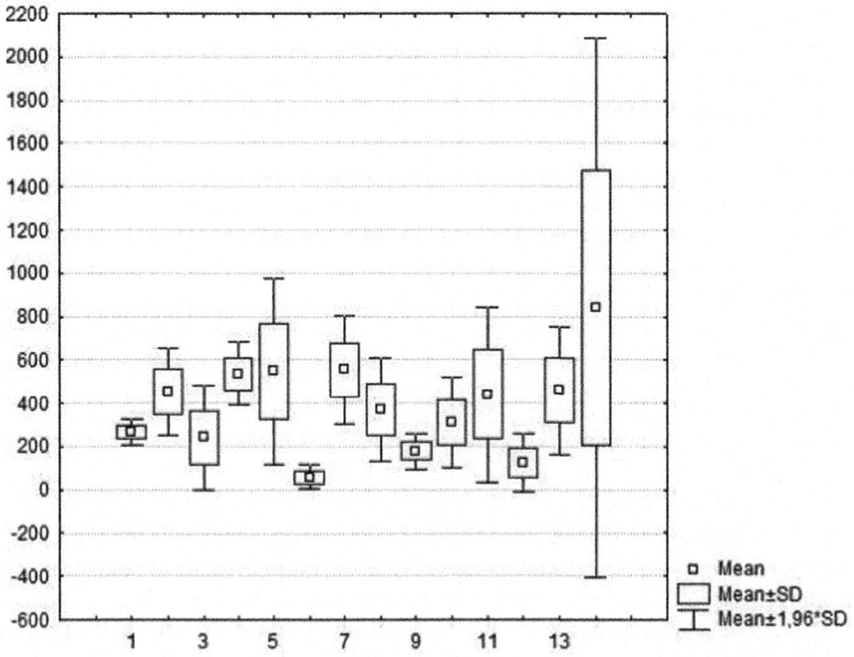

Fig. 1 Productivity of aboveground phytomass of plant communities of coastal marshes, g/m^2

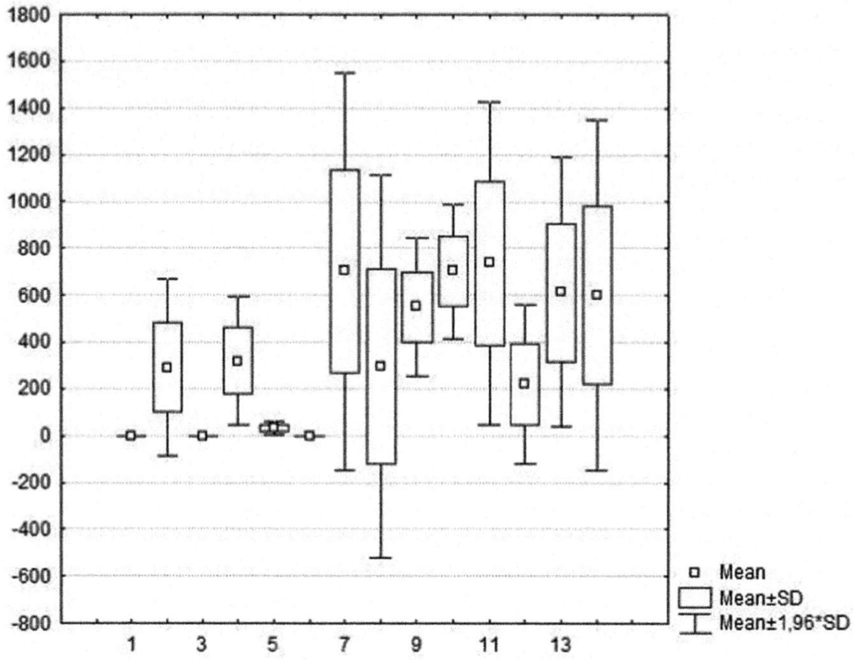

Fig. 2 Aboveground mortmass of plant communities of coastal marshes, g/m^2

Above-ground mortmass values also showed no dependence on phytocoenosis composition and location relative to the sea (Fig. 2). In some locations above ground morphic mass is absent or very low, which is probably due to regular tidal activity.

Consistent with the above, the magnitude of belowground phytomass also showed no dependence on the distance of the phytocoenosis from the sea (Fig. 3).

As can be seen from the figures, the productivity of even the same types of meadow communities can differ in different parts of the study area. The large value of the standard deviation at many points indicates uneven grass density. In addition, there is no tendency for grassland productivity to change with distance from the sea line.

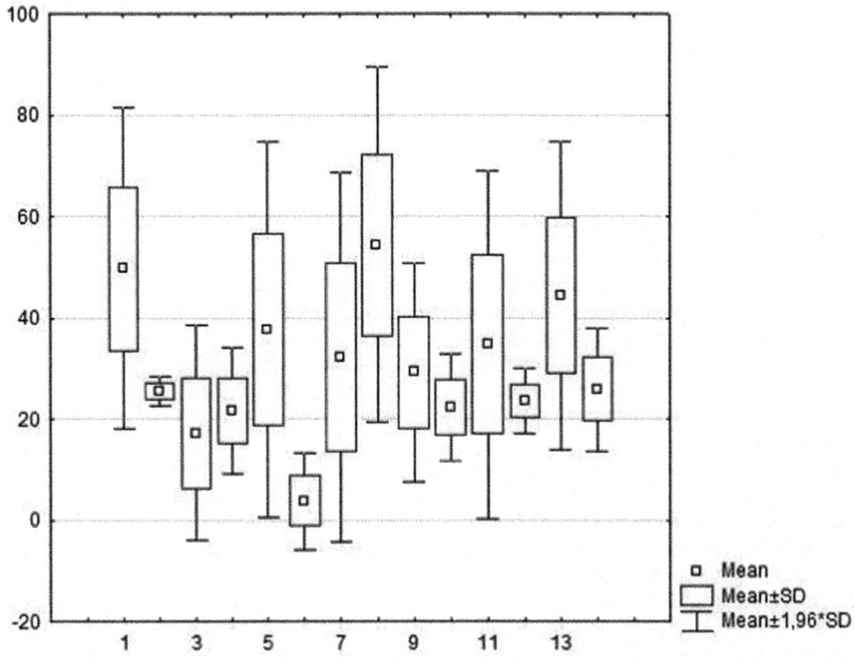

Fig. 3 Magnitude of underground phytomass of plant communities of coastal marshes, g/m^2

4 Conclusion

The productivity of marsh grasslands in the southern part of Sakhalin Island was analysed. On average, the productivity of aboveground phytomass was 100–800 g/m^2, aboveground mortmass—from 0 to 800 g/m^2, root mass—5–60 g/m^2. There were no trends in the productivity of phytocoenoses with distance from the seashore. No relationship between productivity and meadow community structure was also revealed.

Acknowledgements The work was carried out within the framework of the State Task FEFF-2024-0004 "Creation of scientific bases for management of carbon uptake and storage processes in biomorpholithosystems of coastal-marine wetlands".

References

Campbell A, Fatoyinbo L, Goldberg L, Lagomasino D (2022) Global hotspots of salt marsh change and carbon emissions. Nature 612(7941):1–6. https://doi.org/10.1038/s41586-022-05355-z

Hossain L, Kabir H, Nila US, Rubaiyat A (2021) Response of grassland net primary productivity to dry and wet climatic events in four grassland types in Inner Mongolia. Plant-Environ Interact 2(5):250–262. https://doi.org/10.1002/pei3.10064

Lapina AMM, Ivanova KV (2021) Syntaxonomic composition and spatial structure of territorial units of vegetation of seaside marshes of the Malozemelskaya tundra. Phytodiversity Eastern Euro 15(3):52–66. https://doi.org/10.24412/2072-8816-2021-15-3-52-66

Lavrinenko IA (2020) Typology and syntaxonomic composition of territorial vegetation units! A new approach on the example of studying arctic marshes. Vegetation Russ 39:100–148. https://doi.org/10.31111/vegrus/2020.39.100

Moseev DS, Sergienko LA, Leshchev AV, Bragin AVV, Romanov RE, Churakova EY (2021) Rare plant communities of lakes and river estuaries of the Arkhangelsk region. Turczaninowia 24(3):138–162. https://doi.org/10.14258/turczaninowia.24.3.11

Moseev DS, Sergienko LA, Parinova TA, Volkov AG (2021) Halophytic vegetation of the southern coast of Pechora Bay. Bot J 106(11):1050–1065. https://doi.org/10.31857/S0006813621110053

Mucina L, Bültmann H et al (2016) Vegetation of Europe: hierarchical floristic classification system of vascular plant, bryophyte, lichen, and algal communities. Appl Veg Sci 19:3–264. https://doi.org/10.1111/avsc.12257

Neshataeva VY, Neshataev VY, Korablev AP, Kuzmina EY (2014) Vegetation of seaside marshes of the coast of Korfa Bay (Olyutorsky district of Kamchatka Krai). Bot J 99(8):868–894

Quan Q, Zhang F, Meng C, Ma F, Zhou Q, Sun F, Niu S (2020) Shifting biomass allocation determines community water use efficiency under climate warming. Environ Res Lett 15. https://doi.org/10.1088/1748-9326/aba472

Rozhkova-Timina IO (2023) Vegetation of coastal-marine wetland marshes (Carboniferous polygon in Salmon Bay, South Sakhalin). Processes Geomedia 4(38):2215–2220

Sergienko LA, Moseev DS (2015) Taxonomic structure and ecological and geographical characteristics of the florocenotic complex of the coasts of the Russian Arctic. Sci Notes Petrozavodsk State Univ 6(151):28–33

Sergienko LA, Dyachkova TY, Androsova VI (2020) Floristic richness and saturation of seaside plant communities dominated by *Triglochin maritima* L. (Juncaginaceae) on the coast of the northern seas. Nat Assoc Sci 53(1):17–20

Titlyanova AA, Bazilevich NI, Snytko VA et al (1988) Biological productivity of grass ecosystems. In: Geographical regularities and ecological features. Nauka, Novosibirsk, 134 p

Zhang B, Cadotte M, Chen S, Tan X, You C, Ren T, Chen M, Wang S, Li W, Chu C, Jiang L, Bai Y, Huang J, Han X-G (2019) Plants alter their vertical root distribution rather than biomass allocation in response to changing precipitation. Ecology 100. https://doi.org/10.1002/ecy.2828

Zhang L, Gao J, Tang Z, Jiao K (2021) Quantifying the ecosystem vulnerability to drought based on data integration and processes coupling. Agric For Meteorol 301–302. https://doi.org/10.1016/j.agrformet.2021.1083

Zhang W, Ge Z, Li S, Tan L, Zhou K, Li Y, Xie L, Dai Z (2022) The role of seasonal vegetation properties in determining the wave attenuation capacity of coastal marshes: Implications for building natural defenses. Ecol Eng 175:106494. https://doi.org/10.1016/j.ecoleng.2021.106494

Geospatial Position of Elements of Coastal Marine Wetland Geosystems (Carbon Landfill, Sakhalin Island)

Victor V. Afanas'ev, A. V. Uba, A. B. Faustova, and O. A. Korablev

Abstract A digital morphometric model of a coastal marine wetland polygon within the cryptolagoon of Salmon Bay, Aniva Bay is presented. The elevation model is derived from a combination of laser airborne marshland survey and field GNSS RTK survey of the drying area. The distribution of plant communities and their productivity by hypsometric levels in a system of periodic sea level changes and surges is considered. Possibilities of application of the results in the technological scheme of carbon sequestration by managed coastal wetlands of the Sakhalin region are shown. The possibilities of applying the results in a technological scheme for carbon sequestration by managed coastal wetlands of the Sakhalin are shown.

Keywords Estuary · Salt marsh · Intertidal mudflat · Coastal sedimentation · Biomorpholithosystem · Pioneer vegetation · Bioproductivity

1 Introduction

Geospatial parameters of marshes and mudflats formed on the coast of Sakhalin Island were determined as a result of the research conducted in 2021–2022 in the territory of Sakhalin Oblast. The basis was the analysis of aerial photo arrays of 1952, 1971, 1977 and space images of 2014–2020, which was performed in the Quantum GIS geoinformation system. The calculations were performed on the ellipse WGS84 EPSG:7030. At the next stage of the study in the same QuantumGIS environment using standard procedures for analysing remotely sensed materials, we reconstructed changes in marshes areas for the period 1952–2019 of the Piltun, Chaivo, Nyivo, Nabil, Lunskoye, Nevskoye, Ainskoye lagoons and Salmon Bay of Aniva Bay (Afanas'ev et al. 2023).

V. V. Afanas'ev (✉) · A. V. Uba · A. B. Faustova · O. A. Korablev
Sakhalin State University, Yuzhno-Sakhalinsk, Russia
e-mail: vvasand@mail.ru

V. V. Afanas'ev · A. V. Uba
Institute of Marine Geology and Geophysics, FEB RAS, Yuzhno-Sakhalinsk, Russia

T. Chaplina (ed.), *Processes in GeoMedia—Volume VIII*, Springer Geology,
https://doi.org/10.1007/978-981-97-6627-7_6

The age of sediments was determined by reconstructing the change in marshes areas over the period 1952–2019, by dating methods relative to known natural events and by industrial period artefacts in sediment sections. The results allowed us to characterise the mean annual marshes accretion rates and, consequently, the mean annual carbon increment in marshes soils for more than 75% of the total area of lagoons on Sakhalin Island. Sakhalin Island. Thus, for the first time, quantitative parameters of organogenic sedimentation of coastal-marine biomorpholithosystems over a period of 68 years have been estimated based on direct field observations and analysis of remote sensing data (Afanas'ev and Faustova 2023a, b; Faustova and Afanas'ev 2023).

Measurement of carbon stocks in coastal-marine wetland ecosystems (CMWE) covered mainly vegetation carbon pools (including above- and below-ground parts), soil carbon pools and muddy drainage sediments (Afanas'ev 2023).

Programmed formation of coastal blue carbon ecosystems that can be reliably used for climate change mitigation is not possible without an understanding of coastal sedimentation conditions and morpholithodynamically based scenarios for their development.

Geomorphological approaches have already shown their effectiveness in analysing the balance of greenhouse gas producers and sinks, especially in relation to the dynamics of marshes and mudflats (Hu et al. 2015; Curado et al. 2014; Rosentreter et al. 2021).

Natural processes and phenomena, anthropogenic morpholithogenesis in all their diversity and interaction are implemented in the coastal zone mainly through changes in the geomorphological position of biomorpholithosystems and coastal morpholithodynamic processes (Hydro-bio-geomorphological systems: 2024; Afanas'ev and Faustova 2023a, b).

Determination of the geospatial position and sequestration capacity of coastal biomorpholithosystems completes the studies of the first stage. A general conservative estimate of carbon storage by lagoonal biomorpholithosystems alone in terms of CO_2 equivalent is at least 500,000 tonnes/year. Given the sequestration properties of the marshes and drylands of north-west Sakhalin, this figure is likely to increase substantially.

Retrospective assessments of changes in the main sequestration characteristics of biomorpholithosystems along the coast of Sakhalin Island, in accordance with morpholithodynamic laws, have allowed us to formulate a concept of the sequestration characteristics of biomorpholithosystems. Retrospective assessments of changes in the main sequestration characteristics of the biomorpholithosystems of the Sakhalin coastline in accordance with morpholithodynamic laws have allowed us to formulate a concept for the management of carbon uptake and storage processes in the ASBSU area using nature-based technologies, which includes the following areas:

- Sediment management to optimise sedimentation and create new areas;
- Sediment management for overtopping and burial of march vegetation
- Soil formation in the PMWBU area (formation of peat-gley soil);

- Planting and maintenance of march vegetation;
- Supply of saline water to areas with insufficient salinity to prevent microbial transformation of dead marsh vegetation.
- Harvesting of march vegetation biomass;
- Storage of organic carbon in the form of grown and introduced biomass, collection and disposal/disposal of storm emissions in environments that prevent organic decomposition and methane release;
- Protection from degradation and restoration of coastal ecosystems, including secondary watering with controlled salinity of coastal ecosystems (waterlogging);
- Protection from degradation and restoration of drying marshes and seagrass meadows.

2 Methods and Results

It is planned to create a technological offer that meets market demand in the conditions of a full-scale pilot industrial site, the next stage of development of which will be the creation of carbon farms for carbon sequestration in managed coastal wetlands of the lagoons of the Sakhalin region.

Figure 1 shows the location of a site with the already established fact of the highest sequestration capacity of marshes and drylands in the northern part of Aniva Bay, where work is planned to develop a scientific basis and to test nature-like technological solutions. This area is in fact a crypto-lagoon in terms of morpholithodynamic position, as during maximum low tides the barrier form separating this part of Salmon Bay from the main water area is exposed (Fig. 1).

Figure 2 shows a digital high-precision surface model of the polygon where geobotanical, geomorphological and soil surveys were carried out in 2023. The data were processed in QGIS 3.34 and Python 3.12 (Gdal, numpy). The elevation model was derived from a combination of laser airborne marshland survey and field GNSS RTK drying survey. Dense cloud processing of the laser survey was done in Agisoft Metashape 2.1.0. Height correction was performed relative to the base station, which height above the WGS84 ellipsoid was determined by PPP method and converted to the EGM 2008 geoid level. The combined height model was converted into a raster with pixel size 1×1 m in WGS84 UTM Zone 54N coordinate system. In further calculations it is assumed that the pixel area of the model is equal to 1 m^2. When calculating histograms, the model heights below minus 1 m and above 4 m were not taken into account.

The most important result is the determination of the boundaries of plant community types of the territory, as well as the parameters of the potential pioneer zone, above which plants have a high probability of establishment ((Rozhkova-Timina 2023; Rozhkova-Timina et al. in press). A brief description of plant communities in the height intervals of the morphometric model is given below. It should be noted that the minimum observed level is $-$ 0.25 m and the maximum + 1.57 m.

Fig. 1 Location, morphometry and landscape indication elements of the terrestrial Carboniferous polygon within the margins of marshes and mudflats in the northern part of Salmon Bay. [Modified Copernicus Sentinel data 2021/Sentinel Hub (09.09.2021)]

In the height interval from 0.2 to 0.3 m surfaces with sparse sedge vegetation are observed, these are the areas of pioneer vegetation.

Triostrene-sedge vegetation occupies the surfaces of overgrown muddy drainage in the height range of 0.3–0.4 m (Fig. 3). The sedge-triostrene meadows are located in the 0.4–0.55 m elevation interval (Fig. 4). They have lower content of sedge and higher content of triostrene. However, these communities are also developed on silty-clay sediments and are affected by daily tides. In the sisigian tidal zone, arctomyatlice and woolly-leaved meadows are common on sandy substrates. Woolly starflower meadows are close to intra-island psammophyte meadows and are influenced by sisygian tides and surges. Species of reed-brood reed meadow communities do not belong to halophytes, but in case of superposition of sisygian tides and surges of different types their growing areas may be flooded by sea water.

High bioproductivity of plant communities at the polygon should be noted (Figs. 4, 5 and 6).

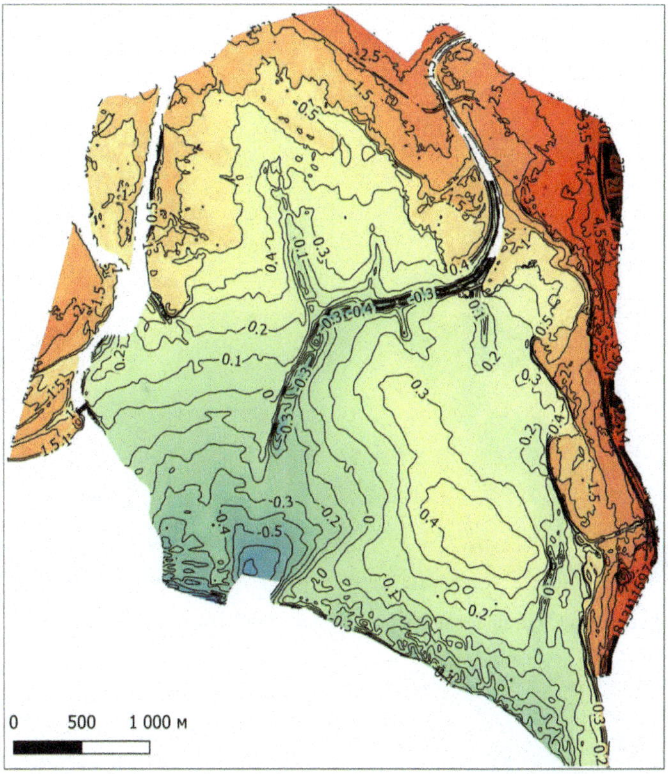

Fig. 2 Isohypsum visualisation of a digital high-precision polygon surface model

No less important result of the performed research is the determination of areas occupied by this or that plant community with the prospect of developing technological schemes of anthropogenic increase of sequestration properties of coastal marine wetland geosystems (Fig. 7).

So the area of muddy drainage in the height range of 0.2–0.3 m with sparse sedge vegetation is about 2 km^2, in the height range of 0.3–0.4 m about 1.5 km^2, etc. i.e. without radical geotechnical manipulations on the area of 3.5 km^2 it is possible to start climatic activity on increase of bioproductivity and formation of marshes. On an area of about 7 km it is possible to carry out sediment management activities to optimise sedimentation and create new areas.

Fig. 3 Surface of muddy drainage in the height interval 0.2–0.3 m survey date 05.06.2024

Fig. 4 Image of low march surface from 0.5 m isohypsis and surface of overgrown silt drying in the height interval 0.3–0.4 m, date of survey 05.06.20244

Fig. 5 Image of the surface of the muddy drainage and low march shown in Fig. 4 from the opposite side, date of survey 05.06.2024

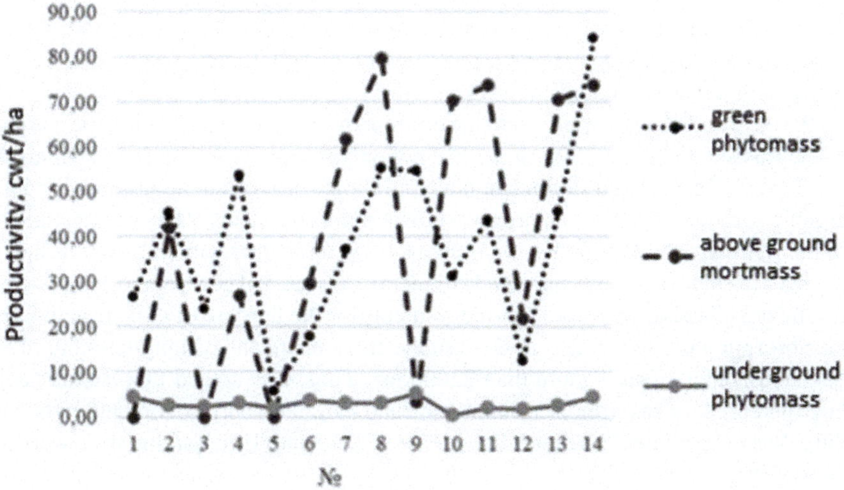

Fig. 6 Vegetation pool productivity in different communities of the carboniferous polygon

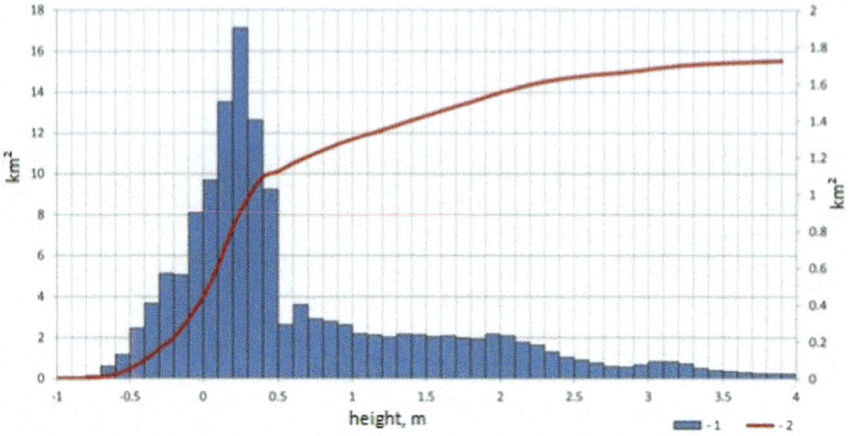

Fig. 7 Histogram of area distribution in 10 cm height intervals 2-cumulative curve

3 Conclusion

Thus, the development of the scientific basis for managing the processes of carbon absorption and accumulation on the territory of the coastal-marine wetlands (CWW) is reaching a new technological level.

We have added a new element to the concept of managed natural-climatic activity of the Russian Federation. We argue that not only mangrove forests are "saviours of the planet", Russia has a significant potential for absorption and long-term burial of organic carbon in the salty environment by march ecosystems. The coastline length of the Russian sector of the Pacific is about 18, and together with the Arctic zone of Russia's northern seas—58 thousand kilometres. The area of geomorpholithosystems for the organisation of coastal-marine wetlands for the purpose of climate action for the absorption of greenhouse gases from the atmosphere and its long-term burial on Sakhalin Island alone is more than 2200 km^2. Thus, it is argued that Russia has a huge potential for absorption, accumulation and long-term burial of organic carbon in the saline environment (prevents the release of methane) by coastal-marine wetland geosystems.

References

Afanas'ev VV (2023) Biogenic morpholitogenesis on sea coasts Sakhalin Island. In: Processes in GeoMedia, vol VII. Springer Nature Singapore, Singapore, pp 221–229

Afanas'ev VV, Faustova AB (2022) Sequestration of organic carbon in salt marsh formations of lagoons of Sakhalin. In: Physical and mathematical modeling of earth and environment processes—2022: proceedings of 8th international scientific conference-school for young scientists. Springer Nature Switzerland, Cham, pp 263–267

Afanas'ev VV, Faustova AB (2023) Carbon sequestration in managed salt wetlands (carbon polygon). In: Processes in GeoMedia, vol VII. Springer Nature Singapore, Singapore, pp 271–277

Afanas'ev VV, Uba AV, Levitsky AI, Korablev OA (2023) Rates of organic carbon accumulation in march formations of lagoons of Sakhalin. In: Chaplina T (eds) Processes in GeoMedia, vol VI. Springer Geology. Springer, Cham. https://doi.org/10.1007/978-3-031-16575-7_23

Anikina NV, Afanasyev VV, Baranova MS et al (2024) Hydro-bio-geomorphological systems: marshes, lakes, lagoons, deltas, and "amber" forest. Media-PRESS Moscow, 228 p. ISBN 978-5-901003-72-5

Curado G, Rubio-Casal AE, Figueroa E et al (2014) Plant zonation in restored, nonrestored, and preserved *Spartina maritima* salt marshes. J Coast Res 30(3):629–634. https://doi.org/10.2112/JCOASTRES-D-12-00089.1

Faustova AB, Afanas'ev VV (2023) Coastal marches and silt drainage of Sakhalin in the context of climate change. In: Conference on physical and mathematical

Hu Z, Van Belzen J, Van Der Wal D et al (2015) Windows of opportunity for salt marsh vegetation establishment on bare tidal flats: the importance of temporal and spatial variability in hydrodynamic forcing. J Geophys Res 120(7):1450–1469. https://doi.org/10.1002/2014JG002870

Rosentreter JA, Al-Haj AN, Fulweiler RW et al (2021) Methane and nitrous oxide emissions complicate coastal blue carbon assessments. Glob Biogeochem Cycles V35:e2020GB006858. https://doi.org/10.1029/2020GB006858

Rozhkova-Timina IO, Korablev OA, Pavlov MV. Structure and productivity of vegetation of coastal marshes of the southern part of Sakhalin Island (carboniferous polygon in Salmon Bay). Sakhalin (Carboniferous polygon in Salmon Bay) (in press)

Rozhkova-Timina IO (2023) Vegetation of coastal-marine wetland marshes (carboniferous polygon in Salmon Bay, South Sakhalin). Process GeoMedia 4(38):2215–2220

Application of Vector Graphics Tools to Increase the Efficiency of Experimental Data Processing

V. P. Pakhnenko

Abstract The paper describes several techniques for automated processing of experimental data. These techniques are developed for the case when the spreading of coloring impurities, solid markers and other objects on a free water surface is studied. And also for the case of penetration of dyeing admixture into the water depth. These methods significantly increase the efficiency of experimental work and subsequent processing of the obtained data sets.

Keywords Automation · Data arrays · Experimental data processing · Flows · Dyeing admixture · Ice · Vector graphics

1 Introduction

In modern science, the processing of experimental data is one of the most important and time-consuming processes. With the increasing volume of data and complexity of experiments, automation of data processing becomes more and more relevant. This paper considers several variants of using automation of experimental data processing with the help of software on the example of experiments with ice rotation, markers and various admixtures on the open water surface, as well as with the penetration of impurities into the current thickness.

The development of software and functionality of various packages, as well as hardware for visualization of the behavior of dynamic environments (capabilities of modern video cameras, laboratory instruments, etc.) open new opportunities in conducting experimental studies and their subsequent processing. Modern video cameras expand the possibilities of video recording of experiments, allowing to shoot more complex experiments from different angles, including directly in the water. This in turn allows to record what is happening in the experiment both on the free water surface and in the water column and even in a container in the water

V. P. Pakhnenko (✉)
A. Yu. Ishlinsky Institute for Problems in Mechanics, Prospect Vernadskogo 101-1, Moscow 119526, Russia
e-mail: pvpmos@gmail.com

T. Chaplina (ed.), *Processes in GeoMedia—Volume VIII*, Springer Geology,
https://doi.org/10.1007/978-981-97-6627-7_7

under the object of observation (for example, in experiments on the spread of oil products under ice). On the other hand, there is a large number of open software libraries and software packages that allow modeling not only the physical process occurring in experiments, but also provide the possibility of animation of various dynamic processes having vector representation.

This paper describes several techniques for processing and visualizing processes occurring on the surface and in the flow depth. Since initially all experiments are recorded on video cameras, the techniques are based on conversion of the original video files into vector representation. Transition from video recording to vector representation and subsequent animation provide the possibility of separate analysis of flow dynamics and processes occurring with ice, markers and admixtures on the surface and in the flow depth.

2 Experiments on the Open Water Surface

The task of time reduction by means of automation of the processing process is important and urgent. Its solution will improve the quality of the experimental work, as well as the efficiency of subsequent processing of the obtained data. One of the possible ways to solve the problem of automation is the use of vector graphics tools (W3C; Eisenberg 2014).

Let us consider the methodology of automation of experimental data processing on the example of a series of experiments, where solid and soluble markers were used as indicators simultaneously. The markers were placed on a free water surface, and the process of their movement was recorded on a video camera. Figure 1 schematically depicts the distribution of leading and lagging arms of uranyl flowing off the solid marker. In this scheme: R_m—radial coordinate of the marker position relative to the center of the free surface; R_{fi}—radial positions of the end of the leading arm; R_{bi}—radial positions of the end of the lagging arm, φ_m—angle of rotation of the marker relative to the center of the rotating free surface, φ_{fi} and φ_{bi}—angular positions of the ends of the leading and lagging painted arms.

Analysis of a series of experiments to record the movement of a painted marker on the surface of the flow requires processing of a large amount of data. Since the marker movement along the free water surface was recorded using a video camera, the initial data for processing were video files in mp4 format. The video camera shooting frequency was equal to 25 frames per second. For the most detailed processing this video file was sliced frame by frame for further processing of photos. Consequently, such a video file converted by frame-by-frame slicing is converted into a set of photos. It turns out that an experiment lasting only 1 min is converted into an array of 1500 images. Naturally, manual processing of such a large number of images is time-consuming. The volume of input data and the time spent on their processing significantly affect the efficiency of the conducted experimental work.

Each frame obtained from the original video recording of the experiment is a regular raster image. However, in vector representation the same frame will be a text

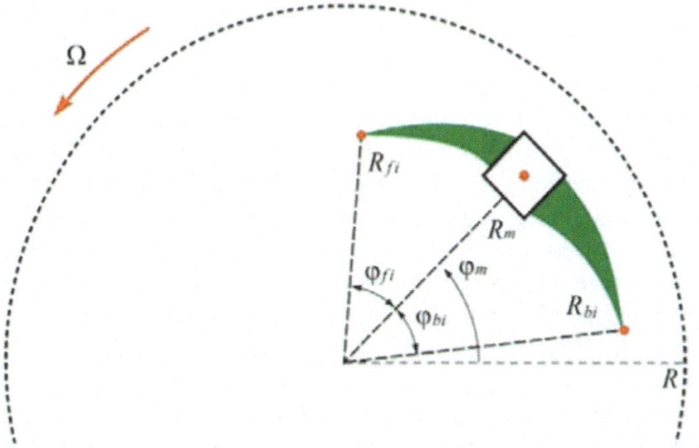

Fig. 1 Schematic of the distribution of the dyeing admixtures relative to the solid-state marker

file. Therefore, if the original image is converted into vector format, we will get a file containing a set of commands in text form, by which this image is displayed on the screen. Such conversion allows us to significantly reduce the amount of source information. In addition, the task of processing text files is much simpler than the task of processing files with two-dimensional raster images.

If instead of using the vector format, the processing was based solely on the original raster images, the estimation of marker movement parameters along the liquid surface required the application of pattern recognition theory (Tu and Gonzalez 1978). It can be said that modern vector graphics is the result of many years of applying pattern recognition theory to the solution of such problems in the creation of geoinformation systems, printing and mapping (Kanasevich 1985).

One of the open sources software tools for converting raster images to vector representation is Potrace software product (Tool to convert raster image to vector representation). Its transformation is based on the approximation algorithm. This algorithm is based on the construction of Bézier curves (W3C; Eisenberg 2014). After applying this product, the output vector files contain the parameter values of these curves calculated from the original raster images from the experiment. However, for the effective application of this package it is necessary that the original raster images have a minimum amount of "noise", while the object under study (marker, admixture or piece of ice) had a continuous clear contour, was contrasted to the water. Therefore, the original photos should be additionally graphically processed beforehand. For example, using filters of the Adobe Photoshop package or analogs, which are selected in such a way that the resulting shape of the studied object was as clear as possible. At the same time, the package also has the function of tracking the center of the object, but for its effective work, the filters should be selected in such a way that in some approximation the object can be approximated by a circle. Then with high enough accuracy the parameters of the obtained Bézier curve can be treated as points lying on the circle. That will allow, using the geometric properties of the circle, to obtain estimates of the position of the center of the object under study in the frame.

The above-described properties of the Potrace program determined the following methodology for automating the processing of experimental data. At the first stage, the video recording is programmatically split into separate frames by seconds and renamed for the convenience of subsequent processing. This program is written in the C# programming language (Richter 2012). It is implemented in such a way that it can process video recordings of any size, which significantly simplifies the work.

At the next stage it is necessary to reduce the amount of graphic "noise" so that only the object under study remains in the frame. For this purpose, a non-linear color filter is used. This filter was applied to a set of graphic files using batch processing (Fig. 2).

Next, using the Potrace program, the resulting raster images are converted to vector form. This leads to the fact that they produce text files in svg format with a set of commands by which these images can be drawn on the screen. These commands are spelled out in the path line (Fig. 3). Figure 3 shows a text file in svg format, which is obtained after conversion from the raster image in Fig. 2a.

Accordingly, after converting each frame from the experiment record in this way, we will get an array of text files, which will contain information about the object under study. The obtained set of svg files from the specified folder is automatically converted into one xml file. This file contains coordinates of marker centers at each moment of time, Fig. 4 shows a graphical representation of this xml file. The xml format provides the possibility of using the file in most common statistical data processing packages.

In addition to information about the displacement of the center of the object under study, which allows tracking its trajectory throughout the experiment, the program makes it possible to record the angle of rotation of the object relative to the center

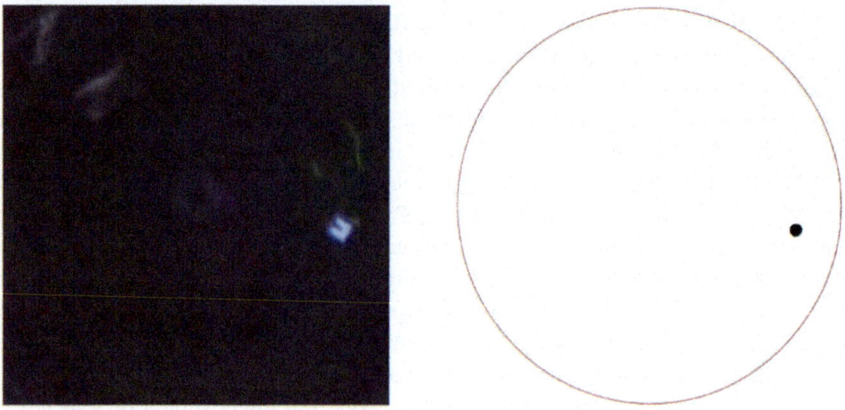

a *b*

Fig. 2 **a** A frame from the experiment without filters applied; **b** images from the experiment after application of filters

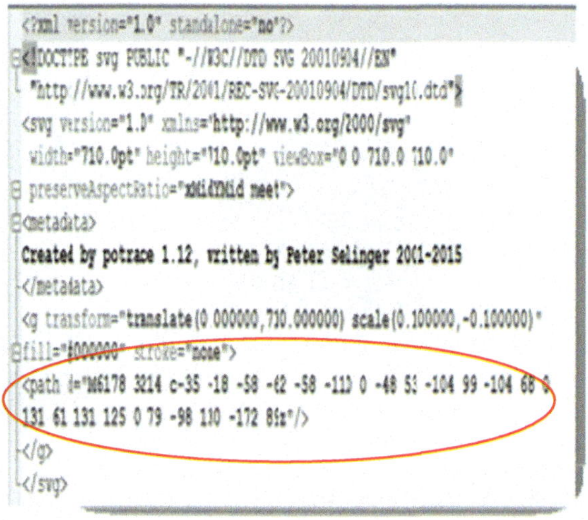

```
<?xml version="1.0" standalone="no"?>
<!DOCTYPE svg PUBLIC "-//W3C//DTD SVG 20010904//EN"
 "http://www.w3.org/TR/2001/REC-SVG-20010904/DTD/svg10.dtd">
<svg version="1.0" xmlns="http://www.w3.org/2000/svg"
 width="710.0pt" height="710.0pt" viewBox="0 0 710.0 710.0"
 preserveAspectRatio="xMidYMid meet">
<metadata>
Created by potrace 1.12, written by Peter Selinger 2001-2015
</metadata>
<g transform="translate(0.000000,710.000000) scale(0.100000,-0.100000)"
fill="#000000" stroke="none">
<path d="M6178 3214 c-35 -18 -58 -62 -58 -110 0 -48 53 -104 99 -104 68 0
131 61 131 125 0 79 -98 110 -172 86z"/>
</g>
</svg>
```

Fig. 3 Textual representation of the information from Fig. 2a

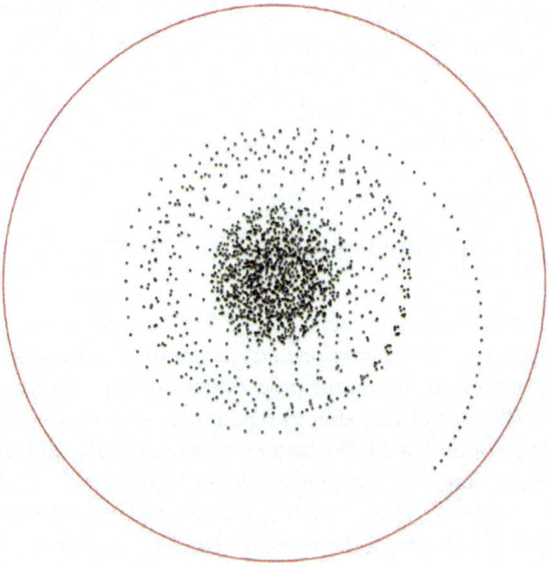

Fig. 4 Results of automatic processing in graphical representation

of its own axis of rotation. This is an important parameter in such experiments as in the study of ice rotation on the free surface of water (Fig. 5) or in the study of hydrocarbon spreading on the water surface (Kistovich and Chaplina 2021).

Accordingly, this technique allows to automate the process of processing experiments in which the processes of movement of markers, admixtures and various investigated objects on a free water surface are studied. That, in turn, allows to obtain the following important regularities: dependence of the change in the radius

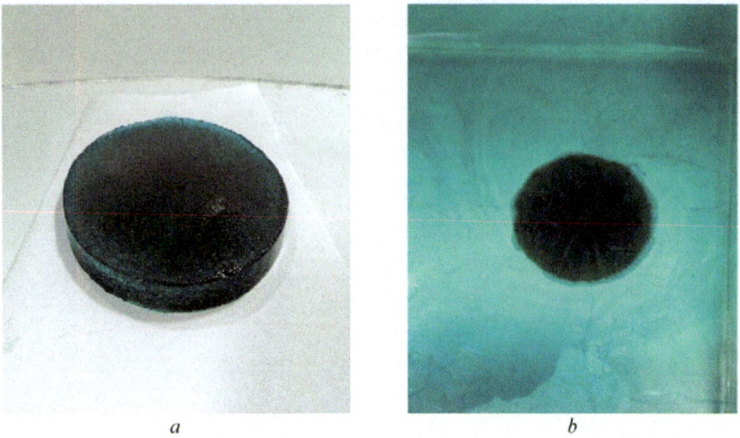

Fig. 5 **a** A sample of tinted ice; **b** images from the experiment with ice rotation on the water surface

of circulation of the investigated object around the center of the rotating free surface; dependence of the change in the frequency of object circulation on time; dependence of the change in the radius of object circulation relative to its own axis of rotation; construction and visualization of the entire trajectory of the object throughout the experiment.

3 Experiments in the Flow Depth

Along with experiments on the free surface of water, no less important are experiments on the penetration of dyeing admixtures into the flow depth. This gives a more complete picture of the processes occurring in water.

The basic methodology remains the same, the original video file of the experiment is converted first into a raster image to preserve all the details of what is happening in the experiment, and then converted into vector format for more efficient processing.

However, further analysis and processing is different. The dyed particles coming downstream from the investigated object are represented as a spatial vector. Visual analysis of such a front shows its high nonstationarity both in time and in the water column, which of course strongly depends on the parameters of the experiment. A characteristic feature of such nonstationarity is the presence of bends in the trajectory of the front of colored particles. This nonstationarity will depend on the parameters of the experiment, whether it is the study of penetration of underpainted particles into the surface of the vortex flow depth or the propagation of particles from the ice surface under the action of its own rotation. Therefore, it is necessary to switch to such a vector representation that will allow us to ensure full preservation of information about the structure of the bends and the trajectory along which the underpainted particles descend. This problem is solved by a spatial vector that consists of sequentially

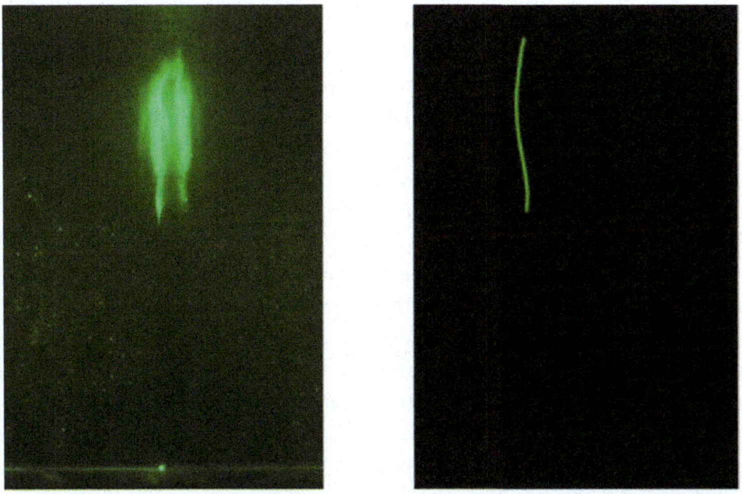

a *b*

Fig. 6 **a** Raster frame from the experiment; **b** vector representation of the left front of the colored column

connected segments. They, in turn, are formed by separate points. Each point of a segment contains information not only about the coordinates of its position in the plane, but also about the direction to the previous and subsequent points, as well as about the distance to them. Figure 6b shows an illustrative example of such a spatial vector from a series of experiments with a dyeing admixture.

The written program allows to manually correct the structure of the generated spatial vector. The program also provides the possibility of building one spatial vector in each raster frame from the experiment. Further, all text information from each frame is converted into a chain of formed vectors and saved as xml files.

This program can be used to generate an xml file containing information on the dynamics of the left and right vortex front behavior in the vortex flow depth. The obtained points of the vector representation of the vortex flow fronts have an arbitrary location on the plane, which is determined by the nature of the front trajectory bends in the experiment itself, and vary in number. Thus, their direct use for comparative analysis of data from different experiments is difficult.

This problem was solved by moving from a non-uniform representation of a spatial vector to its equivalent representation on a given uniform coordinate grid. In this approach, there is some closed contour and a rectangular coordinate sect. The intersection points between the coordinate sect and the given spatial vectors are analyzed. This transition allows us to fully preserve the properties of the spatial vectors, while obtaining points that are located on a uniform coordinate grid. This coordinate grid allows different spatial vectors to be compared to each other. An additional program was written to implement this transition. Further, it was found that the optimal step for a uniform coordinate grid is a step of 20 points (pixels).

Thus, the whole height of the investigated water depth in the experiments was divided into 51 evenly spaced levels. At each of them, the horizontal component of the front trajectory was estimated. This transition provided an opportunity to compare the results of different experiments. The data obtained as a result of this program are much easier to process statistically, since the values of the front trajectories on different frames lie on the same grid. And accordingly they can be studied relative to each other. An example of the program operation throughout the experiment in the presence of a uniform coordinate grid is shown in Fig. 7.

This program can also be used for experiments to study the rotation of ice on the free surface of water relative to the center of its own rotation axis. Figure 8 shows examples of dyeing admixture convergence in the process of ice rotation. By constructing spatial vectors on a uniform coordinate grid, it will be possible to obtain the trajectory and parameters of penetration of the dyeing admixture into the water column and compare the obtained data with each other.

Thus, in the method of automated processing for experiments on the penetration of colored particles into the flow depth, there are three stages of transformation of the source video to vector representation. At the first stage, the transition from video format to a set of raster images is realized. Then from raster images to a representation in the form of spatial vectors. And in the third step, the transition to spatial vectors represented on a uniform coordinate grid is realized. This transition to the uniform sect greatly simplifies the subsequent processing of experimental data and makes it possible to compare the obtained results with each other.

This methodology allows to automate the process of processing of experiments, in which the processes of dyeing admixtures movement into the water and flow depth are studied. That in its turn allows to obtain the following important regularities: dependence of change in penetration of the dye tip on time; dependence of change in

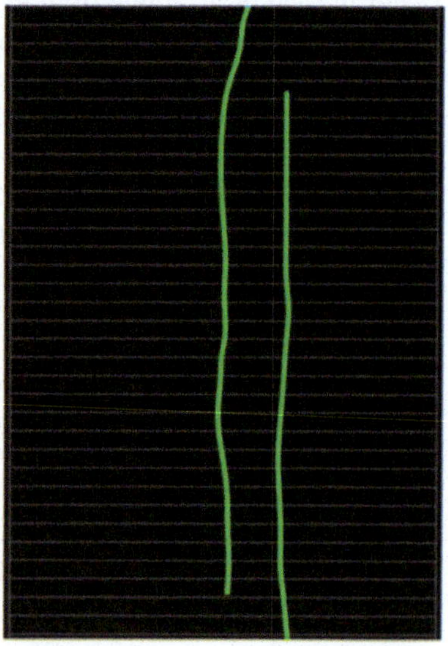

Fig. 7 Propagation of fronts of underpainting particles into the flow depth

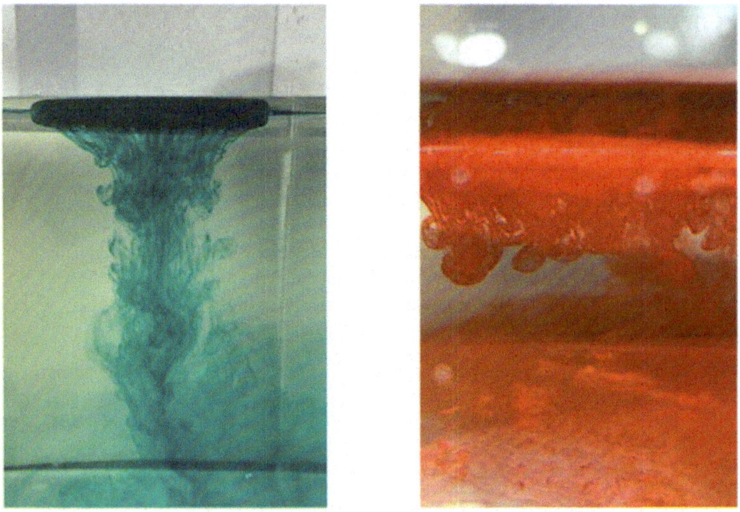

a *b*

Fig. 8 Raster frames from the experiments

the thickness of the colored column on the depth of dye immersion; dependence of deviation of the trajectory of the left and right fronts of the colored column on time; estimation of distributions of wavelengths of oscillations of the fronts of the colored column in the vertical plane.

4 Conclusion

The methods of automated processing of experiments to study the spreading of markers, dyeing admixtures and other objects on the free water surface and in the water depth were presented. These methods proved their operability and efficiency in a large number of laboratory experiments.

It is worth noting that vector representation of the results in turn allows to build an animated visualization of the entire experiment and the processes occurring in it. That provides the possibility of analyzing the dynamics of the events taking place. While a set of raster images can destroy information about the dynamics of the whole process, because there is a loss of dynamic connection between individual frames. Since movements of markers and other investigated objects on the surface and in the water depth are dynamic, it is difficult to study them statically. Therefore, the possibility of trajectory study in dynamics should be emphasized as a significant plus.

Acknowledgements This work was supported by the Russian Science Foundation, project No. 24-27-00029.

References

Eisenberg DJ (2014) SVG essentials, 2nd edn. O'Relly Media, 366 p

Kanasevich ER (1985) Analysis of time sequences in geophysics. Nedra, 300 p

Kistovich AV, Chaplina TO (2021) Analytical and experimental modeling of the hydrocarbon slick form and its spreading on the water surface. Phys Fluids 33

Richter J (2012) CLR via C#. Programming on Microsoft.Net framework 4.0 in C#, 3rd edn. SPB, Piter, p 896

Tool to convert raster image to vector representation. http://potrace.sourceforge.net

Tu J, Gonzalez R (1978) Principles of pattern recognition. Mir, 411 p

W3C. Стандарт Rex-SVG11-20110816 scalable vector graphics (SVG) 1.1. http://www.w3.org/TR/SVG

Non-ferrous and Rare Metals in Wastes from the Blavinskoye Copper and Ore Deposit in the Orenburg Region

G. A. Ponomareva and V. P. Petrishchev

Abstract The article presents data on the content of non-ferrous and rare metals in the wastes of the Blyavinskoye copper-coal deposit of the Orenburg region (Russia). The wastes of the mothballed Blyavinskoye deposit are represented by overburden dumps formed during the construction of the open pit. The depletion of mineral resources with ever-increasing volumes of consumption necessitates deeper processing of minerals extracted from the subsoil. The purpose of this work is to determine the level of content of the most important non-ferrous and rare metals and to identify paragenesis of typomorphic elements for the western dump and quarry lake of the Blyavinskoye deposit. To achieve this goal, samples of soil from the dump, silt and water from the quarry lake, as well as samples of vegetation of the dump were taken. Elemental analysis was performed by atomic-emission method with inductively coupled plasma. The analyses determined Cu, Fe, Al, Ca, K, Mg, Mn, Cr, V, Ni, Co, Ba, Sr, Na, Rb, Cs, Be, Tl, Cd, Ga, Se, As, B, S, P. Industrial Cu content in the soil (~ 1.3%) of the western dump was detected. The contents of S (3.8%), Cd (0.41%), Rb (0.53%), Cs (0.13%), Ga (0.02%), Tl - 0.05%, the sum of Pb and Zn (0.12%) are sufficient in case of their associated extraction at complex processing of the mentioned object. Paragenesis of elements Al, Ca, Fe, Cu, Mn, Mg, K, S is typical for the western dump, quarry lake and vegetation of the western dump of Blyavinskoye deposit of Orenburg region. The occurrence of elements, as a rule, is 100%.

Keywords Blyavinskoye copper-cement deposit · Western dump · Open pit lake · Non-ferrous and rare metals · Unconventional deposits

G. A. Ponomareva (✉) · V. P. Petrishchev
Orenburg State University, Orenburg, Russia
e-mail: galy.ponomareva@mail.ru

V. P. Petrishchev
Institute of Steppe of the Ural Branch of the Russian Academy of Sciences, Orenburg, Russia

T. Chaplina (ed.), *Processes in GeoMedia—Volume VIII*, Springer Geology,
https://doi.org/10.1007/978-981-97-6627-7_8

1 Introduction

The relevance of the project is related to the huge role that non-ferrous metals (copper, zinc, etc.) play in the national economy, being strategic types of mineral raw materials. Low availability of explored reserves for mining and processing enterprises, depletion of prospecting and exploration reserves of the past years, decrease in the share of preliminary estimated reserves, unreliability of development of a significant number of explored deposits when they are reevaluated according to the criteria of the market economy of the region, sharp reduction of state-financed geological exploration works, lack of long-term forecasts of consumption of mineral resources along with the already existing developed infrastructure, especially in the old, developed mining areas of the region. These facts led to the adoption of the State Programme of the Russian Federation (Order of the Government of the Russian Federation No. 91-r dated 30.01.2013), which provides for geological exploration of the objects of technogenic systems in order to put them on the balance sheet and further use.

In the Orenburg Region, all commercial scale copper deposits are of the cuprous type. Currently, 9 such deposits and more than 100 ore occurrences are known. Deposits differ by copper reserves. The Blyavinskoye deposit belongs to medium-sized deposits in terms of copper reserves, and it is mothballed. There are also other depleted cuprous ore deposits, along with sites that have been exploited so far. In the eastern part of the Orenburg region for a century of subsoil use accumulated more than 500 dumps of different composition and volume (Opekunova et al. 2020; Guman et al. 2018; Petrishchev 2016).

The total volume of mining waste in the Eastern Orenburg region is estimated at more than 300 million tonnes accumulated at more than 100 technogenic-mineral sites.

The study of surface layers of waste dumps, sub-dump streams and quarry water bodies allowed to identify points with rather high concentration of non-ferrous metal elements. Sub-dump creek at the edge of the Gayskoye deposit (pit No. 2) Zn—11,333 mg/kg (1.1%); Cu—17,360 mg/kg (1.7%). The slope of the dump near the eastern edge of the Gaiskoye deposit Zn—2666 mg/kg (0.26%); Cu—3910 mg/kg (0,39%). At the same time, the volume of dumps of the Gayskoye deposit alone is 150 million cubic metres, of the Blyavinskoye deposit—40 million cubic metres.

Sludges of Gaisky GOK contain many valuable elements. Enrichment by means of screw apparatuses gives concentrate with copper content—0.54% (minimum industrial content—0.4), sulphur—45% (minimum industrial content—35%), gold—2.6 g/t, silver—10.3 g/t, germanium—3.4 g/t, selenium—0.002%, tellurium—0.002% (Petrishchev 2016). The above facts allow us to consider the wastes of the Gaiskoye deposit as a new type of unconventional deposits, which contributes to the expansion of the mineral resource base of non-ferrous and rare metals of the region, the Ural region, and the Russian Federation as a whole.

The Blyavinsky mine is a unique facility. It is located in the Kuvandyk district. It is one of the oldest mines in the Orenburg region, its structure is well studied and has become a textbook, which in its time contributed to the formation of clearer ideas about the geology of effusives, volcanogenic-sedimentary formations, creating a foundation for the development of many of the Ural's sinter deposits in general and the Orenburg part of the Southern Urals in particular. Taking into account that the open pit mining ended in 1971 and, accordingly, the share of non-ferrous metals recovery in the dumps remained at the level of technologies of half a century ago, the waste of the Blyavinskoye deposit is the most promising in terms of studying the content of non-ferrous and rare metals.

A number of researchers raised these questions. Udachin V. N. et al. studied the migration of trace elements and their role in the functioning of natural ecosystems in connection with the growing technogenic impact, especially during the development of large deposits of non-ferrous metals in the Southern Urals (Shafigulina et al. 2015).

With co-authors he studied the distribution of physico-chemical parameters in the quarry lake of Blyavinskoye deposit, established fundamentally different forms of elements in the dissolved and suspended forms in aerobic and anaerobic zones, determined the position of the chemocline—the area of sharp changes in the concentration of chemical elements (Udachin 2009).

Theoretical developments in combination with the analysis of hydrogeochemical situation of the area of the Gaisky copper-coal deposit, formed under the influence of technogenesis, consideration of the formation of the chemical composition of natural waters under anthropogenic metamorphisation with the assessment of the ecological and hydrochemical state of all components of the natural environment are given in the work of Samarina (1999).

Foreign publications (Buckley et al. 2017; Hu et al. 2018; Mua et al. 2016) are also characterised by a high level of development of issues related to the study of field operation wastes.

The analysis of the conducted studies on the Blyavinskoye deposit allows us to state the available separate, scattered information, mainly concerning natural, open-pit waters and soils. The given review of literature data proves that no systematic work on studying the distribution of non-ferrous and rare metals in the waste dumps of the Blyavinskoye deposit has been carried out.

It is also necessary to take into account that worked-out deposits, their dumps, overburden rocks, remaining sulphide ores are a global environmental problem, polluting vast territories, surface and ground waters. Metal sulphides are oxidised to sulphates under near-surface conditions. Sulfates of copper and iron and other associated metals, in contact with the environment, in turn, further aggravate the environmental problems of the specific territory of the Kuvandyk district and the region as a whole (Petrishchev 2016; Opekunova et al. 2017).

2 Materials and Methods of the Study

The Southern Urals lode deposits are fairly well studied. The Middle Devonian volcanogenic rocks of the island-arc basalt-rhyolite formation are host rocks for cuprous ores. The host rocks have undergone alteration to the greenschist phase of metamorphism. Ore deposits most often gravitate to the upper part of the effusive acidic sequence, which is overlapped by andesite-basalt or sedimentary rocks of low thickness and then by the andesite-basalt sequence. The lying side of the sulphide ores is confined to acidic lavas and tuffs, where the halo of hydrothermal-metasomatic transformations of rocks is intensely manifested, expressed in sericitisation, oxy-quartzisation, pyritisation, chloritisation, as well as in raslacisation and brecciation (Prokin and Necheukhin 1985; Vikentyev et al. 2006).

Sulphide deposits are composed mainly of pyrite, the content of which reaches 95%, as well as chalcopyrite, sphalerite and, to a lesser extent, other sulphides.

The ores of sinter deposits, as a rule, are complex, they contain valuable associated elements: gold, silver, selenium, cadmium, bismuth, tellurium and others (Prokin and Necheukhin 1985; Vikentyev et al. 2006; Pshenichny 1975).

The Blyavinskoye deposit is located in the north-eastern part of the volcano-depression structure of the same name. It is confined to the Blyavinsko-Komsomolsky subvolcanic body of acidic composition, which lies in accordance with the host rocks in the contact zone of the Lower and Upper Blyavinskaya sub-formations. Extension of rocks and ore bodies is submeridional, dip is westerly at an angle of 40–70°. The deposit is represented by 4 lenses of massive cuprous ores: Northern and Southern lenses in the western band, confined to the hanging side of the subvolcanic liparite body; Eastern and Minor lenses in the eastern band, located in the lying side of this body. The Northern lens is the largest in size, and it was mined by open pit, among other things. Interlayers of quartz-hematite rocks are found in the hanging wall of this lens. The deposit has an oxidation zone to a depth of 30–60 m and a zone of secondary sulfide enrichment to a depth of 300 m (Prokin and Necheukhin 1985).

Ores, on average, have the following mineralogical composition: pyrite, marcasite, melnikovite, chalcopyrite are the predominant ore minerals, covellin is widely distributed among the secondary ones. Galena, pale ore, sphalerite, chalcosine, wurzite, enargite are very rare. Non-ore is mainly barite, gypsum, quartz. The secondary sulphide-sulphate enrichment subzone contains melanterite, chalcantite, goslarite and other sulphates.

The sampling scheme took into account the confinement of samples to different levels of the west dump. Samples of pit water and silt were also taken. Vegetation samples from the western dump were taken to study the cure of various metals by above-ground and underground parts of plants. In total, more than 100 samples were taken from the sulphide and copper ore, as well as soils, bulk materials, selected minerals and water.

The contents of ferrous, non-ferrous and a number of rare metals were determined by atomic emission analysis at the Engineering Centre of the Orenburg State University. From the group of ferrous metals concentrations of Fe, Mn, Cr, V were determined, from non-ferrous (conventional term) metals—those that determine the basis of non-ferrous metallurgy—Al, Mg, Ni, Co, Cu, Zn, Pb, Ca, Sr, Na, K. From rare metals, rare alkaline elements Rb, Cs were determined; from light rare—Be; from diffuse metals those that accumulate, as a rule, in sulphide ores—Tl, Ga, Cd, Se, As. From non-metals B, S, P were determined.

Minerals, rocks, ores were ground to a particle diameter of less than 0.1 mm. A 10 g sample was ozolised, dissolved in acids, converted to chlorides, filtered and the final hydrochloric acid solutions were transferred to measuring flasks.

Water samples were evaporated to a wet precipitate, treated with nitric acid and hydrogen peroxide and again evaporated to wet salts. The resulting precipitate was dissolved in distilled water, transferred to a measuring flask and brought to the mark with water for analysis.

For chemical evaluation of plant material for total element content, the plant samples were air-dried to air-dry condition in air. The dried plant samples were ground and then subjected to wet mineralisation.

The method of wet mineralisation is based on complete destruction of organic substances of the product sample by heating with sulphuric and nitric concentrated acids with the addition of perchloric acid or hydrogen peroxide or by heating only with hydrogen peroxide.

The following normative documents for measurement methods were used in the analysis: GOST 31870-2012 Water. Determination of elements content by atomic spectrometry methods; GOST R 53198-2008 Ores and concentrates of non-ferrous metals. General requirements for methods of analysis; PND F 16.2.2:2.3.71–2011. Methodology of measuring mass fractions of metals in sewage sludge, bottom sediments, samples of plant origin by spectral methods.

The obtained sample solutions were analysed by atomic emission spectrometry (AES) on an ICP-5000 spectrometer (Focused Photonics Inc., manufacturer China).

In order to determine the main elemental composition of the samples by NPP method, the concentration of the determined elements was measured three times for each sample. For the analysis we made the construction of calibration dependences for the determined elements of the periodic system in the concentration range from 0.1 to 10 mg/l. Selection of lines for measurement of working solutions was carried out according to the smallest values of relative standard deviation.

3 Results and Discussion

The result of the analyses was the determination of concentration Cu, Fe, Al, Ca, K, Mg, Mn, Cr, Cr, V, Ni, Co, Ba, Sr, Na, Rb, Cs, Be, Tl, Cd, Ga, Se, As, B, S, P (27 chemical elements) by the atomic emission method. The results of determination the concentration of chemical elements in soil samples, in water and vegetation samples are given in Tables 1, 2 and 3.

To establish the peculiarities of distribution of non-ferrous and rare metals in the soil of the western dump, in the vegetation and in the water of the quarry lake of the Blyavinskoye deposit, diagrams were constructed (Figs. 1 and 2).

The diagram shown in Fig. 1 demonstrates the close distribution of elements in the dump soil, in vegetation and in the water of the quarry lake. The highest concentrations of all elements are found in the water of the quarry lake, while the lowest concentrations are found in the vegetation. The highest contents are observed

Table 1 Content (mg/g), occurrence (%) of ferrous and non-ferrous metals in anthropogenic objects of the Blyavinskoye field, Orenburg region

Metals	Soil (21)		Flora (9)		Quarry water μg/l	Eel (2)	
	mg/t	Occurrence, %	mg/t	Occurrence, %		mg/t	Occurrence, %
Fe	*3.11–667.9* 252.92	100	*0.37–41.33* 18.20	100	422.01	0.19	100
Mn	*0.03–19.04* 2.64	100	*0.26–0.76* 0.55	100	164,629.1	4.15	100
Cr	*0.004–0.67* 0.22	100	*0.001–0.10* 0.16	100	1.51	0.48	100
V	*0.002–1.65* 0.31	95	*0.0001–0.09* 0.04	100	200.35	1.41	100
Cu	*0.49–65.43* 12.76	100	*0.13–1.26* 0.46	100	15,197.25	4.38	100
Zn	*0.01–2.39* 0.58	100	*0.06–0.28* 0.16	100	35,917.79	0.89	100
Pb	*0.004–4.32* 0.64	100	*0.003–0.13* 0.06	100	21.73	0.90	100
Al	*0.59–130.43* 29.65	100	*0.09–18.93* 7.26	100	956.51	100.58	100
Mg	*0.09–103.44* 17.80	100	*0.78–6.88* 4.03	100	415,967.6	78.78	100
Ni	*0.01–0.28* 0.06	100	*0.004–0.03* 0.01	100	2501.68	0.12	100
Co	*0.01–1.07* 0.19	100	*0.001–0.01* 0.01	100	866.20	0.06	100
Ba	*0.002–1.17* 0.16	100	*0.006–0.06* 0.03	100	19.74	0.25	100

(continued)

Table 1 (continued)

Metals	Soil (21)		Flora (9)		Quarry water μg/l	Eel (2)	
	mg/t	Occurrence, %	mg/t	Occurrence, %		mg/t	Occurrence, %
Ca	*0.09–111.55* 21.61	100	*4.08–30.97* 10.17	100	292,347.9	28.97	100
Sr	*0.003–0.94* 0.29	100	*0.03–0.15* 0.08	100	2918.87	0.52	100
Na	*0.03–20.44* 3.77	100	*0.07–0.53* 0.25	100	70,809.2	4.50	100
K	*0.09–40.09* 9.21	76	*3.90–17.55* 6.40	100	29,335.1	13.30	100

Minimum and maximum of the concentration are shown in italics, below–average value

Table 2 Contents of rare, disseminated metals, as well as non-metals in technogenic objects of the Blyavinskoye field in the Orenburg region (mg/g)

Elements	Soil (21)		Flora (9)		Quarry water	Eel	
	mg/t	Occurrence, %	mg/t	Occurrence, %	μg/l	mg/t	Occurrence, %
Rb	*0.08–18.15* 5.26	95	*0.02–0.15* 0.06	100	27,284.58	3.20	100
Cs	*0.21–5.89* 1.30	90	*0.002–0.42* 0.20	78	–	4.81	100
Be	*0.00001–0.01* 0.001	24	*0.00004–0.0001* 0.00003	22	2.50	–	0
Tl	*0.0004–6.58* 0.51	62	*0.0005–0.02* 0.01	67	2.50	–	0
Cd	*0.0004–7.95* 4.09	100	*0.003–0.02* 0.01	100	1495.269	1.28	100
Ga	*0.0006–0.75* 0.18	100	*0.0002–0.01* 0.007	78	3.5394	0.189015	100
Se	*0.001–0.34* 0.10	100	*0.00002–0.005* 0.003	89	79.11507	0.07649	100
As	*0.0004–2.27* 0.58	90	*0.003–0.08* 0.03	100	27.84	0.09	100
B	*0.007–4.00* 0.95	100	*0.02–0.11* 0.07	100	19.74	0.89	100
S	*5.22–0.34* 38.20	100	*8.29–52.03* 15.04	100	588124.3	34.43	100
P	*0.001–24.88* 2.76	76	*0.94–3.10* 1.70	100	–	1.26	100

Minimum and maximum of the concentration are shown in italics, below–average value

Table 3 Contents of elements in plants composing plant communities in waste dumps of Blyavinskoye deposit (mg/g)

No. n/a	Copper content in soil	Copper content in herbaceous plants	
		Stems, leaves and flowers	Roots
1	3.62	0.8159	0.4107
2	3.62	1.2626	0.5452
3	3.62	0.1678	0.1434
Average	3.62	0.7488	0.3664

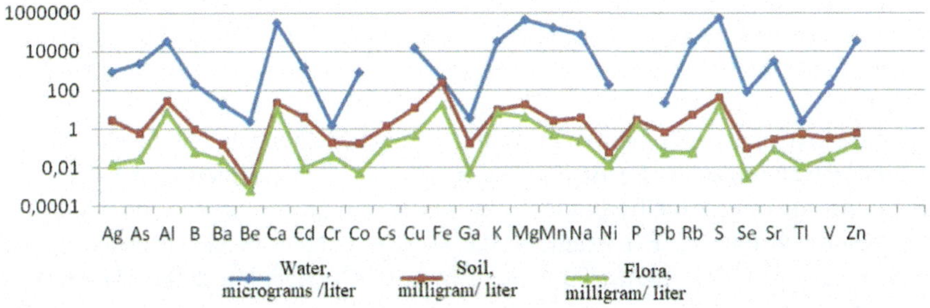

Fig. 1 Distribution of elements (average in dump soil (21 samples), in plants (9 samples), in quarry water) in technogenic wastes of Blyavinskoye deposit (element contents are given in logarithmic scale)

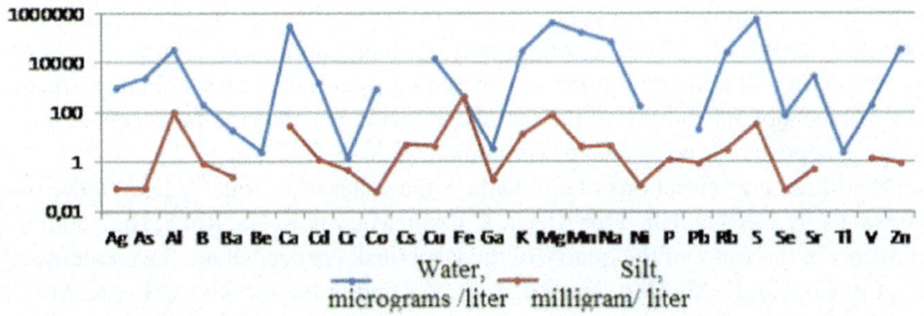

Fig. 2 Distribution of elements: average in silt of the quarry lake (2 samples) and in the water of the quarry of the Blyavinskoye deposit (elemental contents are given in logarithmic scale)

in all study sites for the following elements: Al, Ca, Cu, Fe, K, Mg, Mn, Na, Rb, S. The highest contents of elements in the soil are characteristic for the elements Al, Ca, Cd, Cd, Cu, Fe, K, Mg, Mn, Na, Rb, P, S. The highest concentrations of elements in the soil (by mean values greater than 10 mg/g) are established for Al, Ca, Cu, Fe, Mg, S. The absence of alkali metals in this series is explained by the fact that they are active migrants and eventually accumulate in the water of the quarry lake. Such elements as Co, S, Zn show high content only for soils. This can be considered quite natural, as these elements are common and typical for copper-coal deposits of the Southern Urals. The occurrence of all elements, as a rule, is 100%. According to Glazovskaya (1988), the development of sinter deposits is characterised by "ferritization", an increase in the relative role of copper compared to zinc and other elements, because chemical elements accumulated in deep deposits are "pumped" to the earth's surface, i.e. to exogenous conditions that are not characteristic of metal sulphides (chalcophilic elements) (Glazovskaya 1988). As a result of this "ferritization" the environment is enriched with chalcophiles (Tables 1, 2 and 3). According to Emlin (1993), the day surface can act as a kind of filter for elements in ionic form, because it is on the day surface that the action of various geochemical barriers (oxidative, cryogenic, evaporative) increases, which leads to the separation of ore elements: Fe, Cu, Zn: Cd and protons enrich downward flows that break down silicates and release lithophilic elements: Al, Mg, to a lesser extent Na, Ca, K (Emlin 1993).

Minima of element content in the dump soil, in vegetation and in the water of the quarry lake are observed for Be, Cr, Ga, Ni, Pb, Se, Tl. Nevertheless, their contents are sufficient in case of by-product extraction in case of complex processing of the mentioned objects. For example, the average content of Ga in the ground is approximately 0.02% (industrial content from 0.5% (Avdonin 2005), Tl—0.05%, etc.). The industrial significance of Pb is estimated, as a rule, in the sum with Zn. According to our data, the sum of Pb and Zn is approximately 0.12% on average in the soil. Taking into account the average content of Cu in the ground of about 1.3%, S (3.8%), Cd (0.41%), Rb (0.53%), Cs (0.13%), the western dump is of interest as an unconventional source of a number of non-ferrous and rare metals. It should be noted that according to the copper content of 1–2%, the ores are classified as poor (Avdonin 2005). The discovery of platinum, palladium in association with gold and silver in ores and individual minerals (including pyrite) of copper-coal deposits and other objects of subsoil use in the Orenburg region expands the prospects of waste of these deposits for noble metals (Ponomareva 2023, 2021).

Maximum concentrations of elements in the water of the quarry lake (more than 10,000 μg/l) are established for Al, Ca, Cu, Mn, Mg, Zn, K, Na, Rb, S. High contents elements in the water of the quarry of the Blyavinskoye deposit are the elements Al, Ca, Cd, Co, Cu, K, Mg, Mn, Na, Rb, S, Sr, Zn, and in the silt Al, Ca, Fe, K, Mg, S.

The contents of all the contents of all mentioned elements in the quarry water are higher than in the sludge, except for Cs, P, these elements are absent in the water. Tl and Ve are absent in the silt, but also in the water of the quarry lake their contents are not levelled in the water of the quarry lake their contents are not high (Fig. 2; Table 2). The water in the quarry lake belongs to acid sulphide type, has pH ~ 4, [according to Udachin (2009) pH = 2.92 (Udachin 2009)].

During the development of the Blyavinskoye deposit, as well as all copper and copper-ore deposits deposits, the structure of groundwater changes. Water from the oxidation zone is pumped out, being partially diluted by weakly mineralised fracture waters.

This leads to dissolution of sulphates (melanterite, chalcanthite, goslarite and other soluble sulphates Cu, Zn, Fe, etc.) and oxidation of copper and iron sulphides.

According to Emlin (1993), the mine water flow rate at the initial stages of development of the Blyavinskoye deposit reached 25 m^3/h. The concentration of Cu in mine streams was 6 g/l, and 1314 tonnes of Cu, not counting other elements, were carried out per year. Intense migration of chalcophiles at the Blyavinskoye deposit can be explained by the presence of an oxidation zone enriched with minerals oxidation zone enriched with protosulphates. It is the subzone of secondary sulphide-sulphate enrichment subzone determined the main industrial value of the Blyavinskoye deposit and contributed to the mobilisation of chalcophiles that are in easily soluble form (Emlin 1993).

Different plant species living on the same soil within the Blyavinskoye deposit accumulate different amounts of chemical elements. Let us compare the copper content in plants and the soil on which they grow (western dump of the Blyavinskoye deposit) (Table 3).

Along with soil and water sampling within the western dump of the Blyavinskoye deposit, all types of widely distributed plants (herbs) in the area were collected and analysed. Analyses of the results of the above-ground and underground parts of the plants show that, on average, the above-ground part accumulates more copper (0.7488 mg/g) compared to the underground part (0.3664 mg/g). In stems, leaves and flowers, 0.1678 mg/g copper to 1.2626 mg/g was found. In the underground part of the plants—from 0.1407 to 0.5452 mg/g. The analysis of Table 3 shows that the herbaceous vegetation has less copper than the soil on which they grow. All plants are perennial grasses. No external changes were found in the studied plants.

Comparatively high contents in plants are observed for the following elements: Al, Ca, Cu, Fe, K, Mg, Mn, P, S. Maximum contents occur in the paragenesis of elements Al, Ca, Fe, K, Mg, P, S (more than 1 mg/g).

Comparison of element paragenesis by comparatively high contents in vegetation (Al, Ca, Fe, Cu, K, Mg, Mn, P, S), soils (Al, Ca, Cd, Cu, Fe, K, Mg, Mn, Na, Rb, P, S) and quarry water (Al, Ca, Cd, Co, Cu, K, Mg, Mn, Na, Rb, S, Sr, Zn) shows their almost complete coincidence. Consequently, the paragenesis of elements Al, Ca, Fe, Cu, Mn, Mg, K, S is characteristic for the western dump, quarry lake and plant community of the western dump of the Blyavinskoye deposit of the Orenburg region.

In general, the relatively low concentration of valuable components in technogenic deposits is compensated for by large volumes of mineral raw materials concentrated on the surface (i.e. extraction from the subsurface is not required). The raw materials are mainly of fragmented mechanical composition and are located within accessible territories with developed infrastructure. The environmental and socio-economic effects of the possible use of waste from the Blyavinskoye deposit as mineral raw materials should also be emphasised. Conventional non-ferrous metal ores are usually very poor (metal concentration division), but the other metals accompanying the main component in them can be much more valuable than the main element of the ore. Taking into account the fact that more than 13 components are extracted from copper-coal ores, the share of which in the total value of raw materials is 40–50% (Avdonin 2005), the western dump can be considered quite a promising object.

The main limiting factor for the extraction of scarce metals from waste minerals is the lack of industrial technologies and high costs of scientific research.

The wastes of the Blyavinskoye deposit represent a new type of deposits—technogenic, unconventional deposits formed in recent decades (Makarov et al. 2019). Taking into account the revealed contents of non-ferrous and rare metals in the wastes of the deposit, it is necessary to continue the started work on the study of dumps, the possibility of extraction of non-ferrous and rare metals in the dumps of the Blyavinskoye deposit. This is very promising not only for the Orenburg region, but also for the Urals as a whole.

4 Conclusion

As a result of the conducted research 27 elements were determined in the wastes of Blyavinskoye copper-coal deposit of the Orenburg region: Cu, Fe, Al, Ca, K, Mg, Mn, Cr, V, Ni, Co, Ba, Sr, Na, Rb, Cs, Be, Tl, Cd, Ga, Se, As, B, S, P.

Paragenesis of elements Al, Ca, Fe, Cu, Mn, Mg, K, S is characteristic for the western dump, quarry lake and vegetation community of the western dump of the Blyavinskoye deposit.

The copper content in the soil (~ 1.3%) is commercial. The contents of a number of components are sufficient in case of their associated extraction during complex processing of the mentioned object: S (3.8%), Cd (0.41%), Rb (0.53%), Cs (0.13%), Ga (0.02%), Tl—0.05%, the sum of Pb and Zn (0.12%).

The wastes of the Blyavinskoye deposit are an unconventional source of non-ferrous and rare metals, so there is a need for a more detailed study of technogenic deposits, as well as the definition of possible technologies for the development and processing of wastes.

Acknowledgements The article was prepared with the support of RSF grant №. 23-27-10006 "Analysis of the formation of technogenic geosystems as a result of the development of ore deposits in the Orenburg region in order to rationalise reclamation measures".

References

Avdonin VV, Boitsov VE, Grigoriev VM et al (2005) Deposits of metallic minerals, 2nd edn., revised and additional. Academic Project, Triksta, 720 p

Buckley AN, Hope GA, Parker GK, Steyn J, Woods R (2017) Mechanism of mixed dithiophosphate and mercaptobenzothiazole collectors for Cu sulfide ore minerals. Min Eng 109:80–97

Emlin EF (1993) Technogenesis—the newest stage of the geological history of ore deposits of the Urals. Mining J 1:43–126

Glazovskaya MA (1988) Natural landscape-geochemical complexes and their manifestation on the territory of the USSR. Ser Geogr 5:3–9

Guman OM, Makarov AB, Antonova IA, Zaharov AV (2018) Transformation of geological environment in copper pyrite mining in the Urals. Min Inf Anal Bull (sci Techn J) 6:98–106

Hu Y, Ye L, Li Z, Huang Z, Zhang J (2018) Genesis of Fahlore in the Tianbaoshan lead–zinc deposit, Sichuan Province, China: a scanning electron microscopy–energy dispersive spectroscopy study. Acta Geochimica 37(6):842–853

Makarov AB, Khasanova GG, Talalay AG (2019) Technogenic deposits: peculiarities of research. Izvestia Ural State Min Univ 3(55):58–62. https://doi.org/10.21440/2307-2091-2019-3-58-62

Mua Y, Peng Y, Lauten RA (2016) The depression of pyrite in selective flotation by different reagent systems—a literature review. Min Eng 96–97:143–156

Opekunova MG, Opekunov AY, Papyan EE, Somov VV (2017) Phytoindicational properties of the vegetation in landscape transformation studies on Sibay chalcopyrite deposits (Southern Urals). Contemp Probl Ecol 10(3):301–314. https://doi.org/10.1134/S1995425517030076

Opekunova M, Opekunov A, Somov V et al (2020) Transformation of metals migration and biogeochemical cycling under the influence of copper mining production (the Southern Urals). Catena 189:104512. https://doi.org/10.1016/j.catena.2020.104512

Petrishchev VP, Ryakhov RV, Dubrovskaya SA, Noreika SY (2016) Assessment of the dynamics of recovery processes and reclamation efficiency of technogeosystems of copper and cauldron deposits of the Southern Urals. Probl Reg Ecol 5:85–91

Ponomareva GA, Ponomarev AA (2021) Regularities of platinoids distribution in halogen formations of the Southern Urals. Min J 12:10–14. https://doi.org/10.17580/gzh.2021.12.02

Ponomareva GA, Petrishchev VP (2023) Substantial composition of bitumen of the Orenburg Urals. Ser Geol 1:80–88. https://doi.org/10.17308/geology/1609-0691/2023/1/80-88

Prokin VA, Necheukhin VM, Sopko PF et al (1985) Copper-coal deposits of the Urals: geological conditions of location. UNC AS USSR, Sverdlovsk, 288 p

Pshenichny GN (1975) Gayskoye copper-coke deposit of the Southern Urals (in Russian). Nauka, Moscow, 187 p

Samarina VS (1999) Technogenic metamorphisation of natural water chemistry [Text]: on the example of ecol. Hydrochemical mapping of the Ural River basin, Orenburg region. In: Samarina VS et al (eds) Russian Academy of Sciences, Ural Branch, Perm Scientific Centre, Gore Institute. Ural Branch of RAS, Ekaterinburg, 444 p. Illus. Bibliography, p 435

Shafigulina GT, Udachin VN, Filippova KA, Aminov PG (2015) Geochemical characteristics of technogenic soils of mining and industrial landscapes of the Southern Urals. Bull Acad Sci Republ Bashkortostan 20(4):93–101

Udachin VN, Aminov PG, Williamson BD (2009) Geochemistry of the environment of geotechnical systems of the Southern Urals. Nat Techn Sci 6(44):298–306

Vikentyev VI, Moloshag VP, Yudovskaya MA (2006) Forms of occurrence and conditions of concentration of noble metals in the Urals sinter ores. Geol Ore Deposits 48(2):91–125

Meteorite Impact Craters in the Territory of Central Chernozem Region

A. V. Zhabin and V. G. Gadiyatov

Abstract Circular depressions of various depths have been known in the Central Chernozem region for decades. Their diameter ranges from the first tens of meters to two kilometers. They are surrounded by a shaft up to 50 m wide. The interior of the depressions is flat, often swampy, and sometimes lake-type reservoirs of varying depths are developed there. According to the conclusion of various researchers, the rocks in the areas of these depressions are volcanic ash or impactites. However, despite the presence of a number of signs indicating the cosmic origin of the depressions, various opinions were expressed, but the main hypotheses were related to karst and suffusion manifestations, as a result of which funnels appeared on the earth's surface. We have established that the rounded depressions observed on the Earth's surface are impact craters formed during explosions of celestial bodies and during their contacts with the earth and water surfaces. Rocks formed during explosions are impactites. Confirmation of the impact nature of rocks arising from impact metamorphism is the specific fracturing, polymorphic modifications of quartz group minerals (stishovite, tridymite, cristobalite) contained in the rocks, the genesis of which is associated with high temperatures and pressures, as well as abnormally high concentrations of platinum group metals, which are millions of times higher than the clark of the Earth's crust and from several thousand to the first tens of thousands of times higher than their content in chondrites. Among them, the maximum concentrations are iridium, osmium and rhenium. At such high concentrations, impactites are of industrial interest and deserve exploration work. Given the modern geological processes taking place on the earth's surface, it can be assumed that the territory of the Central Chernozem region was subjected to intense bombardment by meteoritic material at the end of the Pleistocene—the beginning of the Holocene.

Keywords Central Chernozem region · Depressions · Impact craters · Cosmic bodies · Impactites · Platinum group metals · Noble metals

A. V. Zhabin
Voronezh State University, 1 Universitetskaya Pl., Voronezh, Russia

V. G. Gadiyatov (✉)
Voronezh State Technical University, 84 20-Letiya Octyabrya Str., Voronezh, Russia
e-mail: gadiatovvg@mail.ru

T. Chaplina (ed.), *Processes in GeoMedia—Volume VIII*, Springer Geology,
https://doi.org/10.1007/978-981-97-6627-7_9

1 Introduction

Depressions have been identified in the sedimentary rocks of the Central Chernozem region, which stand out in the form of funnels or ring structures on satellite images. The "rings" are located at different distances from each other, forming both separate rather large objects and small clusters of rounded shapes, similar to craters that occur when a large meteorite, asteroid, comet or other celestial body collides with the Earth. Shallow depressions are more often several dozen unevenly scattered over an area of up to 100 km^2, and the total area on which they are scattered is 5 times larger. The diameter of the depressions ranges from 20 to 2 km, most depressions have a diameter of 100–500 m. The depth reaches up to several meters. The bottom of the depressions is often swampy or covered by natural water reservoirs (Grishchenko 1976; Tselmovich et al. 2015). Uplifts are noted in the center of a number of large depressions, their slopes are rounded and gentle. On all sides, some depressions are surrounded by shafts or fragments of them have remained, which are clearly visible on satellite images (Fig. 1).

According to the opinion of various experts, the rocks in the areas of these depressions are volcanic ash or impactites that are not characteristic of sedimentary rocks. Various opinions were expressed, but the main hypotheses were related to karst and suffusion manifestations, as a result of which funnels appeared on the earth's surface. Despite the presence of a number of signs indicating the cosmic origin of the depressions, almost all authors do not associate these rocks with impact craters that arose from the fall of cosmic bodies. Depressions formed in the sedimentary rocks of the Central Chernozem region in the Late Pleistocene and Holocene (Zhabin and Dmitriev 2019). Meteorites fell everywhere on Earth at different times. Thus, when a comet collided with the Earth, a cataclysm occurred within 8–15 thousand years, which led to catastrophic consequences (Firestone et al. 2007).

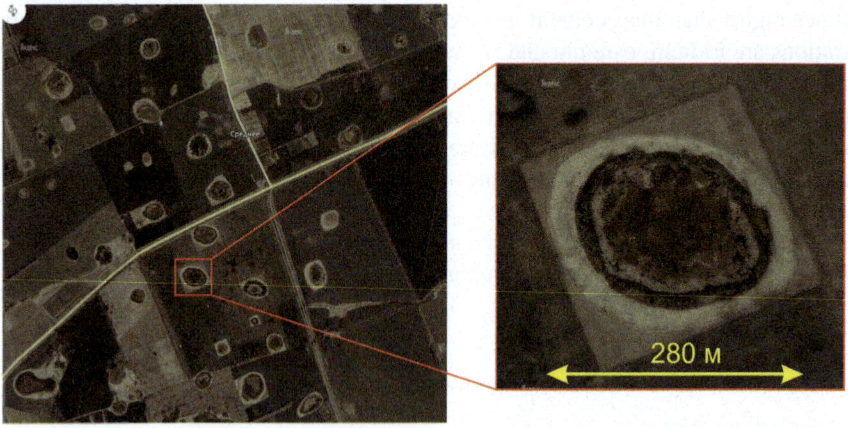

Fig. 1 General view of the depressions in the satellite image (on the left). The right image shows a medium-sized funnel (Zhabin and Dmitriev 2019)

The mechanism of depressions formation is well studied, so we will only give a general scheme of the occurrence of impact and explosion craters. When a cosmic body moves, a volume of gas moves in the atmosphere equal to its cross-section multiplied by the length of the trajectory. When the mass of a gas is 10 times the mass of a cosmic body, then passing through the atmosphere, its velocity decreases significantly and when falling to the surface of the earth is no more than 10% of the original, i.e. up to 3 km/s (Zhabin et al. 2020a; Zhabin and Zhabina 2019; Gerasimov 1984). Upon impact with the earth's surface, the destruction of the cosmic body occurs and fragments of the parent rock and the cosmic body are ejected from the resulting depression. This is how an impact meteorite crater is formed. In the event that the velocity of a cosmic body is higher, an explosion occurs upon impact with the earth. At the same time, the diameter of the resulting depression significantly exceeds the diameter of the fallen cosmic body. An explosive crater is formed. If a cosmic body explodes before reaching the Earth, an impact crater also forms on the earth's surface, and the material in the form of small fragments is scattered over a significant area.

The signs that the depression is a crater are the following: the presence of surrounding shafts; the presence of minerals uncharacteristic of host rocks that occur at high pressures and temperatures; fracturing of rocks, as well as the formation of impactites; abnormal concentrations of trace elements in host rocks, characteristic of space aliens, magmatic or metamorphic formations; debris (fragments) cosmic bodies, etc. Fragments of cosmic bodies are extremely rare, so in this case we do not consider them signs.

During our research, we adopted the following working hypothesis: impact craters arose from different cosmic bodies with different mechanisms of impact on the earth's surface; their fall and crater formation occurred at different historical times; in different conditions that existed then (dry land, sea). As a result, the material testifying to the cataclysms may not have been preserved or few have been preserved.

Therefore, we conducted studies in areas with large craters, where rocks attributed by researchers to volcanic ash are described, as well as in places with known depressions (Iosifova 1992; Zhabin et al. 2008). In the course of field work, rocks of Neogene and Quaternary ages were studied. They are located in the immediate vicinity of two large craters located in the areas of the villages of Gorelka (Borisoglebsky administrative district) and Alexandrovka Donskaya (Pavlovsky administrative district). districts), which administratively belong to the Voronezh Region, as well as rocks in the areas of five small craters located in the settlements of Khleborob (Talovsky district), Gnezdilovo (Oryol region), Novoshevchenkovskoye (Olkhovatsky district), Yablochnoye (Khokholsky district) and Lipetsk (Fig. 2).

Fig. 2 Research locations in Voronezh oblast triangles show the settlements of Duvanka and Gorelka

2 Purpose of the Study

The objective of the research was to study the depressions and rocks that make up their interior and lie beyond them. The purpose of the research is to prove that the depressions that appeared on the surface of the earth in the sedimentary rocks of the Central Chernozem region are impact craters formed when cosmic bodies fell onto the surface of the Earth.

3 History of the Study

The history of studying depressions (ring structures) for their formation associated with the fall of cosmic bodies to the surface of the Earth has not been previously conducted. At various times, researchers were interested only in rocks lying among the sediments of the Quaternary and Neogene periods, represented, according to their data, by volcanic ash.

These formations were first described and studied by Dubyansky (1935). He suggested that the layer of volcanic ash described by him in the area of the Duvanka settlement and attributed to the Upper Pleistocene was "brought by air currents from the Caucasus during catastrophic volcanic eruptions." He also linked the formation of this ash layer "with the rise in the level of the Sea of Azov during the melting of glaciers, which caused the flow of waters up the Don valley." Ash deposits were eroded off the shores of Azov and Caspian Seas and, because of transport, accumulated among the sandy formations of the Don floodplain terrace.

In the formation of an ash layer in the area of the village of Gorelka, researchers (Milkov 1977; Ermolov 2008) find a connection with volcanic activity that occurred in the Carpathians or in Italy. That is, when volcanoes erupt, the ash is supposedly transported by air currents and settled on the earth's surface in that area. Ash deposits belong to the Lower Neogene.

Later, deposits composed of volcanic ash were studied by other researchers. According to the conclusion of some (Savko et al. 2001), these are fine–grained and silty sands forming lenticular interlayers among rocks of the Quaternary and Neogeonic periods. According to other researchers (Grishchenko 1976; Kholmovoy 2008, 2013), these deposits, studied in the area of Alexandrovka Donskaya village, are volcanic ash. All determinations of the mineral composition of rocks were carried out in immersion liquids under a microscope, the chemical composition was studied by analytical methods.

4 Research Methods

During the field work on the study of rocks containing "ashes", furrow samples from 30 to 50 cm long were selected throughout the rock section. The mineral and elemental compositions of the samples were studied by X-ray diffraction and electron microscopic analyses using the ICDD PDF-2 Siroguant databases and Sietronics Pty Ltd software. Powder and oriented preparations were used in the X-ray diffraction study.

In addition to the gross samples, the mineral composition of the light and heavy fractions of all samples was studied. To do this, a fraction of more than 0.01 mm was extracted, and then separated in a bromoform. Minerals were determined in immersion liquids under an optical microscope. In addition, the chemical composition was checked during a microprobe study. The results of X-ray diffraction analysis were compared with X-ray spectral analysis data for randomly selected grains. The elemental composition of the ashes was determined from three samples taken from the upper and lower parts of the formation, as well as from the contact zone. Ten grains were analyzed in each of them, after which the obtained data were averaged.

Samples weighing 1 kg were taken from rocks in the area of small craters (Khleborob, Gnezdilovo, Yablochnoye, Olkhovatka, Lipetsk), which after crushing were worn down to a fraction of 0.07 mm. The elemental composition was determined using an electron microscope TM-1000, Hitachi, Japan with an EDS-SWIFT energy dispersion prefix. The mineral composition was determined using an ARL X'tra X-ray diffractometer (Switzerland). Qualitative phase analysis was performed using the

ICDD PDF-2 database. Quantitative X-ray phase analysis using Siroquant Sitronics Pty Ltd software.

5 Research Results and Discussion

In the area of the village of Gorelki, in an outcrop located on the northern outskirts of the village in the northern side of the gully of the sublatitudinal strike, a section of rocks with a thickness of 11.9 m was studied (Zhabin et al. 2020b). From top to bottom, it is represented by sand, clay, light green siltstone of glauconite-quartz composition, aleuropelite and siltstone of greenish-gray quartz composition. Aleuropelites, previously described as volcanic ash, lie at a depth of about 9 m, the thickness of the formation is 2.5 m. In the dry state, the rock is white, in the wet state it is gray. The structure is fine-grained, the texture is layered, in the sole with a thickness of up to 0.3 m, the rock is compacted to the state of siltstone, the texture is massive.

The clay fraction of all rocks is represented by montmorillonite, illite and kaolinite. The content of illite is about 10%. The percentage ratio of kaolinite and montmorillonite in rocks containing ash is 30:60. In aleuropelites (ashes), the content of kaolinite prevails over montmorillonite (60:30). According to X-ray diffraction analysis, about 90% of the substance of aleuropelites (ashes) is represented by an X-ray amorphous phase. The rest is quartz and clay component. According to the Energy-dispersive X-ray spectroscopy (EDX) of the lower and upper parts of the formation, hurricane concentrations of platinum group metals (PGMs), gold and silver were found in the samples (Table 1).

Judging by the composition, aleuropelites have nothing to do with volcanic ash. And the abnormal contents of PGMs and precious metals, which are not found in other sedimentary deposits, indicate that the formation of these rocks is directly related to the fall of a cosmic body to Earth, the depression from which is 2 km away. According to the content of clay minerals (kaolinite, montmorillonite, etc.),

Table 1 The content of platinum group metals and precious metals in aleuropelites (volcanic ash) in the Gorelka village area (according to the EDX)

Object name	Elemental composition, %								
	Ru	Rh	Pd	Ag	Re	Os	Ir	Pt	Au
Upper part of the formation	0.4	0.6	0.7	0.3	3.9	2.1	4.6	1.0	0.7
Lower part of the formation	0.4	0.6	1.4	0.7	2.5	4.2	2.9	1.7	2.9
Lower formation contact	1.6	0.9	0.4	0.2	1.8	5.2	2.1	2.5	3.0
Average	0.8	1.0	0.83	0.4	2.73	3.83	3.2	1.73	2.2

it is highly likely that the rocks are of marine origin. Since the presence of such an amount of these minerals in the composition of the rock is inherent only in formations of marine genesis.

In the area of the village of Alexandrovka Donskaya, in the ledge of a sand quarry located on the northern outskirts of the village, a 21 m thick rock section was studied (Zhabin et al. 2020b). From top to bottom, it is represented by loam, fine-grained quartz sand of light yellow color, siltstone (ash), sand (similar the above mentioned), siltstone (ash), sand (similar the above mentioned).

Two layers of siltstone, previously mistaken for volcanic ash, are distinguished in the section. At the top, the siltstones are light gray to white, loose in a dry state. The contacts with the above and below sands are uneven, the reservoir thickness is 0.5–1.2 m. There are similar siltstones below. The middle part of this formation contains lenticular sand layers with a thickness of up to 0.4 m. The thickness of the formation is from 0.8 to 1.0 m. The composition of siltstones is up to 70% montmorillonite, about 5% clay fraction, glauconite, etc. Energy dispersion analysis revealed a significant content of PGMs and precious metals (Table 2). The results of energy dispersion analysis do not contradict the data of X-ray diffraction analysis. Unlike the "ashes" of the Gorelka section, no X-ray amorphous mineral phases were found in similar rocks of the Duvanka.

Thus, abnormal concentrations of PGMs and precious metals in siltstones indicate their impact origin. It is highly likely that two layers of identical rocks were formed during the fall of a cosmic body. At the same time, the "alien" fell into a marine reservoir, which is confirmed by the high content of illite (Table. 3), montmorillonite and glauconite grains in siltstones.

To the south of the village of Khleborob, located in Talovsky district of Voronezh Oblast, rocks similar to slag were found at an ancient human site (Medvedev and Zhabin 2019). According to X-ray diffraction analysis, in addition to quartz, the rock composition is mainly occupied by diopside and yelimite, the formation of which is associated with high-temperature metamorphism of calcareous rocks with a high silica content, in addition, a high-temperature polymer modification of quartz, cristobalite. The rock also contains wollastonite, bismuth minerals, etc. (Table 3). According to energy-dispersive determinations, anomalous contents of PGMs, gold and silver were established in the samples (Table 4). Based on the above analysis results, the rock was identified as impactites.

In the area of the village of Gnezdilovo (Oryol Oblast), a quarry uncovered quaternary deposits with a capacity of a little more than 7 m (Zhabin et al. 2020a). At the base there are yellowish-gray, fine-grained, clay sands. The apparent power of the pack is 5.0 m. A layer of quartzite-like porous (cellular) rock of light yellowish-gray color lies on them with sharp disagreement. The layer is aged, traced for a distance of about 100 m along the wall of the quarry, the thickness of the layer is from 10 to 20 cm. The rock is fine-grained, massive. The pore size is up to 10 mm, cavities up to 6 cm are rare, the pore surfaces are smooth, without signs of dissolution. According to the results of X-ray diffraction analysis, the rock is represented by minerals of the silica mineral association, which includes quartz, high-temperature polymorphic

Table 2 The content of platinum group metals and precious metals in aleurites (volcanic ash) in the Duvanka village area (according to the EDX)

Object name	Elemental composition, %								
	Ru	Rh	Ag	Os	Ir	Pt	Au	Pd	Re
Upper formation	0.6	0.7	1.1	9.4	0.8	4.1	2.0	–	–
Lowerformation	0.4	0.8	–	3.4	3.9	1.4	3.3	0.5	2.2
Average	0.5	0.75	0.55	6.4	2.35	2.75	2.5	0.25	1.1

Table 3 Mineral composition (%) of impact rocks in the area of the settlements of Khleborob, Gnezdilovo, Novoshevchenkovskoye, Lipetsk (according to X-ray diffraction analysis)

Minerals	Khleborob	Gnezdilovo quartzite rock	Gnezdilovo sandstones	Novoshevchenkovskoye	Lipetsk
Quartz	45.6	65.8	47.3	0.9	0.6
Tridymite		5.8	11.8		
Diopside	17.3	0.9			
Cristobalite	4.6	5.7	12.8		
Helenite (melilite)					85.4
Fayalite				51.7	
Wustit				15.3	
Titanomagnetite				1.0	
Kubanite	2.2				
Wollastonite	4.3	4.1			
Ielimite	8.0				
Bismuth	4.0				
Cobalt	0.8		2.6		
Illit		3.5			
Spodumen		1.1		1.5	
Sanidin		1.1	4.2		
Chromite		2.5			
Sodalite	0.7				
Galena	0.6				
Forsterite				2.6	
Edenite				2.4	
Orthoclase				2.3	
Stishovit					1.7
Calcite					1.3
Siderite					1.3
Chlorite	1.3		4.1	1.7	
Chloritoid			1.2	2.3	
Platinum					
Grossular	1.6			2.3	
Marcasite	1.6				0.3
Chalcopyrite	0.6				
Antimonite	0.4	2.2		1.4 (?)	
Pyroxene					
Pyrite			0.5		0.1

(continued)

Table 3 (continued)

Minerals	Khleborob	Gnezdilovo quartzite rock	Gnezdilovo sandstones	Novoshevchenkovskoye	Lipetsk
Arsenopyrite				1.1	
Hesterite				1.5	
Silver	1.3	1.5		1.6	0.8
Gold	1.0				
Dolomite				1.4	
Rhodium		1.6			
Osmium			0.2		
Zircon		0.5			

Table 4 Average content of platinum group metals and precious metals in impact rocks in the area of the settlements of Khleborob, Gnezdilovo, Novoshevchenkovskoe, Lipetsk, Gorelka, Duvanka (according to energy-dispersive determinations)

Item	Chemical element, weight %								
	Ru	Rh	Pd	Ag	Re	Os	Ir	Pt	Au
Khleborob	0.7	0.1	0.6	0.6	2.7	2.4	3.6	2.6	0.6
Gnezdilovo (quartzite rock)	0.4	1.2	0.8	0.4	5.8	2.2	9.5	1.2	4.3
Gnezdilovo (sandstones)	0.4	2.0	1.0	0.4	1.8	7.6	3.5	3.7	3.6
Novosevchenkovskoe	0.2	0.1	0.2		0.7	0.4	1.2	0.1	0.3
Yablochnoye	0.6	0.3	1.5	0.6	6.9		10.6	0.5	4.5
Lipetsk	0.6	1.3	0.5	1.1	1.9	4.9	1.7	3.5	2.1
Gorelka	0.8	1.0	0.8	0.4	2.7	3.8	3.2	1.7	2.2
Duvanka	0.5	0.8	0.3	0.6	1.1	6.4	2.4	2.8	1.1

modifications of quartz tridymite and cristobalite. In addition, it contains spodumene, sanidine, diopside, compounds of rhodium, platinum, silver, etc. (Table 3).

Above the quartzite-like porous rock is a pack of blocky Aleuropean sandstones, broken by cracks. The rock is white with a grayish tinge, finely grained, massive. The average size of the blocks is 40 × 30 × 20 cm. The thickness of the rock pack is about 2.0 m. The rock also belongs to a siliceous mineral association with a higher content of tridymite, cristobalite. In addition, chlorite, microcline, sanidine, compounds of cobalt, platinum, vanadium, silver, osmium, etc. have been identified in the rock. (Table 3). According to the data of energy dispersion analysis, hurricane concentrations of PGMs, gold and silver were found in the samples (Table. 4).

Judging by the presence of tridimitacrystabalite in the rock, as well as PGMs and precious metals, quartzite-like porous and fractured Aleuropean sandstones are also impacted.

On the northern outskirts of the village of Novoshevchenkovskoye, a cluster of poorly rounded, often sharp-angled fragments of dark gray rocks was uncovered under the soil and vegetation layer. A rock with a cellular texture, magnetic to varying degrees. Among the fragments of the rock there are rounded "blanks" up to 8 cm in size, of which about 93.0% is iron. In the rock itself, the main place is occupied by high–lying minerals—fayalite, wustite, titanomagnetite, etc. In addition, abnormal concentrations of MPG and precious metals were determined (Table 4).

According to the above analysis results, these rocks are classified as impactites.

To the east of the village of Yablochnoye, under the soil and vegetation layer, as in the previous case, an accumulation of similar fragments of rocks was found. In this case, the rocks are not magnetic. Their composition is dominated by the X-ray amorphous phase, which is about 90%. Minerals with a crystalline structure are represented by cristobalite (6.8%), quartz (2.1%), and eckermanite (1.1%). In addition, MPG, silver and gold were determined in the rock (Table 4).

It can be assumed that the rock was formed under conditions corresponding to the highest temperature values and very rapid cooling, as a result of which the molten material did not have time to crystallize. Perhaps this place was the epicenter of the explosion. The rock formation belongs to the impactites.

On the northwestern outskirts of the city of Lipetsk, in an abandoned quarry at a depth of 14 m, a layer of fine-celled (porous) rock 15–20 cm thick was discovered. In appearance, the rock is similar to the previously described rock from the area of the village of Gnezdilovo. In the mineral composition of the rock, the main place belongs to gehlenite (with point inclusions of melilite) (86.1%) and carbonates (2.6%). In addition, stishovite, ruthenium, mullite, silver, etc. were identified (Table 3).

Helenite, which predominates in the composition of the rock, is included in the isomorphic series of minerals of the melilite group. It is found in carbonaceous chondrites, it is also found in the Allende meteorite and on the comet 81P/Wild. In nature, galena occurs in ultrabasic alkaline and contact-metasomatic rocks. According to energy dispersion definitions, abnormal concentrations of PGMs, silver and gold were also determined in the rock. According to the above analysis results, the rocks are impactites.

6 Conclusion

During the field study, we found that the outer shaft near a number of rounded depressions was well preserved or fragments of it remained, distinguished by light stripes on satellite images. The mineral composition of the rocks inside the depressions and the surrounding shafts is identical. Given the modern geological processes taking place on the earth's surface, it can be assumed that the territory of the Central Chernozem region was subjected to intense bombardment by meteoritic material. There is reason to believe that this happened at least twice in the Late Pleistocene—early Holocene.

The rounded depressions observed on the Earth's surface are depressions formed during an explosion in celestial bodies and during their contacts with the earth and water surface. Rocks formed during explosions are impactites. Confirmation of the impact nature arising from impact metamorphism is the following: specific fracturing of rocks, which is described in the Gnezdilovo area; polymorphic modifications of quartz group minerals (stishovite, tridimite, cristobalite) contained in the rocks, characterized by formation conditions. Their genesis is associated with high temperatures and pressures arising from explosions of cosmic bodies and their impact on the Earth's surface. Tridymite and cristobalite have been identified in rocks from the Gnezdilovo area, and stishovite has been established in rocks from the Khleborob district and in the Lipetsk Oblast. The latter is a mineral of impact meteorite craters, a superdense type of silica (4.35 g/cm^3, synthesized at a pressure of 135 thousand atm.).

The formation of impact rocks is also confirmed by abnormally high concentrations of platinum group metals, which are millions of times higher than the clarke of the Earth's crust (Beus 1972; Ovchinnikov 1990; Gadiyatov and Biryulkin 1992; Gadiyatov 1992), and from several thousand to the first tens of thousands of times higher than their contents in chondrites (Palme et al. 2004). Among PGMs, the maximum concentrations are iridium, osmium and rhenium. For example, in the quartzite rock of Gnezdilovo the clarke of concentration (Cc) of iridium in relation to the clarke of chondrites is 202,128, in the rocks of Yablochnoe the Cc = 225,532. All rocks also have abnormally high concentrations of gold (Cc from 750,000, relative to the clark of the Earth's crust). With such high concentrations of PGMs and gold, the rocks are of industrial interest and deserve exploration work.

Thus, rounded depressions on the territory of the Central Chernozem region represent impact craters formed when cosmic bodies fell to the Earth's surface.

References

Beus AA (1972) Geochemistry of the lithosphere. Moscow, 296 p (in Russian)

Dubyansky AA (1935) Preliminary information about volcanic ash located in Pavlovsk (Voronezh Oblast). Proc Central Sci Res Geol Exploration Inst 39:3–18 (in Russian)

Ermolov VA (2008) Fundamentals of geology. Moscow State Mining University Publishing House, 598 p (in Russian)

Firestone RB, West A, Kennett JR et al (2007) Evidence for an extraterrestrial impact 12900 years ago that contributed to the megafaunal extinctions and the Younger Dryas cooling. In: Firestone RB, West A, Kennett JR (eds). PNAS, vol 104, pp 16016–16021

Gadiyatov VG (1992) Geochemical methods of searching for gold mineralization in the greenstone strata of the west of the Aldan shield. In: Abstract of the dissertation for the degree of D of Geological and mineralogical Sciences. Irkutsk, 17 p (in Russian)

Gadiyatov VG, Biryulkin GV (1992) Background gold content of Precambrian greenstone formations of the Temulyakite-Tungurchinsk folded zone of the west Aldan shield. Geochem Int 29:138–141

Gerasimov MV (1984) Shock degassing of the earth and terrestrial planets. In: Abstract for the degree of Ph.D.-Mathematical. Moscow, 20 p (in Russian)

Grishchenko MN (1976) Pleistocene and Holocene of the Upper Don basin. Moscow, 229 p (in Russian)

Iosifova YI (1992) On the age of the ash-containing strata in the section near the village of Gorelka, Voronezh Oblast. Phanerozoic Stratigraphy of the Center of the East European Platform, pp 36–59 (in Russian)

Kholmovoy GV (2008) Morphoscopic features of neogene and quaternary volcanic ash in the Voronezh region. Bull Voronezh State Univ Ser Geol 1:19–22 (in Russian)

Kholmovoy GV (2013) Geological map of pre-Quaternary sediments. In: The book: ecological-geographical atlas-book of the Voronezh Oblast. Voronezh State University Publishing House, 40–42 (in Russian)

Medvedev AP, Zhabin AV (2019) Archaeological discovery of an accumulation of impactites in a Late Bronze Age building on the Elan River. Bull Voronezh State Univ Ser History Polit Sci Sociol 3:5–20 (in Russian)

Milkov FN (1977) Genesis and genetic series of landscape complexes. In: Geoscience, vol 12. Moscow Society of Natural Scientists, College of Works, New Series. Moscow University Publishing House, pp 5–11 (in Russian)

Ovchinnikov L.N. 1990. Applied geochemistry. Moscow, 246 p (in Russian)

Palme H, O'Neill HSC (2004) Cosmochemical estimates of mantle composition, 2. In: Holland HD, Turrekian KK (eds) Treatise on geochemistry. Elsevier, Amsterdam, The Netherlands, 1–38

Savko AD, Manukovsky SV, Mizin AI (2001) Lithology and facies of pre-Neogene deposits of the Voronezh anteclise. Proc Sci Res Inst Geol Voronezh State Univ 3:1–2001 (in Russian)

Tselmovich VA, Sungatullin RK, Vafin RA (2015) Impact origin of Lake Rabiga Kul. In: All-Russian annual seminar on experimental mineralogy, petrology and geochemistry VESEMPG. Abstract. Reports. Moscow, p 132 (in Russian)

Zhabin AV, Savko AD, Sirotin VI (2008) Clay minerals of the sedimentary cover of the Voronezh Anteclise. Proc Sci Res Inst Geol Voronezh State Univ 51:1–92 (in Russian)

Zhabin AV, Dubkov AA, Zolotareva GS, Kuznetsov YuA (2020a) Impactites in quaternary deposits of the Voronezh anteclise. Subsoil Volga Caspian Regions 102:18–31 (in Russian). https://doi.org/10.24411/1997-8316-2020-11023

Zhabin AV, Dmitriev DA (2019) Impact events in the Late Pleistocene and Holocene. Facies analysis in lithology. Moscow State University (in Russian)

Zhabin AV, Zhabina AA (2019) The influence of impact events on global changes in the natural environment in the Late Pleistocene and Holocene. In: Global climate changes: regional effects, models, forecasts. Mathematical international scientific and practical conference, vol 1. Voronezh, pp 40–43 (in Russian)

Zhabin AV, Ignatenko IM, Donetsky SV (2020) On the origin of woodcanic ash deposits on the territory of the Voronezh region. In: Subsoil of the Volga and Caspian regions, vol 102, pp 39–52 (in Russian)

Variations of Multifractal Scaling of Spatial Distributions of Seismic Energy Before the Preci Earthquake (Central Italy, 2016, Mw = 6.6)

I. R. Stakhovsky

Abstract The results of multifractal analysis of spatial distributions of seismic energy (seismic energy fields) in the epicentral area of the 30.10.2016 Central Italy earthquake ($M \approx 6.6$) are presented. It is shown that before the main shock of this event the singularity spectrum of seismic energy field got significantly widen and changed the form of asymmetry. From the physical point of view these variations may be explained as a consequence of seismogenerating system transition in strongly non-equilibrium state. The quantitative characteristics of the observed effects are proposed. These characteristics may be used for the purposes of the seismogenerating system current state monitoring.

Keywords Earthquake · Seismic energy · Multifractal · $f(a)$-spectrum asymmetry · $f(a)$-spectrum widening

1 Introduction

The modern paradigm of seismogenesis explicitly takes into account the nonequilibrium nature of seismicity, which makes it possible to consider the process of lithospheric earthquake preparation as the transition of the seismogenerating medium in the focal area of the forthcoming earthquake from a weakly nonequilibrium state into a strongly nonequilibrium one. The theory of nonequilibrium systems (Prigogine 1980) asserts that the transition of the physicochemical system into a strongly nonequilibrium state leads to the growth of fluctuations of its parameters and to the formation of scale-invariant dissipative structures, which can be identified by methods of structural, for example, multifractal analysis of the system (Mandelbrot 1989; Takens and Verbitski 2000; Ouadfeul 2012).

I. R. Stakhovsky (✉)
Schmidt Institute of Physics of the Earth, Russian Academy of Sciences, Moscow, Russian Federation
e-mail: stakhov@ifz.ru

T. Chaplina (ed.), *Processes in GeoMedia—Volume VIII*, Springer Geology,
https://doi.org/10.1007/978-981-97-6627-7_10

In recent decades it has been shown that spatial distributions of seismic epicenters can be well modeled by multifractal measures (Geilikman et al. 1990; Hooge et al. 1994; Stakhovsky and Belousov 1996). In the following, multifractal measures modeling the spatial distributions of seismic epicenters will be denoted by the term "seismic fields" for brevity. In the paper (Stakhovsky 2002) it was found that the transition of the seismogenerating system into a strongly nonequilibrium state is accompanied by characteristic changes in the structure of the seismic field in the focal area of the forthcoming earthquake, which were called the effect of a seismic field $f(a)$-spectrum widening (it is described below).

The effect of a seismic field $f(a)$-spectrum widening precedes the main shock of a strong earthquake and can be considered as a method of monitoring the current state of the seismogenerating medium, but its accurate determination requires very representative seismic data. In the analysis of seismic fields, the lack of data leads in calculations to artificial scaling break, as well as to the loss of weak perturbations of the $f(a)$-spectrum, which are the subject of the search, which actualizes the development of other methods of structural analysis of the seismic process.

The present work was aimed at further investigation of structural variations of seismicity in the process of preparation of a strong earthquake, i.e., investigation of the structural evolution of the seismogenerating system in the strongly nonequilibrium state preceding the main shock of a strong earthquake. It is shown below that the period of transition of a seismogenerating system into a strongly nonequilibrium state can be detected by analyzing the spatial distribution of seismic energy (instead of the distribution of seismic epicenters). In the following, multifractal measures modeling the spatial distributions of seismic energy will be denoted by the term "seismic energy fields" for brevity. Seismic energy field analysis provides additional advantages when it is necessary to distinguish a seismogenerating system that is in the process of preparing for a large earthquake from a system in which such preparation has not yet begun.

2 Seismic and Seismic Energy Fields

In (Stakhovsky 2002), it was found that in the vicinity of strong earthquakes in the last ~ 2.5 years before their main shock, the spatial distributions of epicenters of weak seismicity show an increase in fluctuations of seismic activity, i.e., local short-term bursts or decays in the generation of seismic events. Because of the high intermittency inherent in the seismic process as a whole, these fluctuations cannot be singled out as an independent object of study; however, in multifractal measures modeling the distributions of seismic epicenters, the growth of seismic activity fluctuations in cumulative form manifests itself as an increase in the width of the $f(a)$-spectrum of the seismic field (i.e., as an increase in the difference between the maximum and minimum values of the seismic field singularity indices) compared to an earlier field. Note that in accordance with the theory of nonequilibrium systems (Prigogine 1980), the described result can be regarded as theoretically expected.

This effect allows to distinguish the steady seismic regime, i.e., the weakly nonequilibrium state of the seismogenerating system, from the transient regime, i.e., the strongly nonequilibrium state preceding the global destruction. In the steady regime, numerous but "weak" (in the scale of the system under study) events do not destroy the global integrity of the system, do not change the boundary conditions that determine the evolution of the system, and do not lead to changes in its dynamics. The properties of the seismic process in the steady regime are described by the topological characteristics of its attractor (Konstantinou and Lin 2004; Stakhovsky 2016). In the transient regime, the entire system under consideration is covered by the process of preparation of an event that will further lead to the destruction of the system, i.e., the process of preparation of a "strong" (in the scale of the system under study) earthquake. In the steady regime, the $f(a)$-spectra of seismic fields are subject to only weak variations; in the transient regime, before the global rupture, the singularity spectra experience a significant widening.

The study of the effect of seismic fields $f(a)$-spectra widening requires large amounts of data. The theoretical consideration of the construction of the $f(a)$-spectrum for a scale-invariant distribution of single-type events (spatial distribution of epicenters) leads to estimates of the required data volumes, which, as a rule, exceed the capabilities of modern seismic catalogs (Eckmann and Ruelle 1992). At the same time, there is a firmly established power-law dependence between seismic energy and the number of events inducing this energy [the Gutenberg-Richter law (Gutenberg and Richter 1956)]. Thus, if the measures modeling the spatial distributions of seismic epicenters (i.e., seismic fields) are multifractals, then the measures modeling the spatial distributions of seismic energy (i.e., seismic energy fields) are also subject to multifractal organization. The action of the Gutenberg-Richter law leads to the fact that when spatial distributions of seismic energy are covered by the renormalizable scale grid used in multifractal analysis, the total energy of seismic events falling into the box of the scale grid is no longer determined by the number of events, but by the energy of only a few events with maximum magnitudes. The other events, regardless of their number, contribute to the total sum only corrections, predominantly small compared to the absolute value of the sum.

Thus, the analysis of seismic energy fields has several advantages over the analysis of seismic fields. First, the total numbers of events in the boxes of the scale grid usually differ by no more than one or two orders of magnitude, while the total seismic energy of these events may differ by three or four orders of magnitude and more, which serves as a kind of multiplier of the variations under study. Second, in the analysis of seismic energy fields, the dependence of the calculation accuracy in constructing $f(a)$-spectra on the number of events available for study is qualitatively reduced. In accordance with the physics of the process, the variations of the seismic and seismic energy fields related to each other should be synchronous. Variations of the seismic process associated with the transition of the seismogenerating system into a strongly nonequilibrium state preceding the main shock of a strong earthquake can be detected in seismic energy fields with a higher probability than in seismic fields. This is illustrated below by analyzing the seismic process leading up to the

October 30, 2016 Central Italy earthquake, which was the strongest earthquake in Italy in the previous 36 years.

3 Initial Data

The earthquake of Central Italy on October 30, 2016 has happened at 06:40:19 (UTC) in Central Apennines near the administrative border between Umbria and Marche regions (Galli et al. 2016; Zhu et al. 2016) (Fig. 1). Its magnitude was $M \cong 6.6$ and the hypocenter depth was $H \approx 9.6$ km. The epicenter had coordinates 42.83° N, 13.10° E and was located about 60 km to the north–north-west from the town of L'Aquila, destroyed in 2009 by the earthquake with magnitude $M = 6.3$. The nearest settlement to the epicenter of the earthquake in 2016 was the agricultural community of Preci, so this event is often called the Preci earthquake. It also caused significant damages in nearby villages, particularly in Arquata del Tronto. In the neighboring town of Norcia, medieval buildings such as the Basilica of St. Benedict and the Church of San Salvatore collapsed, that was not without the local population casualties.

In seismic terms, the region is controlled by the Italian National Seismic Network, which consists of more than 400 modern seismic stations, of which at least 30 are located in or near the epicentral area of the Preci earthquake (Michelini et al. 2016). Seismic information is transmitted to the International Seismological Centre (ISC) in Great Britain (International Seismological Centre, Summary of the Bulletin of the

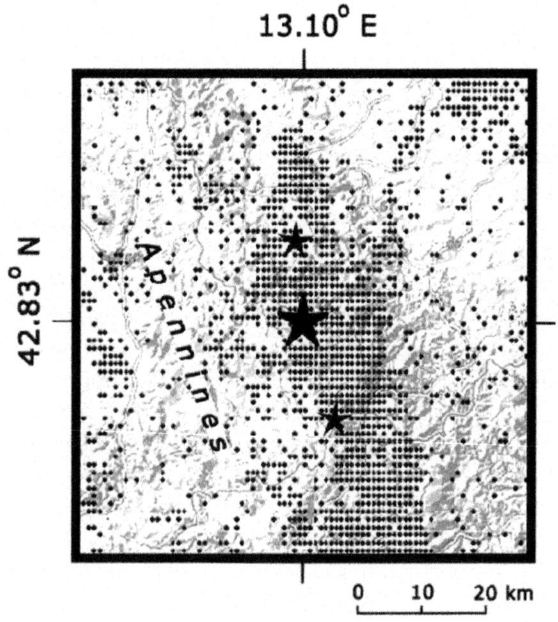

Fig. 1 Spatial distribution of seismicity in the vicinity of the epicenter of the Preci earthquake. Dots indicate the locations of epicenters of earthquakes with $M \geq 2$ for the period from 01.04.2006 to 30.10.2016. Small asterisks marked the locations of epicenters of two strongest foreshocks—24.08.2016 with $M = 6.0$ and 26.10.2016 with $M = 6.1$. Large asterisk in the center—epicenter of the Preci earthquake 30.10.2016 with $M = 6.6$

International Seismological Centre 2019), where it (together with data from other seismic networks) is systematized in the ISC catalog (International Seismological Centre 2000).

Despite the fact that the Preci earthquake had not very high magnitude value even by the standards of European seismicity, it was preceded by a rare phenomenon—a long period of extremely high seismic activity in a relatively small area with a radius of ~ 30 km in the vicinity of the epicenter of the forthcoming event. So, the seismicity preceding this earthquake will be investigated further within the boundaries of the polygon with the size of 70 × 70 km with sublatitudinal and submeridional orientation of the sides and the center coinciding with the epicenter of the Preci earthquake. The size of the investigated polygon was chosen as the minimum possible one, provided that it covers the whole area of high seismic activity before this earthquake. For the territory within the boundaries of the polygon, and for the period of time 1.08.2016 ÷ 30.10.2016 ISC-catalogue provides information about more than 20,000 earthquakes with $M > 0$, which corresponds to the intensity of seismic events generation on average ~ 10 events per hour during 3 months (at that, the real distribution of events in time is quite heterogeneous).

The continuous flow of seismic energy radiated by the seismogenerating system during the mentioned period allows to say that during all this time before the main shock of the Preci earthquake the seismogenerating system was in a strongly nonequilibrium state, not passing to the unloading state. The strongest foreshocks of the Preci earthquake occurred on August 24 ($M = 6.0$) and October 26 ($M = 6.1$) (see Fig. 1). The period of strong nonequilibrium of the lithosphere material ended with the main shock of the Preci earthquake on October 30 ($M = 6.6$), followed by a series of aftershocks, which completed the unloading of the seismogenerating system from excessive tectonic stresses.

Below we compare structural characteristics of spatial seismic energy distributions within the designated polygon under the steady seismic regime (i.e., long before the main shock of the Preci earthquake under conditions of weak nonequilibrium of the medium) and under the transient regime (i.e., just before the main shock of the Preci earthquake under conditions of strong nonequilibrium of the medium). The ISC-catalog data were used in the calculations. The verification showed that for the study area and for the time period 2006 ÷ 2015 (in the steady seismic regime) the completeness magnitude of the catalog is $M \cong 2.0$ (Fig. 2). The ISC-catalog data satisfactorily follow the law of earthquake recurrence (Gutenberg-Richter law) in the range of magnitudes $M \geq 2.0$, while in the range of lower magnitudes a lack of data becomes significant. Accordingly, the information on events with $M < 2.0$ that were included in the catalog was not used in the calculations as unrepresentative.

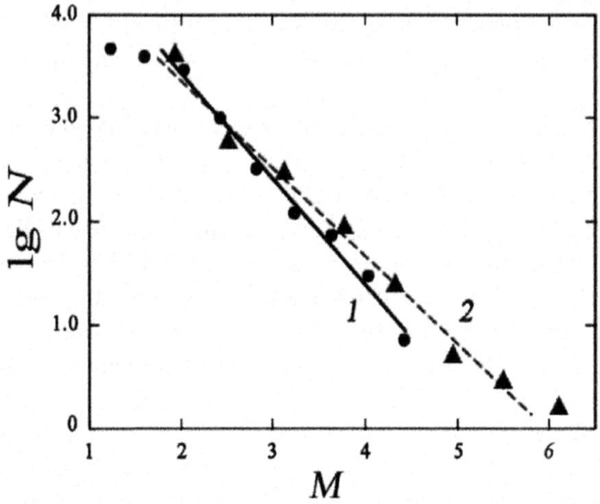

Fig. 2 Recurrence plots for the seismic data used in the calculations. Graph 1 uses data for 2006 ÷ 2015 (period of steady seismic regime). The points (circles) are calculated from catalog data, the straight solid line is a linear regression plot based on data with M ≥ 2.0. In the range M < 2.0, the catalog data lose representativeness. Graph 2 uses data for 01.01.2016 ÷ 30.10.2016 (period of transient seismic regime). The points (triangles) are calculated from the catalog data, the straight dashed line is a linear regression plot based on these data

4 Calculation Techniques

The magnitudes of seismic events systematized in catalogs are related to the seismic wave energy by empirical relationships, among which one of the popular ones is the Bath formula (Bath 1973):

$$\lg E = 5.24 + 1.44 M_S \quad (1)$$

where E—seismic energy (J), M_S—surface-waves magnitude. A number of alternative formulas for relating magnitudes to energies are known, which lead to slightly different absolute values of energies, but further we will calculate only the parameters of the energy scaling independent of the absolute values of energy.

During the scaling analysis, the polygon under study (the system under study) was covered by a two-dimensional renormalizable scaling grid of nonoverlapping square boxes. The grid boxes were numbered by index $i (i = 1,2, 3 \ldots)$. The spatial distribution of seismic energy was modeled by the measure P, the substance of which in the i-th grid box p_i was estimated by normalization:

$$p_i = E_i / E_0, \quad (2)$$

where E_i—the total value of seismic energy in the i-th box, i.e., the sum of seismic energies determined in accordance with expression (1) for all events falling into the

i-th box, E_0—the total value of seismic energy of all events in the investigated catalog data sample, i—is the ordinal index of the box. Let us use the algorithm of direct calculation of the singularity spectrum of the measure P (Chhabra et al. 1989). Let us introduce a new local characteristic of the measure:

$$\tilde{p}_i(q, r) = p_i^q(r) / \sum_j p_j^q(r), \tag{3}$$

where q—the order of the measure moment, r—grid box size (scale). Then the singularity spectrum of the measure P in parametric form can be calculated using the equations:

$$a(q) = \lim_{r \to 0} (\ln r)^{-1} \sum_{i=1}^{N} \tilde{p}_i(q, r) \ln p_i(r), \tag{4}$$

$$f(a(q)) = \lim_{r \to 0} (\ln r)^{-1} \sum_{i=1}^{N} \tilde{p}_i(q, r) \ln \tilde{p}_i(q, r), \tag{5}$$

where $a(q)$—singularity index characterizing the local self-similarity of the measure, $f(a(q))$—singularity spectrum characterizing the global self-similarity of the measure, N—the total number of non-empty grid boxes.

In the present work $f(a)$-spectra of seismic energy fields were constructed using Eqs. (4) and (5). The limits in (4) and (5) for calculating the parameters a and $f(a)$ were estimated as coefficients of linear regressions $\sum_{i=1}^{N} \tilde{p}_i(q, r) \ln p_i(r)$ and $\sum_{i=1}^{N} \tilde{p}_i(q, r) \ln \tilde{p}_i(q, r)$ on $\ln r$. The changes in the scale $r\,(r = r_1, r_2, r_3 \ldots)$ were carried out by renormalizing the grid. In numerical calculations, the values of q were sampled discretely with step $\Delta q = 0.05$ in the range of values $-10 \leq q \leq 10$.

Let us also introduce quantitative characteristics of $f(a)$–spectra:

$$\Delta a = a_{\max} - a_{\min}, \tag{6}$$

where Δa—the opening (width) of the $f(a)$-spectrum, $a_{\max}$ and $a_{\min}$—are respectively the maximum and minimum values of singularity indices of the studied fields, and

$$\mathbb{A} = \frac{2a_{ex} - a_{\max} - a_{\min}}{a_{\max} - a_{\min}}, \tag{7}$$

where $\mathbb{A}$—asymmetry parameter of the $f(a)$-spectrum, a_{ex}—value of the singularity index corresponding to the position of the extremum of the $f(a)$-spectrum. The asymmetry parameter $\mathbb{A}$ specified by formula (7) can accept values in the range $-1 < \mathbb{A} < 1$. The signature of $\mathbb{A}$ indicates the direction of spectrum extremum displacement relative to the center of the range Δa, negative values of $\mathbb{A}$ correspond to the displacement of the extremum towards low values of a, positive values of $\mathbb{A}$ correspond to the displacement of the extremum towards high values of a. For a symmetric spectrum $\mathbb{A} = 0$.

5 Results of the Calculations

The ISC-catalog data used in the calculations were divided into two samples: the 1st sample covered the time period from 01.04.2006 to 31.12.2015 and contained 3441 events with $M \geq 2.0$, the 2nd sample covered the time period from 01.01.2016 to 30.10.2016 and contained 3427 events with $M \geq 2.0$. The hypocenter depths of these events did not exceed 40 km (with single exceptions). The second sample characterized the seismic process immediately preceding the main shock of the Preci earthquake (the Preci earthquake itself was not included in the sample). Due to the large-scale swarm of seismic activity observed in the territory of the investigated polygon during the period of strongly non-equilibrium state of the medium before the main shock of the Preci earthquake, the number of events in the 2nd sample proved to be relatively large, although the 2nd sample covers a time interval of only 10 months. To avoid errors in calculations caused by different representativity of data, the 1st sample contained a similar number of events, but during the period of the steady seismic regime this number of events within the polygon boundaries occurred within 9.5 years. The recurrence plots constructed from the data of the 1st and 2nd samples are shown in Fig. 2 with the numbering corresponding to the sample number.

Figure 3 shows $f(a)$-spectra of seismic energy fields constructed from the data of the 1st and 2nd samples. When analyzing the spatial distributions of seismic energy, the $f(a)$-spectra can be constructed relatively complete. Further, for brevity, the subscripts under parameters $\mathbb{A}$, Δa and $f(a)$ will be used to identify them with the corresponding data samples.

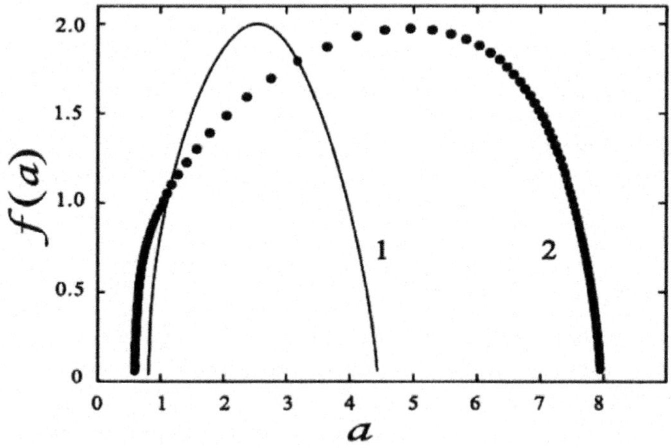

Fig. 3 Singularity spectra of seismic energy fields plotted using data from both samples. Numeral 1 (solid line) denotes the $f(a)$-spectrum constructed from the data of the first sample, numeral 2 (dots) denotes the $f(a)$-spectrum constructed from the data of the second sample

The event sets in the 1st and 2nd samples have no intersections, but the values of the extrema of $f(a)$-spectra (monofractal dimensions of seismic energy fields), as shown in Fig. 3, differ slightly $f(a_{ex})_1 \approx f(a_{ex})_2$. The differences are explained by inevitable boundary effects that artificially distort the distribution of seismic events (the studied samples are always limited in space, time, and magnitude). At the same time the $f(a)$-spectra widths are $\Delta a_1 = 3.62\ldots$ and $\Delta a_2 = 7.36\ldots$, i.e., they differ very significantly.

The transition of the seismogenerating system from a weakly nonequilibrium state (steady regime) to a strongly nonequilibrium state (transition regime), which preceded the main shock of the Preci earthquake, led to a change not only in the intensity, but first of all in the structure of the seismic process. Figure 3 demonstrates the effect of the multifractal field $f(a)$-spectrum widening, detected not in the seismic but in the seismic energy field. The transition of the seismogenerating system into a strongly nonequilibrium state has fundamentally changed the shape of the $f(a)$-spectrum of the seismic energy field.

The changes refer not only to the widths of the spectra $f(a)_1$ and $f(a)_2$ but also to their asymmetry. The extremum of the spectrum $f(a)_1$, constructed from the data of the first sample and characterizing the seismic process in the steady regime, is shifted towards low values of singularity indices. For the spectrum $f(a)_1$, the asymmetry parameter has a negative signature, $\mathbb{A}_1 = -0.044\ldots$. The extremum of the spectrum $f(a)_2$, constructed from the second sample data and characterizing the seismic process in the transient regime is shifted towards high values of singularity indices. The parameter $\mathbb{A}_2$ of the spectrum $f(a)_2$ has already a positive signature, $\mathbb{A}_2 = 0.17\ldots$.

The asymmetry of the $f(a)_1$-spectrum in the studied case is not large, but the fact that the asymmetry of the $f(a)$-spectrum of the seismic energy field changes during the preparation of a strong earthquake is apparently not accidental. The asymmetry of the $f(a)$-spectrum is caused by the ratio (proportion) between events with relatively low and relatively high values of magnitudes in the seismic flux. Long before the main shock of a strong earthquake, the part of relatively large-scale events in the total flux is small, but as we approach the moment when the system loses global stability, this part gradually increases. As can be seen from Fig. 2, a change in the proportion between events with relatively low and relatively high magnitudes in the seismic flux leads not only to a change in the $f(a)$-spectrum of the seismic energy field, but also to a change in the slope of the recurrence plot. When the system enters a strongly nonequilibrium state, the amplitudes of energy fluctuations increase, which, in accordance with the theory (Prigogine 1980), makes the change in the asymmetry of the $f(a)$-spectrum of the seismic energy field a natural phenomenon, apparently inherent in the process of preparation of any lithospheric earthquake.

Thus, during the period of strong nonequilibrium in the course of preparation of the Preci earthquake, fluctuation bursts and decay of seismic activity led both to an increase in the maximum values and a decrease in the minimum values of the singularity indices of the seismic energy field, which was manifested in a significant $f(a)$-spectrum widening. The relative increase in share of strong energy fluctuations (i.e., events of increased magnitude) in the seismic flux, also led to a change in the asymmetry parameter $\mathbb{A}$. The seismic process in the transient regime had rightly ended with a strong earthquake.

As it was proposed in Stakhovsky (2002, 2020), the effect of $f(a)$-spectra widening can be quantitatively estimated by the help of the functional δS:

$$\delta S = S^{**} - S^{*} = \int_{a^{**}_{\min}}^{a^{**}_{\max}} f^{**}(a)\mathrm{d}a - \int_{a^{*}_{\min}}^{a^{*}_{\max}} f^{*}(a)\mathrm{d}a, \tag{8}$$

where S is an area of the coordinate space enveloped by the $f(a)$-spectrum (and another characteristic of the $f(a)$-spectrum opening), the superscript in the form of two asterisks refers to the $f(a)$-spectrum of the seismic energy field that preceded the strong earthquake, the superscript in the form of one asterisk refers to the $f(a)$-spectrum of the earlier seismic energy field. The functions $f(a)$ in the integration domain are continuous, smooth and limited, i.e., the widening parameter δS can always be estimated numerically. In our case the widening parameter is $\delta S = 6.57\ldots$. The functional δS is convenient for quantitative evaluation of the effect, first of all, because it takes into account not only the width, but also the shape of the $f(a)$-spectra.

As we can see, the effect of $f(a)$-spectrum widening can be considered as a method of obtaining a quantitative characteristic of the process of preparation of a strong earthquake. However, the study of seismic energy fields also provides a qualitatively new possibility compared to seismic fields—the analysis of the asymmetry of $f(a)$-spectra.

6 Discussion of the Results and Conclusions

Detection of the period of strong nonequilibrium preceding a strong earthquake requires parameterization of the seismic process and reduction of its complex physical manifestations to quantitative parameters that allow comparison of the preparation processes of different seismic events. The quantitative characteristics of the effect of seismic energy field $f(a)$-spectrum widening can be considered suitable for this purpose. Although the construction of the distribution function of such a random variable as seismic energy is always associated with well-known uncertainties, many aspects of the effect of seismic energy field $f(a)$-spectrum widening can be discussed and interpreted quite specifically today.

The results of the study show that the monofractal dimensions of seismic energy fields are practically insensitive to the preparation of the critical transition (destruction) of the seismogenerating system, i.e., the sets of energy dissipation points in the lithosphere are predominantly stationary. Dissipation points (weakened points of the lithosphere, fault nodes, focal centers) appear to induce seismic events repeatedly without spatial migration. In other words, seismic energy fields (as well as seismic fields) are assumed to be multifractal fields with localized singularities. This aspect is important both in the mathematical modeling of the seismic process and in the modeling of the processes of crystalline rock fracture preparation at smaller scale levels (Berkowitz and Hadad 1997; Stakhovsky 1995), which requires an adequate choice of mathematical modeling procedures. But it must be recognized that today we have too weak seismic statistics to draw final conclusions about the evolution of the monofractal dimensions of seismic energy fields in time.

The effect of seismic energy field $f(a)$-spectrum widening is determined by the behavior of the higher moments of the multifractal measure modeling the spatial distribution of seismic energy. Physically, this means that the multifractal structure of the spatial distribution of seismic energy is created due to the energy activity of presumably localized dissipation points in the seismogenerating system. Changes in the state of the system, i.e., the transition from a weakly nonequilibrium state to a strongly nonequilibrium state, are reflected precisely in the energy activity of the dissipation points, which is manifested primarily in the growth of seismic energy fluctuations.

The effect of seismic energy field $f(a)$-spectrum widening combines into a single integral effect the details of a rather complex process of evolution of a seismogenerating system in a strongly nonequilibrium state. Nevertheless, the possibility of describing the effect by scalar characteristics (e.g., δS, $\mathbb{A}$) allows to speak about seismic scaling as a macroparameter of the seismogenerating system, providing an opportunity to measure, in a certain sense, the process of approaching the moment of a loss of system global stability. In this context, not only the numerical value of the widening parameter δS, but also the numerical value of the asymmetry parameter $\mathbb{A}$ can be useful.

As shown above, the change of seismic regimes can be accompanied by a change in the form of asymmetry of the seismic energy field $f(a)$-spectrum. The asymmetry of the $f(a)$-spectrum of the seismic energy field is associated with the absence of strong energy fluctuations in the steady regime and with the growth of the fluctuation range in the transient regime, which leads to a change in the value or even an inversion of the signature of parameter $\mathbb{A}$. Previously, the asymmetry of $f(a)$-spectra could not be studied in the analysis of seismic fields, because the lack of data usually leads to the impossibility of constructing full right branches of $f(a)$-spectra of seismic fields.

Thus, the effect of seismic energy field $f(a)$-spectrum widening can be used both to detect the very fact of the existence of strong earthquakes preparation processes in the seismogenerating medium and to quantitatively describe these processes. It is closely related to the physics of the lithospheric earthquake preparation process and, apparently, is more convenient than the effect of seismic field $f(a)$-spectrum widening as a method of monitoring the current state of the seismogenerating medium. This is explained by the fact that the $f(a)$-spectrum of the seismic energy field quite weakly depends on the number and energies of events with low values of magnitudes, which are cut off from the studied data by its representativeness threshold. Variations in the scaling structure of seismic energy fields can be considered as a tool for analyzing the fine structure of the seismic process, which is not yet available for analysis by other methods.

It should also be noted that the effect of $f(a)$-spectrum widening before strong earthquakes for both seismic and seismic energy fields is determined by the common laws of evolution of nonequilibrium systems and is not directly related to the tectonic aspects of the seismogenerating system. Within the scope of the well-known concept of universality classes (Wilson 1979; Odor 2004), it can be reasonably assumed that this effect is also inherent in nonequilibrium systems of other physical nature. Thus, in the paper (Macek et al. 2011), devoted to astrophysics, it is argued that the variations of the magnetic field of solar wind can be successfully modeled in the framework of the multifractal formalism, while the "degree of multifractality" of the magnetic field "seems to be modulated by the solar activity", i.e., associated with solar flares (the term "degree of multifractality" is used in Macek et al. (2011) to denote the parameter Δa of the multifractal field, i.e., the width of its $f(a)$-spectrum). Further studies should show how this effect, detecting with the help of functional tools of various private fields of physics, is universal for nonequilibrium processes preceding critical transitions in complex systems.

References

Bath M (1973) Introduction to seismology. Basel and Stuttgart. Birkhauser Verlag, p 350

Berkowitz B, Hadad A (1997) Fractal and multifractal measures of natural and synthetic fracture networks. J Geophys Res Sol Earth 102(B6):12205–12218. https://doi.org/10.1029/97JB00304

Chhabra AB, Meneveau C, Jensen RV, Sreenivasan KR (1989) Direct determination of the $f(a)$-singularity spectrum and its application to fully developed turbulence. Phys Rev A 40(9):5284–5294

Eckmann JP, Ruelle D (1992) Fundamental limitations for estimating dimensions and Lyapunov exponents in dynamical systems. Physica D 56:185–191

Galli P, Castenetto S, Peronace E (2017) The macroseismic intensity distribution of the 30 October 2016 earthquake in central Italy (Mw 6.6): seismotectonic implications. Tectonics 36:2179–2191. https://doi.org/10.1002/2017TC004583

Geilikman MB, Golubeva TV, Pisarenko VF (1990) Multifractal patterns of seismicity. Earth Planetary Sci Lett 99(1/2):127–132

Gutenberg B, Richter CF (1956) Magnitude and energy of earthquakes. Ann Geofis 9(1):15–20

Hooge C, Lovejoy S, Pecknold S, Malouin F, Schertzer D (1994) Universal multifractals in seismicity. Fractals 2(3):445–449

International Seismological Centre (2000). On-line Bulletin. https://doi.org/10.31905/D808B830

International Seismological Centre, Summary of the Bulletin of the International Seismological Centre (2019), vol 53, no II. https://doi.org/10.31905/M8L1R7WI

Konstantinou KI, Lin CH (2004) Nonlinear time series analysis of volcanic tremor events recorded at Sangay Volcano Ecuador. Pure Appl Geophys 161:145–163. https://doi.org/10.1007/s00024-003-2432-y

Macek WM, Wawrzaszek A, Carbone V (2011) Observation of the multifractal spectrum at the termination shock by Voyager 1. Geophys Res Lett 38:L19103. https://doi.org/10.1029/2011GL049261

Mandelbrot B (1989) Multifractal measures, especially for geophysicist. PAGEOPH 131(1–2):5–42. https://link.springer.com/journal/24/volumes-and-issues/131-1

Michelini A, Margheriti L, Cattaneo M, Cecere G, D'Anna G, Delladio A, Moretti M, Pintore S, Amato A, Basili A, Bono A, Casale P, Danecek P, Demartin M, Faenza L, Lauciani V, Mandiello AG, Marchetti A, Marcocci C, Mazza S, Mele FM, Nardi A, Nostro C, Pignone M, Quintiliani

M, Rao S, Scognamiglio L, Selvaggi G (2016) The Italian national seismic network and the earthquake and tsunami monitoring and surveillance systems. Adv Geosci 43:31–38. https://doi.org/10.5194/adgeo-43-31-2016

Odor G (2004) Universality classes in nonequilibrium lattice systems. Rev Mod Phys 76:663–724. https://doi.org/10.1103/RevModPhys.76.663

Ouadfeul S (ed) (2012) Fractal analysis and chaos in geosciences. IntechOpen, London, 184 p. https://doi.org/10.5772/3309

Prigogine I (1980) From being to becoming. W.H. Freeman and Co, San Francisco, p 200

Stakhovsky IR (1995) Modeling of crack aggregation in a nonequilibrium medium. Math Model 7(6):54–64 (in Russian)

Stakhovsky IR (2002) Widening of the $f(a)$-spectra of seismic fields in preparation zones of strong earthquakes. Izvestiya Phys Solid Earth 38(2):156–160

Stakhovsky IR (2016) Attractor reconstruction from the time series of information entropy of seismic kinetics process. Izvestiya Phys Solid Earth 52(5):740–753. https://doi.org/10.1134/S1069351316040091

Stakhovsky IR (2020) Variations of seismic scaling before strong earthquakes. In: Processes in GeoMedia, vol 1. Springer Geology. Springer Nature, Switzerland, Cham, pp 271–277. https://doi.org/10.1007/978-3-030-38177-6_29

Stakhovsky IR, Belousov TP (1996) Scale invariants in seismotectonics. Trans (Doklady) Acad Sci 347(2):252–255

Takens F, Verbitski E (2000) Multifractal analysis of dimensions and entropies. Reg Chaotic Dyn 5(4):361–382. https://doi.org/10.1070/RD2000v005n04ABEH000154

Wilson KG (1979) Problems in physics with many scales of length. Sci Am 241:158–179

Zhu C, Wang C, Shan X, Zhang G, Li Q, Zhu J, Zhang B, Liu P (2022) Rupture models of the 2016 central Italy earthquake sequence from joint inversion of strong-motion and InSAR datasets: implications for fault behavior. Remote Sens 14:1819–1841. https://doi.org/10.3390/rs14081819

Quantum Kinetic Theory and the Image Particle Method for Kinetic Simulation of Wakes in Plasma

Maxim Glebovich Ponomarev

Abstract This paper shows availability of the quantum mechanical model in addition to the previous classical physics, classical mechanics models. Quantum kinetic theory and the Image Particle Method are introduced to simulate charged particle fluxes, which allow to include effects of generated electric and magnetic fields in addition to all previous results for charged particle flows simulations in external electromagnetic fields. A gauge invariant Wigner function has been introduced in this paper for kinetic simulation of wakes, disturbances in plasma with formulation in terms of general electro-magnetic fields. This Wigner formulation is the natural language to describe quantum transport, and to monitor decoherence of macroscopic quantum states in interaction with the environment. According to this new "Image Particle Method", the collision integrals, corresponding simulated collision operators for these "image particles" describe the effects from the generated electric and magnetic fields.

Keywords Quantum transport · Interaction with the environment · Gauge invariant Wigner function · Wigner model · Quantum kinetic theory · Wakes · Disturbances · Weak coupling · Intermediate coupling · Strong coupling · Weak magnetization · Strong magnetization · Extreme magnetization · Geomagnetic fields · Magnetic fields · Charged particle flows · Magnetic field effect on ionized flows · Plasma · Geomagnetism · Charged particle sources · Explosive charged particle sources · Charged particle fluxes · Electric and magnetic field effects · Effect of induced electric and magnetic fields · Electromagnetic fields

1 Introduction

This article introduces yet another example from extensive number of innovative physical models with applications to real world problems published in Ponomarev (2022a, b, c, 2023a, b, c, d), Carati and Ponomarjov (2003), Gun'ko and Ponomarev

M. G. Ponomarev (✉)
Cambridge Research and Consultancy, Cambridge, UK
e-mail: maxim.pv@gmail.com

T. Chaplina (ed.), *Processes in GeoMedia—Volume VIII*, Springer Geology,
https://doi.org/10.1007/978-981-97-6627-7_11

(1993, 1995a, b), May et al. (2006), Iovea et al. (2012, 2013), Ponomarev et al. (2010, 2011, 2012a, b, 2013a, b, 2014, 2022), Ponomarjov (1995, 1996, 1997a, b, 1998, 1999, 2000a, b, c, d, e, f, 2001a, b, c, d, 2002), Ponomarjov and Gunko (1995), Ponomarjov and Carati (2003, 2004a, b, c, d, 2006, 2004), Webb and Ponomarev (2007).

Specifically, this paper extends previous results regarding non-quantum "Image Particle Method" (Ponomarev 2023a) to the "Image Particle Method" for quantum cases. This "Image Particle Method" reduces the problem of arbitrary generated (induced) electric and magnetic fields to consideration of interactions with image particles through the corresponding simulated collision operators.

These kinetic simulations extend directly previous results by Ponomarev MG, published in Ponomarev (2023a, b), Carati and Ponomarjov (2003), Gun'ko and Ponomarev (1993, 1995a, b), Ponomarev et al. (2011); Ponomarjov (1995, 1996, 1997a, b, 1998, 1999, 2000a, b, c, d, e, f, 2001a, b, c, d, 2002), Ponomarjov and Carati (2003, 2004a, b, c, d, 2006, 2004), Ponomarjov and Gunko (1995), Webb and Ponomarev (2007).

It is worth to note that this method is another new one in addition to previous "Imaginary Emission Method". In real problems these two methods can be used in the same models. Especially, the "Imaginary Emission Method" reduced the problem of wakes in plasma to consideration of charged particle flows from the "Imaginary Emission" sources (Ponomarjov 1995, 1996, 1998, 1999, 2000a, c, e, 2001a, b, c, d, 2002).

2 General Corresponding Points of Classical Mechanics and Quantum Mechanics

To describe a physical system there is a need to formulate the Hamiltonian function or operator, that is usually an expression for the energy of this physical system.

To formulate a complete quantum theory, there is a need in a definition of the Hamiltonian operator. This Hamiltonian operator defines the system dynamics. Due to an ambiguity in quantizing the classical Hamiltonian function, different ways can be explored. The explorations of these different ways generated different formulations of quantum mechanics theories.

Currently a number of different formulations of quantum mechanics are suggested including formulations by Schrödinger (1926), Wigner (1932), Feynman (1948), Keldysh (1965), Husimi (1940), Bohm (1952a, b).

Regarding the more recent advances, there are reviews in the more recent papers (Weinbub and Ferry 2018; Curtright et al. 2018; Brodin and Zamanian 2022).

A system can be described by the Hamiltonian function $H = H\left(\overrightarrow{r_j}, \overrightarrow{p_j}, t\right)$ or corresponding operator $\hat{H} = \hat{H}\left(\widehat{\overrightarrow{r_j}}, \widehat{\overrightarrow{p_i}}, t\right)$ for the quantum cases. These Hamiltonians usually define the energy of the corresponding systems. The usual explicit time dependence is due to time dependent external forces acting on the system.

Initial state (at time t = 0) is required to mark the start of the simulation. In classical case it is named as the initial position in a multidimensional phase space for the radius vector defined point $\left(\overrightarrow{r_j}(t), \overrightarrow{p_j}(t)\right)_{t=0}$, which represent the system. For the quantum cases there is a need for the state function (at time t = 0) $\psi\left(\overrightarrow{r_j}, t\right)_{t=0}$.

Furthermore, we have the first order (in time) equations of motion in both classical, and quantum models. These equations can be evaluated to model the state of the system for any lager time t > 0 both in classical $\left(\overrightarrow{r_j}(t), \overrightarrow{p_j}(t)\right)$, and in quantum models $\psi\left(\overrightarrow{r_j}, t\right)$.

Finally, the magnitude of the appropriate phase function $Q\left(\overrightarrow{r_j}(t), \overrightarrow{p_j}(t)\right)$ can be used to get the evolved value of an observable in classical mechanics. In quantum mechanics, corresponding quantity is calculated as expectation value $\langle\psi(t) \mid \hat{Q} \mid \psi(t)\rangle$ of an operator using the evolved final state-vector. So, the Table 1 (that is similar to the Table 1 from Buck and Merchant 2015) summarises this comparison of the two frameworks of classical mechanics and quantum mechanics.

Another table summarize the more general mathematical physics comparison of the two frameworks. This Table 2 corresponds to Table 2.1: "Comparison of classical and quantum framework" from Lein (2018).

Table 1 Comparison of the two frameworks of classical mechanics and quantum mechanics

	Classical model	Quantum model
Specification of system	Hamiltonian function $H = H\left(\overrightarrow{r_j}, \overrightarrow{p_j}, t\right)$	Hamiltonian operator $\hat{H} = \hat{H}\left(\widehat{\overrightarrow{r_j}}, \widehat{\overrightarrow{p_i}}, t\right)$
Initial state (at time t = 0)	Canonical variables $\left(\overrightarrow{r_j}(t), \overrightarrow{p_j}(t)\right)_{t=0}$	State function $\psi\left(\overrightarrow{r_j}, t\right)_{t=0}$
Equations of motion	Hamilton's equations $\dot{\vec{r}}_j = \frac{\partial H}{\partial \vec{p}_j}, \dot{\vec{p}}_j = -\frac{\partial H}{\partial \vec{r}_j}$	Schr¨odinger equation $\hat{H}\psi = i\hbar\frac{\partial\psi}{\partial t}$
Final state (at time t)	Evolved variables $\left(\overrightarrow{r_j}(t), \overrightarrow{p_j}(t)\right)$	Evolved state $\psi\left(\overrightarrow{r_j}, t\right)$
Observable property	Phase function $Q\left(\overrightarrow{r_j}(t), \overrightarrow{p_j}(t)\right)$	Expectation value $\langle\psi(t) \mid \hat{Q} \mid \psi(t)\rangle$

Table 2 Major general corresponding points of classical mechanics and quantum mechanics

	Classical model	Quantum model
Observables, the spectrum is the set of all possible outcomes of a measurement	Real valued functions	Selfadjoint operators acting on Hilbert space
Building block observables	Position $\vec{r}$ and momentum $\vec{p}$	Operators: Position $\hat{\vec{r}}$ and momentum $\hat{\vec{p}}$
States	Probability measures	Density operators
Pure states	Points in phase space	Wave functions $\psi(\vec{r}, t)$

3 Gauge Invariant Wigner Function for Kinetic Simulation of Disturbances and Wakes of Bodies in Plasma

The time-dependent gauge invariant Wigner function have already been investigated and published in (Stratonovich 1956; Serimaa et al. 1986; Levanda and Fleurov 2001).

First, a correct definition of the gauge invariant one-particle Wigner function $f_W(\vec{r}, \vec{v}, t)$ was published in Stratonovich (1956).

Because in this work we do not consider relativistic phenomena, we write Wigner function $f_W(\vec{r}, \vec{v}, t)$ in a non-covariant form,

$$f_W(\vec{r}, \vec{v}, t) = \left(\frac{m_i}{2\pi\hbar}\right)^3 \int d\vec{\xi} \exp\left(\frac{i\vec{\xi}}{\hbar} \cdot \left(m_i\vec{v} + e_i \int_{-1/2}^{1/2} d\theta \vec{A}\left(\vec{r} + \vec{\xi}\theta, t\right)\right)\right)\overline{\psi}\big(\vec{r} + \vec{\xi}/2, t\big)\psi\left(\vec{r} - \vec{\xi}/2, t\right) \quad (1)$$

Here, in Eq. 1, we have $\vec{r}$ corresponds to position and $\vec{v}$ is a velocity at time t, e_i is electric charge and m_i is a mass of a particle in a pure state (however, the properties to be considered here work equally well in the case of mixed states), which defined by the wave-function $\psi(\vec{r}, t)$. The wave-function $\psi(\vec{r}, t)$ supposed to have the probability normalization to unity. The time dependent vector function $\vec{A}(\vec{r}, t)$ describes the vector potential.

Furthermore, we note that here, in Eq. 1, the $f_W(\vec{r}, \vec{v}, t)$ is defined in terms of the kinetic momentum $m\vec{v}$. And this definition is different to the original definition of the Wigner function (Wigner 1932) via the canonical momentum. We see the additional integral in Eq. 1, which integrates the equation containing the vector potential $\vec{A}(\vec{r}, t)$. This additional integral makes compensation for the change in the wave-function $\psi(\vec{r}, t)$ during a local gauge transformation.

For our goals, the use of this specific form according to Eq. 1 is due to efficiency for the further applications, which provide a non-ambiguous way to calculate the moments as averaged values.

For a case of the usual (gauge-dependent) Wigner function $f_W^{gd}(\vec{r}, \vec{p}, t)$ we have

$$f_W^{gd}(\vec{r}, \vec{p}, t) = \left(\frac{1}{2\pi\hbar}\right)^3 \int d\vec{\xi} \exp\left(\frac{i\vec{p} \cdot \vec{\xi}}{\hbar}\right)\overline{\psi}\left(\vec{r} + \vec{\xi}/2, t\right)\psi\left(\vec{r} - \vec{\xi}/2, t\right) \quad (2)$$

This Eq. 2 is formulated in terms of the canonical momentum $\vec{p}$, so that

$$\vec{p} = m_i\vec{v} - e_i\vec{A} \quad (3)$$

We suppose that all the physical objects should be gauge independent.

For time dependent gauge-independent Wigner function, according to Serimaa et al. (1986), we have

$$\frac{\partial f_W(\vec{r},\vec{v},t)}{\partial t} + \left(\vec{v} + \delta\hat{\vec{v}}\right) \cdot \frac{\partial f_W(\vec{r},\vec{v},t)}{\partial \vec{r}} + \frac{e_i}{m_i}\left(\left[\vec{v} + \delta\hat{\vec{v}}, \hat{\vec{B}}\right] + \hat{\vec{E}}\right) \cdot \frac{\partial f_W(\vec{r},\vec{v},t)}{\partial \vec{v}} = Q_i(\vec{r},\vec{v},t) \quad (4)$$

and we have the following operators here:

$$\delta\hat{\vec{v}}(\vec{r},t) = ie_i \frac{\hbar}{m_i^2}\frac{\partial}{\partial \vec{v}} \times \int_{-1/2}^{1/2} \xi d\xi \vec{B}\left(\vec{r} + i\frac{\hbar}{m_i}\xi\frac{\partial}{\partial \vec{v}}, t\right) \quad (5)$$

$$\hat{\vec{E}}(\vec{r},t) = \int_{-1/2}^{1/2} d\xi \vec{E}\left(\vec{r} + i\frac{\hbar}{m_i}\xi\frac{\partial}{\partial \vec{v}}, t\right) \quad (6)$$

$$\hat{\vec{B}}(\vec{r},t) = \int_{-1/2}^{1/2} d\xi \vec{B}\left(\vec{r} + i\frac{\hbar}{m_i}\xi\frac{\partial}{\partial \vec{v}}, t\right) \quad (7)$$

Here $\vec{E} = \vec{E}(\vec{r},t)$ and $\vec{B} = \vec{B}(\vec{r},t)$ are time-dependent electric and magnetic fields (respectively).

$Q_i(\vec{r},\vec{v},t)$ simulates the sources of particles.

This Eq. 4 deducted from the Schrödinger equation for the wave-function or, as an alternative, from the von Neumann equation solved by the density matrix.

Furthermore, Eq. 4 looks very similar to the Boltzmann equation. However, there are three differences:

(1) the operators $\hat{\vec{E}}$ and $\hat{\vec{B}}$ replace the electromagnetic fields $\vec{E}$ and $\vec{B}$ in accordance to Eq. 6 and Eq. 7;
(2) added operator of quantum mechanical perturbation $\delta\hat{\vec{v}}$ as a shift of the velocity vector $\vec{v}$ according to Eq. 5;
(3) the Wigner function can be negative in some regions of phase space, therefore it is not exactly a probability distribution.

It should be noted that the quantum mechanical perturbation $\delta\hat{\vec{v}}$ (according to Eq. 5) disappears in the electrostatic case.

4 Kinetic Simulation of Disturbances and Wakes in Plasma, Image Particle Method

Let's consider induced electric $\overrightarrow{E_{ind}}$ and induced magnetic $\overrightarrow{B_{ind}}$ fields of arbitrary, unlimited values so that

$$\overrightarrow{E} = \overrightarrow{E_0} + \overrightarrow{E_{ind}} \quad (8)$$

$$\overrightarrow{B} = \overrightarrow{B_0} + \overrightarrow{B_{ind}} \tag{9}$$

Then we have the Wigner equation, Eq. 4, in the following form:

$$\frac{\partial f_W(\vec{r}, \vec{v}, t)}{\partial t} + \left(\vec{v} + \delta\hat{\vec{v}}\right) \cdot \frac{\partial f_W(\vec{r}, \vec{v}, t)}{\partial \vec{r}} + \frac{e_i}{m_i}\left(\left[\vec{v} + \delta\hat{\vec{v}}, \hat{\vec{B}}\right] + \hat{\vec{E}}\right) \cdot \frac{\partial f_W(\vec{r}, \vec{v}, t)}{\partial \vec{v}} = Q_i(\vec{r}, \vec{v}, t) \tag{10}$$

$$\delta\hat{\vec{v}}(\vec{r}, t) = ie_i \frac{\hbar}{m_i^2} \frac{\partial}{\partial \vec{v}} \times \int_{-1/2}^{1/2} \xi d\xi \overrightarrow{B_0}\left(\vec{r} + i\frac{\hbar}{m_i}\xi\frac{\partial}{\partial \vec{v}}, t\right) + ie_i \frac{\hbar}{m_i^2} \frac{\partial}{\partial \vec{v}} \times \int_{-1/2}^{1/2} \xi d\xi \overrightarrow{B_{ind}}\left(\vec{r} + i\frac{\hbar}{m_i}\xi\frac{\partial}{\partial \vec{v}}, t\right) \tag{11}$$

$$\hat{\vec{E}}(\vec{r}, t) = \int_{-1/2}^{1/2} d\xi \overrightarrow{E_0}\left(\vec{r} + i\frac{\hbar}{m_i}\xi\frac{\partial}{\partial \vec{v}}, t\right) + \int_{-1/2}^{1/2} d\xi \overrightarrow{E_{ind}}\left(\vec{r} + i\frac{\hbar}{m_i}\xi\frac{\partial}{\partial \vec{v}}, t\right) \tag{12}$$

$$\hat{\vec{B}}(\vec{r}, t) = \int_{-1/2}^{1/2} d\xi \overrightarrow{B_0}\left(\vec{r} + i\frac{\hbar}{m_i}\xi\frac{\partial}{\partial \vec{v}}, t\right) + \int_{-1/2}^{1/2} d\xi \overrightarrow{B_{ind}}\left(\vec{r} + i\frac{\hbar}{m_i}\xi\frac{\partial}{\partial \vec{v}}, t\right) \tag{13}$$

where e_i is electric charge and m_i is mass of the "i" particle.

Let's rewrite Eq. 10 to see the effect of the perturbation operator $\delta\hat{\vec{v}}$

$$\frac{\partial f_W(\vec{r},\vec{v},t)}{\partial t} + \vec{v} \cdot \frac{\partial f_W(\vec{r},\vec{v},t)}{\partial \vec{r}} + \frac{e_i}{m_i}\left(\left[\vec{v}, \hat{\vec{B}}\right] + \hat{\vec{E}}\right) \cdot \frac{\partial f_W(\vec{r},\vec{v},t)}{\partial \vec{v}} = Q_i(\vec{r}, \vec{v}, t) - \delta\hat{\vec{v}} \cdot \frac{\partial f_W(\vec{r},\vec{v},t)}{\partial \vec{r}} - \frac{e_i}{m_i}\left[\delta\hat{\vec{v}}, \hat{\vec{B}}\right] \cdot \frac{\partial f_W(\vec{r},\vec{v},t)}{\partial \vec{v}} \tag{14}$$

The right part of Eq. 14 contains the image particle sources terms, corresponding simulated collision operators, which are similar to non-quantum case outlined in previous paper (Ponomarev 2023a).

It should be noted that we have the excellent correspondence here for the case of extremely small quantum effects. The imaginary particle sources term (in Eq. 7 from the previous paper Ponomarev 2023a) contains e instead of e_i and m instead of m_i because $e = e_i$ and $m = m_i$ for the cases considered in this previous paper

$$C^{image}(\vec{r}, \vec{v}, t) = -\frac{e}{m}\left(\left[\vec{v}, \overrightarrow{B_{ind}}\right] + \overrightarrow{E_{ind}}\right) \cdot \frac{\partial f_i(\vec{r}, \vec{v}, t)}{\partial \vec{v}} \tag{15}$$

This Eq. 15 has been derived there as a part of the Boltzmann equation (Eq. 8 from the previous paper Ponomarev 2023a) for the classical (non-quantum) case

$$\frac{\partial f_i(\vec{r},\vec{v},t)}{\partial t} + \vec{v}\cdot\frac{\partial f_i(\vec{r},\vec{v},t)}{\partial \vec{r}} + \frac{e}{m}\left(\left[\vec{v},\overrightarrow{B_0}\right] + \overrightarrow{E_0}\right)\cdot\frac{\partial f_i(\vec{r},\vec{v},t)}{\partial \vec{v}} = Q_i(\vec{r},\vec{v},t) + C^{image}(\vec{r},\vec{v},t) + \sum_{j=1}^{N} C_{ij}\big(f_i(\vec{r},\vec{v},t), f_j(\vec{r},\vec{v},t)\big) \quad (16)$$

and the corresponding isodensity contours have been plotted as well.

The result for the simulation of magnetic field effect on the wake for the circular disk is presented at the Fig. 1. The boundary condition on the disk surface corresponds to the case of absorbing disk surface.

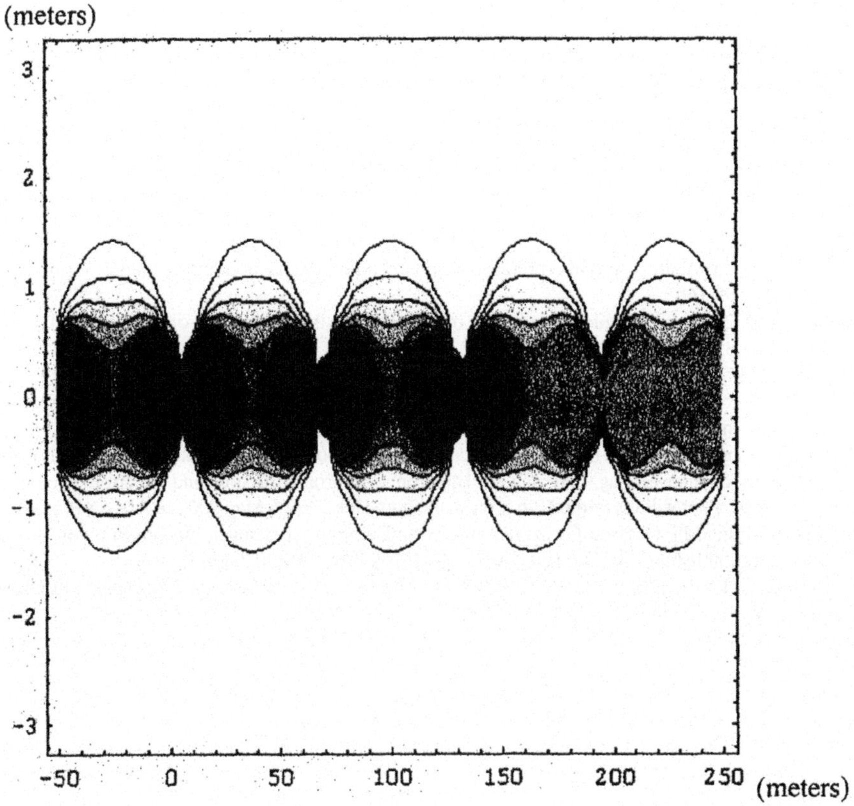

Fig. 1 Cross-sections of isodensity contours (at the plane of magnetic field) for the disturbances of ion concentration by the circular disk. The absorbing disk moves in parallel to the magnetic field. The center of the disk is located at the point 0 at the left side of this Figure. The brighter contours correspond to the higher ion concentrations. The darker contours correspond to the lower ion concentration

5 Conclusions

This paper first adapts Quantum Kinetic Theory and the Image Particle Method to introduce new method to simulate charged particle fluxes that allows to include effects of generated electric and magnetic fields in addition to all previous results for charged particle flows simulations in external electric and magnetic fields. These electric and magnetic fields can be unlimited and non stationary. A Wigner model has been introduced in this paper for kinetic simulation of wakes in plasma with formulation in terms of general electro-magnetic fields. This Wigner formulation is the natural language to describe quantum transport, and to monitor decoherence of macroscopic quantum states in interaction with the environment.

According to this new "Image Particle Method", the corresponding simulated collision operators with these image particles describe the effects from the generated electric and magnetic fields.

This paper shows availability of the quantum mechanical model in addition to the previous classical physics, classical mechanics models.

References

Bohm D (1952a) A suggested interpretation of the quantum theory in terms of hidden variables. I Phys Rev 85:166

Bohm D (1952b) A suggested interpretation of the quantum theory in terms of hidden variables. II Phys Rev 85:180

Brodin G, Zamanian J (2022) Quantum kinetic theory of plasmas. Rev Mod Plasma Phys 6:4

Buck B, Merchant AC (2015) Probabilistic foundations of statistical mechanics: a Bayesian approach. https://arxiv.org/abs/1512.01368

Carati D, Ponomarjov MG (2003) Statistical description of currents induced by two electron cyclotron counter-propagating waves. Electron cyclotron emission and electron cyclotron heating. In: Proceedings of the 12th joint workshop. Held 13–16 May 2002 in Aix-en-Provence, France. Edited by Gerardo Giruzzi (Association Euratom-CEA sur la Fusion, France). World Scientific Publishing Co. Pte. Ltd., pp 77–82. ISBN #9789812705082

Curtright TL, Fairlie DB, Zachos CK (2018) A concise treatise on quantum mechanics in phase space. World Scientific Publishing. ISBN 978-981-4520-43-0

Feynman RP (1948) Space-time approach to non-relativistic quantum mechanics. Rev Mod Phys 20:367

Gun'ko YF, Ponomarev MG (1993) The charged particles emission in a magnetic field. Vestnik Sankt-Peterburgskogo Universiteta Ser 1 Matematika Mekhanika Astronomiya 2:89–94

Gun'ko YF, Ponomarev MG (1995a) The charged particle layer expansion in a magnetic field. Vestnik Sankt-Peterburgskogo Universiteta Ser 1 Matematika Mekhanika Astronomiya 4:105–108

Gun'ko YF, Ponomarev MG (1995b) The charged particle distribution due to emission in a magnetic field. Astronomische Nachrichten 316(1):17–21

Husimi K (1940) Some formal properties of the density matrix. Proc Phys Math Soc Japan 22:264

Iovea M, Neagu M, Stefanescu B, Mateiasi G, Halai H, Amos M, Ponomarev M, Kappatos V, Selcuk C, Gan T-H, Gierl C (2012) Preliminary NDT investigation of sintered powder metallurgy parts by high-resolution TDI based X-ray digital radiography. In: European international powder metallurgy congress and exhibition. Euro PM 2012, vol 1. Basel, Switzerland. Code 105676

Iovea M, Neagu M, Stefanescu B, Mateiasi G, Clarke A, Nicholson I, Ponomarev M, Kappatos V, Selcuk C, Gan T-H (2013) PM parts fast in-line X-ray digital radiography. In: International powder metallurgy congress and exhibition. Euro PM 2013. European PM conference proceedings, vol 1. European Powder Metallurgy Association (EPMA)

Keldysh LV (1965) Diagram technique for nonequilibrium processes. Zh Eksp Teor Fiz Sov Phys JETP 20

Lein M (2018) Weyl quantization and semiclassics (Quantization and Semiclassics). University Lecture Notes. https://arxiv.org/pdf/1009.0444

Levanda M, Fleurov V (2001) A Wigner quasi-distribution function for charged particles in classical electromagnetic fields. Ann Phys 292:199

May J, Ponomarev M, Kuball S, Gallardo J (2006) A case for new statistical software testing models. In: Annual reliability and maintainability symposium. RAMS'06, article number 1677399. Newport Beach, CA, pp 349–353. Category number CH3744; Code 69734. https://doi.org/10.1109/RAMS.2006.1677399

Ponomarev MG (2022a) A novel physical model to enhance precision and performance of 3-dimensional force sensors. Process Geomedia 2(32):1589–1600

Ponomarev MG (2022b) Experimental validation of novel physical model for improvement of sensing 3-dimensional fluid flow loads and responses in real sea conditions with South Western Mooring Test Facility (SWMTF). Process Geomedia 2(32):1579–1589

Ponomarev MG (2022c) A novel physical model to enhance precision and performance of multi-dimensional force sensors. In: Physical and mathematical modeling of earth and environment processes—2022, proceedings of 8th international scientific conference-school. Springer, pp 347–364. ISBN 978-3-031-25961-6. https://doi.org/10.1007/978-3-031-25962-3_35

Ponomarev MG (2023a) New image particle method for kinetic simulation of wakes in plasma with weak and strong coupling. In: Proceedings of the 9th international conference on physical and mathematical modelling of earth and environmental processes. PMMEEP 2023. Springer proceedings in earth and environmental sciences. Springer, Cham. https://doi.org/10.1007/978-3-031-54589-4_59

Ponomarev MG (2023b) Exact analytical expression for maximum heat flux from impulsive sources of charged particles in magnetic field. Process GeoMedia VII. https://doi.org/10.1007/978-981-99-6575-5_31

Ponomarev MG (2023c) Physical foundations of mechanics: part 1: Newtonian mechanics. World of Science LLC, pp 1–83

Ponomarev MG (2023d) Physical foundations of mechanics: part 2: relativistic mechanics. World of Science LLC, pp 1–78

Ponomarev MG, Johanning L, Parish D (2010) Enhancing precision and reliability of tri-axial load cells for mooring load measurements. In: Proceedings of the 3rd international conference on ocean energy (ICOE2010). ISBN 978-84-693-5467-4

Ponomarev MG, Garrison BJ, Vickerman JC, Webb RP (2011) A molecular dynamics study of a 5 keV C60 fullerene impact on a two-component organic molecular sample. Surf Interface Anal 43(1–2):107–111

Ponomarev MG, Kappatos V, Selcuk C, Gan T-H, Amos M, Halai H, Gierl C, Iovea M (2012a) PM dimensional control: development of a digital radiographic inspection technique for production friendly quality assessment of powder metallurgy parts. In: Proceedings of the international euro powder metallurgy congress and exhibition. Euro PM, vol 1. Basel, Switzerland. Code 105676

Ponomarev MG, Selcuk C, Gan T-H (2012b) Construction of pattern recognition system optimized for X-ray inspection of plastic electronics and OLED displays. In: 51st annual conference of the British Institute of Non-Destructive Testing 2012. NDT 2012. BINDT. Northamptonshire, pp 354–360. Code 107048

Ponomarev MG, Kappatos V, Selcuk C, Gan T-H, Amos M, Halai H, Gierl C, Iovea M (2013a) Digital radiographic inspection technique for production friendly quality assessment of PM parts. Powder Metall 56(2):92–95

Ponomarev MG, Selcuk C, Gan T-H, Amos M, Nicholson I, Iovea M, Neagu M, Stefanescu B, Mateiasi G (2013b) Non destructive testing: a defect detection and classification system for automatic analysis of digital radiography images of powder metallurgy parts. In: International powder metallurgy congress and exhibition. Euro PM 2013. Gothenburg, Sweden. Code 110976

Ponomarev MG, Selcuk C, Gan T-H, Amos M, Nicholson I, Iovea M, Neagu M, Stefanescu B, Mateiasi G (2004) Defect detection and classification system for automatic analysis of digital radiography images of PM parts. Powder Metall 57(1):17–20

Ponomarev MG, Johanning L, Parish D (2022) Experimental validation of novel physical model for improvement of sensing multidimensional fluid flow loads and responses in real sea conditions with South Western Mooring Test Facility. In: Physical and mathematical modeling of earth and environment processes—2022, proceedings of 8th international scientific conference-school. Springer, pp 365–376. ISBN 978-3-031-25961-6. https://doi.org/10.1007/978-3-031-25962-3_36

Ponomarjov MG (1995) Disturbances of the ambient magnetoplasma due to its interactions with object surfaces. Imaginary emission method. Far-wake of objects moving through a rarefied plasma at different angles to the ambient magnetic field. Planet Space Sci 43(10–11):1419–1427

Ponomarjov MG (1996) Imaginary-emission method for modeling disturbances of all magnetoplasma species: reflecting and absorbing objects in motion through a rarefied plasma at different angles to the ambient magnetic field. Phys Rev E 54(5):5591–5598

Ponomarjov MG (1997a) Pressure of charged particle flows in ambient magnetic fields. Astron Nachr 318(3):187–192

Ponomarjov MG (1997b) New ways of protecting astronomical equipment and solar batteries of spacecrafts. Bull Am Astron Soc 29:1023

Ponomarjov MG (1998) Outer atmosphere and wake of space objects, kinetic simulation. Disturbances of ambient magnetoplasma due to diffuse reflecting bodies in motion. APS Div Plasma Phys Meet Abstr. Abstract id. U9Q.09

Ponomarjov MG (1999) Kinetic simulation of stratifications and flute structures of charged particle jets and wakes in the ambient magnetic field. Bull Astron Soc 31(4):1157

Ponomarjov MG (2000a) Space flows and disturbances due to bodies in motion through the magnetoplasma. Astrophys Space Sci 274(1):423–429

Ponomarjov MG (2000b) Simulation of oscillations in charged particle systems under the ambient magnetic field control. In: 2nd international conference. Control of oscillations and Chaos (COC 2000), vol 1. St. Petersburg, Russia. Code 57512. Proceedings (Cat. No.00TH8521), pp 167–170. https://doi.org/10.1109/COC.2000.873548

Ponomarjov MG (2000c) 3D collisional kinetic simulation of stratifications and flute structures of plasma flows and wakes in external magnetic fields. Bull Am Astron Soc 32:1022

Ponomarjov MG (2000d) 3D kinetic dynamical models of ionized HII clouds in external magnetic field. In: Ionized Gaseous Nebulae. Mexico City. Meeting abstract id. 44

Ponomarjov MG (2000e) 3D collisional kinetic simulation of stratifications and flute structures of plasma flows and wakes in external magnetic fields. In: 32nd annual meeting of the division for planetary sciences. Session 15. Outer planets IV-aurorae and magnetospheres. Pasadena, CA. Abstract 15.10. 1022

Ponomarjov MG (2000f) 3D time-dependent kinetic simulation of turbulent plasma flows under the effect of external magnetic field. In: American Physical Society, 42nd annual meeting of the APS division of plasma physics combined with the 10th international congress on plasma physics. Québec City. Canada Meeting ID: DPP00. Abstract id. NO1.007. https://ui.adsabs.harvard.edu/abs/2000APS.DPPNO1007P/abstract

Ponomarjov MG (2001a) Acceleration and transport of particles in collisionless plasmas: wakes due to the interaction with moving bodies. Astrophys Space Sci 277(1):39–44

Ponomarjov MG (2001b) Kinetic simulation of magnetic field effects on stratifications and flute structures of space plasma flows and wakes of bodies. In: 2001 joint assembly American geophysical union spring meeting. Boston, MA. Abstracts SM52B-04

Ponomarjov MG (2001c) Kinetic simulation of magnetic field effects on wakes of meteoroids imaginary emission method. European Space Agency (Special Publication) ESA SP 495:295–300

Ponomarjov MG (2001d) Kinetic modeling magnetic field effect on ion flows, disturbances and wakes in space plasma. In: Büchner J, Dum CT, Scholer M (eds), Space plasma simulation: proceedings of the sixth international school/symposium. ISSS-6. Garching, Germany. Berlin: Schaltungsdienst Lange o.H.G, p 328

Ponomarjov MG (2002) 3D time-dependent kinetic simulation of space plasma disturbances due to moving bodies with the ambient magnetic field effect. Adv Space Res 29(9):1397–1402

Ponomarjov MG, Gunko YF (1995) Kinetic modeling of charged particle cloud expansion and emission in magnetic and electric fields. Planet Space Sci 43(10–11):1409–1418

Ponomarjov MG, Carati D (2003) Kinetic simulations of relativistic electron flows in time-dependent electromagnetic fields. In: Electron cyclotron emission and electron cyclotron heating, conference proceedings. World Scientific Publishing Co Pte Ltd, pp 137–142

Ponomarjov MG, Carati D (2004a) Search for optimal 3D wave launching configurations for the acceleration of charged particles in a magnetized plasma: resonant moments method. In: American Physical Society. 46th annual meeting of the division of plasma physics. Savannah, GA. Meeting ID: DPP04. Abstract id. RP1.042

Ponomarjov MG, Carati D (2004b) Acceleration of charged particles by crossed cyclotron waves, resonant moments method. In: 35th COSPAR Scientific Assembly. Paris, France, p 2851

Ponomarjov MG, Carati D (2004c) Acceleration of electron populations by crossing EC waves in an external magnetic field. In: 13th joint workshop on electron cyclotron emission and electron cyclotron resonance heating (conference paper). Nizhny Novgorod, Russia. https://ec13.iapras.ru/papers/Ponomarjov.pdf

Ponomarjov MG, Carati D (2004d) Enhanced acceleration of charged particles by crossing electromagnetic waves in a magnetized plasma, resonant moments method. https://hal.archives-ouvertes.fr/hal-00001926v1

Ponomarjov MG, Carati D (2006) Enhanced acceleration of electrons populations by crossing electron cyclotron waves in an ambient magnetic field, resonant moments method. Adv Space Res 38(8):1576–1581

Ponomarjov MG, Carati D (2004) Search for optimal 2D and 3D wave launching configurations for the largest acceleration of charged particles in a magnetized plasma. Resonant Moments Method. arXiv 2004. https://arxiv.org/format/physics/0411075. arXiv:physics/0411075 [physics.plasm-ph]

Schr¨odinger E (1926) Quantisierung als Eigenwertproblem. Annalen der Physik 385:437–490

Serimaa OT, Javanainen J, Varró S (1986) Gauge-independent Wigner functions: general formulation. Phys Rev A 33:2913

Stratonovich RL (1956) Gauge invariant generalization of Wigner distribution. Dok Akad Nauk SSSR

Webb RP, Ponomarev MG (2007) Molecular dynamics simulation of low energy cluster impacts on carbon nanotubes. Nucl Instrum Methods Phys Res Sect B Beam Interact Mater Atoms 255(1):229–232

Weinbub J, Ferry DK (2018) Recent advances in Wigner function approaches. Appl Phys Rev 5:041104

Wigner E (1932) On the quantum correction for thermodynamic equilibrium. Phys Rev 40:749

Temporal Trends of Persistent Organic Pollutants in Prussian Carp (*Carassius gibelio*) of the Amur (Heilong) River Basin (2019–2021)

Maksim M. Donets and Vasiliy Yu. Tsygankov

Abstract The Amur (Heilong) River is one of the largest in northeast Asia. Even though the Amur River Basin has high economic, transport and environmental importance for the region, its monitoring is very limited. Existing research on the POPs pollution of the Amur River focuses primarily on the study of abiotic components of the ecosystem—water and bottom sediments. Most of the research are carried out in the Songhua River, China (a tributary of the Amur). The work presents the concentrations of POPs (DDT, HCH, PCBs, dieldrin and endrin) in the muscles of Prussian carp (*Carassius gibelio*) in the lower reaches of the Amur River and describes the temporal trends of pollutant levels from 2019 to 2021. Concentrations of pesticides in samples increased significantly from 2019 to 2020 and decreased just as sharply in 2021. Nevertheless, on average levels of the Amur River pollution were less or of the same magnitude as in other regions. Prussian carp are convenient objects for biomonitoring and well reflect trends in changes in pollutant concentrations in the Amur River.

Keywords POPs · OCPs · PCBs · Amur · Heilong · Prussian carp · Temporal trends

1 Introduction

The Amur (Heilong) River is one of the largest in northeast Asia. Its length is 4440 km, and its drainage basin area is 1,855,000 km^2. The Amur has an extremely extensive network of tributaries, among which the most important are the Argun, Zeya, Bureya,

M. M. Donets · V. Yu. Tsygankov (✉)
Pacific Geographical Institute, Far Eastern Branch, Russian Academy of Sciences, 7 Radio Street, 690041 Vladivostok, Russia
e-mail: tsig_90@mail.ru

M. M. Donets
Institute of the World Ocean, Far Eastern Federal University, 10 Ajax Bay, Russky Island, 690922 Vladivostok, Russia

T. Chaplina (ed.), *Processes in GeoMedia—Volume VIII*, Springer Geology,
https://doi.org/10.1007/978-981-97-6627-7_12

Songhua (Sungari) as and Wusuli (Ussuri) Rivers. The Amur basin is located on the territory of three countries: Mongolia, Russia and China. The main water usege is carried out mainly by the Russian and Chinese sides (Kryukov 2008). Obviously, Russia and China are the main contributors to water pollution in this region. For example, according to the Amur Basin Water Administration, the river is influenced by industrial enterprises (primarily—oil refineries, mechanical engineering and agricultural enterprises) from the Russian side and also pulp, cardboard, petrochemical, and chemical plants from the Chinese side. In addition, intensive urban development is underway in the region of Heilongjiang province, and there are also significant agricultural areas (Naumov 2021).

Despite the full-flowing and sufficient feeding of the river, there is quite pronounced competition for water resources in the Amur basin—primarily for unpolluted water. Even in the most extensive Russian part, there is a lack of unpolluted water below the confluence of the Songhua River (passes mainly through China). Here, on November 13, 2005, a series of explosions occurred at the largest petrochemical plant in Jilin province (northeast China). As a result, about 100 tons of benzene and other chemicals were discharged into the Sungari River, and the full list of the resulting mixture is unknown (Kryukov 2008). During the passage of this so-called "benzene spot," nitrobenzene, polycyclic aromatic hydrocarbons, and persistent organic pollutants (POPs) were detected in the water samples. Even after almost 20 years, the results of the accident are reflected in the condition of the main riverbed of the Basin. This is annually confirmed by the persistence of a significant number of water management problems in the Amur and its largest transboundary tributaries—the Wusuli and Songhua Rivers, which have a decisive influence on the entire water management complex of this river basin (Trofimchuk 2022).

Researchers focus on the abiotic components of the ecosystems (water, bottom sediments), and most of the works are carried out on the Songhua River (Donets and Tsygankov 2023). In this regard, the aim of the work was to assess the temporal trends of the accumulation levels of POPs (DDT, HCH, PCBs, dieldrin and endrin) in the body of Prussian carp (*Carassius gibelio*) from the lower reaches of the Amur River (2019–2021).

2 Materials and Methods

Individuals of the Prussian carp (*Carassius gibelio*) were sampled in the lower reaches of the Amur River, in the fishing zone of the Troitsk fish processing plant (the Dzhari village, Khabarovsk Krai) from 2019 to 2021 (Fig. 1).

Fish samples were frozen at – 20 °C and transported to the laboratory. Before analysis, carcasses were thawed and dissected. Fish muscles were analyzed. Lipids were extracted from organ homogenates using a mixture of n-hexane and acetone, followed by destruction of the fatty components with concentrated sulfuric acid. The resulting extract was separated with non-polar (for PCB) and polar (for OCP) solvents on a chromatographic column with Florisil® sorbent.

Fig. 1 The map of the studied region: red point—the location of the fish processing plant; the yellow zone—sampling site

Concentrations of organochlorine pesticides (OCPs) (α-HCH, β-HCH, γ-HCH, δ-HCH, o,p′-DDT, p,p′-DDT, o,p′ -DDD, p,p′-DDD, o,p′ -DDE, p,p′-DDE, dieldrin and endrin) and polychlorinated biphenyls (PCBs) (congeners 28, 52, 155, 101, 118, 143, 153, 138, and 180 congeners) in samples were measured by gas chromatography–mass spectrometry (GC–MS).

The main identification of the organochlorine compounds in the biomaterial was carried out using a Shimadzu GC MS-QP 2010 Ultra gas chromatography-mass spectrometer. Statistical analysis of the results was carried out using IBM SPSS Statistics software for Windows 10. The reliability of the data was assessed using a two-tailed Kruskal-Wallis test with a significance level of $p \leq 0.05$. Results are presented as concentration range (min-max) and mean ± standard deviation (Mean ± SD) (Donets et al. 2022).

3 Results and Discussion

POPs were found in all the samples analyzed. Average levels of organochlorine pesticides (OCPs) exceeded those of polychlorinated biphenyls (PCBs) throughout the study period. Among the OCPs, HCH isomers were dominant with a prevalence of the α- and γ-forms, which indicates both the presence of "fresh" contamination with the original lindane (γ-HCH) and the ongoing decomposition of the compound to more stable forms (α-HCH). Among DDT and its metabolites, DDE isomers were the main compounds to be found, which indicates the predominance of the oxidative pathway of destruction of the original compound (DDT) and long-term circulation of the pollutant in the ecosystem (ATSDR DDT 2022b). In 2019, dieldrin was also determined in the muscles of Pussian carp. This compound is considered to be the one of the most persistent among all POPs (ATSDR Dieldrin 2022a). However, its determination in a single year suggests recent use during agricultural work or leakage of storage locations (during the flood). Endrin was below detection limits in all the samples analyzed. Among polychlorinated biphenyls, low-chlorinated PCB 52 was detected during all sampling periods. In 2020 and 2021, more highly chlorinated congeners (101, 118, 153, 138) were also determined.

The range of $\sum$OCP ($\sum$DDT + $\sum$HCH + dieldrin) in 2019 varied from 0.02 to 1.80 ng/g wet weight (w.w.). In 2019, all the isomers of HCH were detected (Table 1). The total levels of this compound varied within the range of 0.02–1.42 ng/g w.w. α-HCH was dominating isomer, reflecting the use of the technical mixture of HCH, as well as the process of lindane degradation (Mehboob et al. 2013). Among DDT and its metabolites, only degradation products of the maternal compound were detected—DDD and DDE (with the dominance of the latter). Total DDT levels ranged from 0.02 to 0.30 ng/g w.w. and were significantly ($p \leq 0.05$) lower compared to HCH. A wide range of congeners was found among $\sum$PCBs, including both low-chlorinated (PCB 52, PCB 101) and high-chlorinated (PCB 118, PCB 153, PCB 138) with a total range from 0.13 to 0.80 ng/g w.w. (Table 1).

In 2020, the average range of $\sum$OCP concentrations increased significantly and varied from 0.85 to 7.67 ng/g w.w. Among the pesticides, HCH dominated, with concentraions ranging from 0.68 to 6.07 ng/g w.w. As in 2019, the α-form of the compound dominated among the others (Table 1). At the same time, the frequency of detection of β-HCH decreased significantly. This may indicate the decomposition of the most stable b-form and a decrease in "historical" contamination with an increase in "fresh" pesticide input. DDD and DDE were found among DDT and its metabolites. p,p'-DDT and dieldrin were detected in one sample (Table 1). $\sum$PCB levels ranged from 0.04 to 6.75 ng/g w.w. and were represented by 28 and 52 congeners only. Compared to 2019, there is a significant increase in POP concentrations in the muscles of Prussian carp.

In 2021, average $\sum$OCP concentrations were detected in the lowest concentration range—from 0.07 to 0.34 ng/g w.w. As in previous cases, the basis of the pesticides was HCH, varying from 0.07 to 0.21 ng/g w.w. Unlike the two previous periods, the α- and β-forms of the compound were not detected in 2021, and the dominant

Table 1 Concentrations of individual POPs in the fish studied, ng/g w.w*

Pollutants	2019	2020	2021
α-HCH	0.45 ± 0.49	1.78 ± 1.35	< LOD
β-HCH	0.03 ± 0.01	0.20 ± 0.22	< LOD
γ-HCH	0.03 ± 0.02	0.31 ± 0.22	0.07 ± 0.02
δ-HCH	0.04 ± 0.01	0.26 ± 0.16	0.05 ± 0.03
4,4-DDT	< LOD[a]	0.43[b]	< LOD
2,4-DDD	0.06 ± 0.05	< LOD	< LOD
4,4-DDD	0.10 ± 0.05	0.44	0.07 ± 0.04
2,4-DDE	0.04 ± 0.01	0.20 ± 0.04	< LOD
4,4-DDE	0.19	0.72 ± 0.84	0.02 ± 0.01
Dieldrin	0.05 ± 0.01	0.46	< LOD
PCB 28	0.08 ± 0.05	1.11 ± 1.86	< LOD
PCB 52	0.07 ± 0.05	0.22 ± 0.39	0.04 ± 0.01
PCB 101	0.04 ± 0.01	< LOD	0.03 ± 0.01
PCB 118	0.07 ± 0.04	< LOD	0.04 ± 0.03
PCB 143	0.08 ± 0.06	0.09	< LOD
PCB 153	0.13 ± 0.11	< LOD	0.08 ± 0.03
PCB 138	0.12 ± 0.10	< LOD	0.08 ± 0.04

* The table shows only compounds that were detected in the sample
[a]Less, than the limits of detection
[b]Found in one sample

representative of this group of pollutants was lindane (Table 1). This confirms the removal of "old" pollution from the ecosystem and the presence of sources of original pesticides in the Amur River Basin. As in previous years, only DDD and DDE were detected among DDT and its metabolites. Total concentrations of this group of compounds varied from 0.02 to 0.14 ng/g w.w. The range of PCB concentrations varied from 0.02 to 0.36 ng/g w.w. Among the congeners, PCB 52 was dominating. OCP concentrations exceeded the ones for PCBs, indicating a predominant contribution of agricultural activities to pollution of the Amur River throughout the study period.

Levels of α-HCH and o,p'-DDE were significantly ($p \leq 0.05$) higher in 2020 compared to 2019. γ- and δ-HCH were differ significantly in the order 2020 > 2021 > 2019. p,p' -DDE, PCB 52 and $\sum$OCPs differed significantly from year to year in the order 2020 > 2019 > 2021. When comparing the total levels of POPs, from 2019 to 2020 there was a sharp increase of the concentrations in the muscles of Prussian carp (Fig. 2). At the same time, the POP concentrations decreased significantly in 2021. Such a sudden and serious change in concentrations is most likely due to the simultaneous entry of a large volume of pesticides in the environment of the studied region. For example, this could be affected by violations of the hydrological

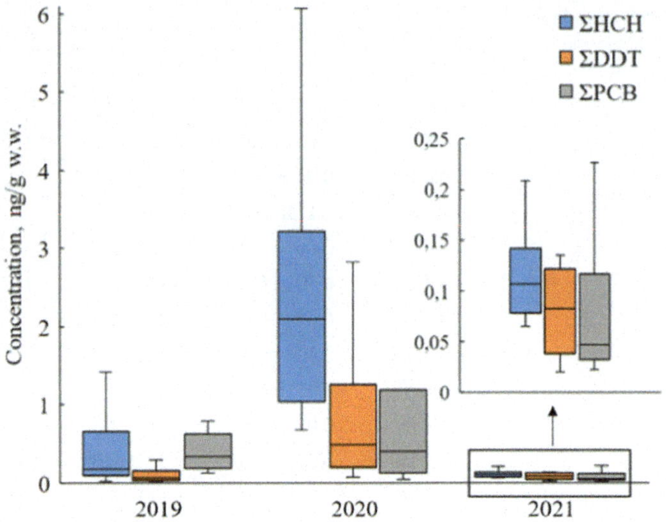

Fig. 2 Comparison of the total concentrations of HCH, DDT and PCBs in the muscles of Prussian carp

regime (flood), which was recorded in the lower reaches of the Amur River in 2020 (Shesterkin and Shesterkina 2022).

PCB concentrations differ significantly in the order 2019 > 2020 > 2021 (Fig. 2). At the same time, the qualitative composition of biphenyls shortened significantly from 2019 to 2020 and expanded again by 2021.

Existing research on the POPs pollution of the Amur River focuses primarily on the study of abiotic components of the ecosystem—water and bottom sediments, which allows us to compare the resulting concentrations in fish with their habitat. Some authors reported a significant increase in the levels of DDT, HCH and aldrin in the Amur and Wusuli Rivers in 2020 (Mishchenko et al. 2022). In summer, the increase in concentrations was more than several hundred times for DDT and HCH and about two times for aldrin; in winter—up to 7 times (Fig. 3).

Thus, our results are consistent with those of other authors and confirm the increase in pesticide levels from 2019 to 2020. The most likely explanation for this spike and subsequent decrease in toxicant levels in fish is heavy flooding, including in agricultural areas—potential sources of pesticides. At the same time atmospheric precipitation causes an increase in terrigenous runoff, which increases the potential removal of toxicants and the content of suspended particles in water. It is known that suspended organic particles are a kind of "sorbent" for POPs in natural ecosystems (Böhm et al. 2016). In addition, rising water levels and precipitation provoke resuspension of bottom sediments, which are the main reservoirs of POPs in aquatic ecosystems (Sapota 2006). As a result, the concentrations of pollutants in the water increase significantly, and bentho- and detritivorous organisms (in our case, Prussian carp) have access to more contaminated "deep" layers of bottom sediments. At the

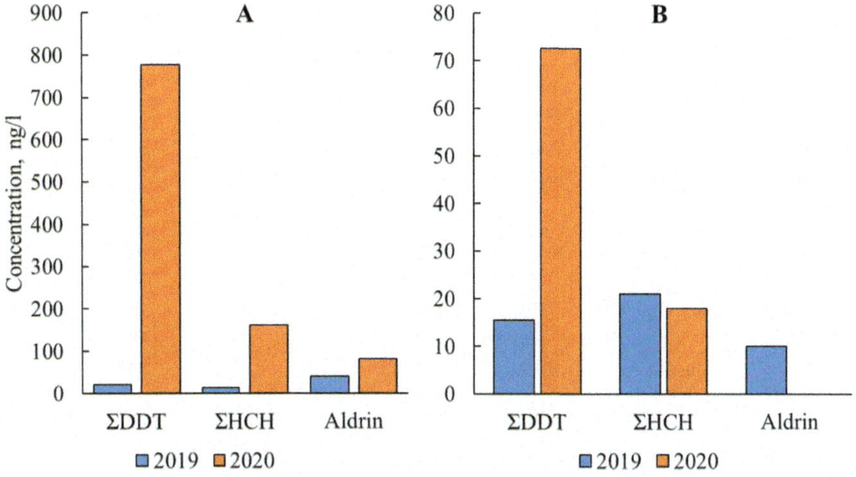

Fig. 3 Comparison of OCP concentrations in water of the Amur River in the summer (**a**) and winter (**b**) of 2019 and 2020 (according to Mishchenko et al. 2022)

same time, a sharp decrease in pollutant concentrations is most likely due to the fact that the fish selected in 2020 were influenced by floods and an increase in the overall level of pollution (which means they were living directly in the most contaminated site), while those caught in 2021 were not, since such a rapid removal of such significant levels of lipophilic toxicants is impossible.

To assess the degree of pollution of the river, the obtained POP concentrations were compared with different regions of the world (Table 2).

The HCH concentrations in Prussian carp from the Amur River were lower than in lakes Taihu and Dianshan during the entire observation period (2019–2021) (Ke et al. 2007; Yang et al. 2019). In 2020, pollutant levels are comparable to those in Lake Taihu and exceed the concentrations of Lakes Vransko and Nansi (Zhang et al. 2014; Romanić et al. 2018). Determined levels of DDT are significantly lower than in all compared regions. Especially with the lakes of China, where active use of OCPs was previously carried out (Ke et al. 2007; Zhang et al. 2014; Yang et al. 2019). PCB concentrations in the fish we studied were significantly lower than in Lake Dianshan during the entire sampling period and were higher than those for fish from Lake Vransko (about twice) (Romanić et al. 2018; Yang et al. 2019). Thus, even despite a significant increase in OCP concentrations in 2020, fish from the Amur River remains quite "clean" in terms of POPs. At the same time, unlike most regions of the world, the levels of HCH exceed DDT here, as in the entire Russian Far East.

Table 2 Average concentrations of POPs in crucian and Prussian carp from different regions of the world, ng/g w.w.

Species	Area, year	Year of sampling	∑HCH	∑DDT	∑PCB	Reference
Carassius auratus	Lake Taihu,	2004	2.5[a]	4.6[b]	–[c]	Ke et al. (2007)
	Nansi Lake	2011	1.90[a]	2.61[b]	–[c]	Zhang et al. (2014)
	Lake Dianshan	2016	4.04[a]	4.24[b]	6.54[d]	Yang et al. (2019)
Carassius gibelio	Lake Vransko	2014	0.20[e]	0.87[b]	0.5[f]	Romanić et al. (2018)
	Amur River	2019	0.38[a]	0.10	0.39[f]	This work
		2020	2.34[a]	0.79	1.23[f]	
		2021	0.11[a]	0.08	0.10[f]	

[a]As a sum of α-, β-, γ- and δ-isomers
[b]As a sum of p,p'-DDT, DDD and DDE
[c]No data
[d]As a sum of 28, 52, 77, 81, 101, 105, 114, 118, 123, 126, 138, 153, 156, 157, 167, 169, 180, and 189 congeners
[e]As a sum of α-, β- and γ-isomers
[f]As a sum of 28, 52, 101, 138,153 and 180 congeners

4 Conclusion

Thus, during 2019–2021 a significant increase in OCP concentrations in the muscles of Prussian carp of the Amur River was recorded (2019–2020), followed by an equally sharp decrease (2020–2021). A sharp spike in pollutant concentrations is a reflection of changes in their levels in the environment as a result of natural and anthropogenic factors. Despite the fluctuations of toxicants, the levels of POPs in Prussian carp, on average, is lower or comparable to other regions of the world. It is necessary to expand the biomonitoring network of the Amur River for more detailed monitoring of POP concentrations and their fluctuations.

Financial Support The work was supported by the Russian Science foundation (agreement No. 23-74-10032).

References

ATSDR (2022a) Toxicological profile for Aldrin/Dieldrin. U.S. Department of Health and Human Services, ATSDR, Atlanta, GA

ATSDR (2022b) Toxicological profile for DDT, DDE, and DDD. U.S. Department of Health and Human Services, ATSDR, Atlanta, GA

Böhm L, Schlechtriem C, Düring R-A (2016) Sorption of highly hydrophobic organic chemicals to organic matter relevant for fish bioconcentration studies. Environ Sci Technol 50:8316–8323. https://doi.org/10.1021/acs.est.6b01778

Donets MM, Tsygankov VY (2023) Organochlorine compounds in the Amur (Heilong) River basin (2000–2020): a review. JoX 13:439–462. https://doi.org/10.3390/jox13030028

Donets MM, Tsygankov VY, Gumovskiy AN et al (2022) Fish as a risk source for human health: OCPs and PCBs in Pacific salmon. Food Control 134:108696. https://doi.org/10.1016/j.foodcont.2021.108696

Ke R, Xu Y, Huang S et al (2007) Comparison of the uptake of polycyclic aromatic hydrocarbons and organochlorine pesticides by semipermeable membrane devices and caged fish (*Carassius carassius*) in Taihu Lake, China. Environ Toxicol Chem 26:1258–1264. https://doi.org/10.1897/06-454R1.1

Kryukov V (2008) Chapter 5. Possibility of sustainable development of the basin of Amur River from ecological viewpoints (in Russian). In: Energy and environment in slavic eurasia: toward the establishment of the network of environmental studies in the Pan-Okhotsk region, Slavic Research Center, Sapporo, Japan, pp 127–156

Mehboob F, Langenhoff AAM, Schraa G, Stams AJM (2013) Anaerobic degradation of lindane and other HCH isomers. In: Malik A, Grohmann E, Alves M (eds) Management of microbial resources in the environment. Springer Netherlands, Dordrecht, pp 495–521

Mishchenko OA, Gladun IV, Volosnikova GA (2022) Monitoring of pesticide pollution of surface waters of the Amur River in the area of the city of Khabarovsk. Materials of the XI scientific-practical conference with international participation. Pacific State University, Khabarovsk, pp 98–103

Naumov YA (2021) On the features of surface water pollution in the territory of the Russian Far East. Ecumene Regional Stud, 102–112. https://doi.org/10.24866/1998-6785/2021-3/102-112

Romanić SH, Vuković G, Klinčić D et al (2018) Organochlorine pesticides (OCPs) and polychlorinated biphenyls (PCBs) in Cyprinidae fish: towards hints of their arrangements using advanced classification methods. Environ Res 165:349–357. https://doi.org/10.1016/j.envres.2018.05.003

Sapota G (2006) Persistent organic pollutants (POPs) in bottom sediments from the Baltic Sea. Oceanol Hydrobiol Stud 35(4):295–306

Shesterkin VP, Shesterkina NM (2022) Hydro/chemical features of the Amur River waters near Khabarovsk during a very high flood in 2020. Water Sect Russ Probl Technol Manag, 97–110. https://doi.org/10.35567/19994508_2022_1_7

Trofimchuk MM (ed) (2022) Surface water quality in the Russian Federation: yearbook 2021. Roshydromet, Rostov-on-Don

Yang J, Qadeer A, Liu M et al (2019) Occurrence, source, and partition of PAHs, PCBs, and OCPs in the multiphase system of an urban lake, Shanghai. Appl Geochem 106:17–25. https://doi.org/10.1016/j.apgeochem.2019.04.023

Zhang G, Pan Z, Bai A et al (2014) Distribution and bioaccumulation of organochlorine pesticides (OCPs) in food web of Nansi Lake, China. Environ Monit Assess 186:2039–2051. https://doi.org/10.1007/s10661-013-3516-5

Geoecological Problems in the Ural Basin Under a Changing Climate

I. V. Kudelina, T. V. Leontyeva, M. V. Fatyunina, and V. P. Petrishchev

Abstract Geoecological problems in the Ural basin are caused by intense and growing technogenic load on the environment over time. It is created mainly by mining agricultural enterprises that pollute the environment and natural waters, including water intakes for domestic, drinking and technical purposes. To protect them from pollution in the river basin Ural research was carried out, mainly within the Orenburg region to hydrogeological studies of the territory. It made it possible to establish patterns of changes in the dynamics and chemical composition of water at existing water intakes. The previously identified processes of depletion of water resources due to the uneven manifestation of seasonal and long-term changes in the elements of the water balance confirm the need for the use of new technologies. Their implementation will help overcome the shortage of water resources in the Orenburg region. When using water replenishment technology, the productivity of water intake wells increases and water quality improves, which can be supported by the introduction of barrier technologies. New water management technologies open up the possibility of providing the population of the Orenburg region with drinking water in the future.

Keywords Mining areas · Water management technologies · Pollutants · Groundwater replenishment

1 Introductory Part

The research is focused on the problem of protecting water resources of drinking quality in the Ural basins within the Orenburg region and its water-scarce areas. These areas include Central and Eastern Orenburg. The state of the Ural River basin was perfectly illustrated by A.A. Chibilev (Sivohip 2021; Sokolov 2020; Sivokhip 2019), which greatly facilitates the disclosure of this topic. The main attention is paid to the nature of the relationship between groundwater and surface waters in

I. V. Kudelina (✉) · T. V. Leontyeva · M. V. Fatyunina · V. P. Petrishchev
Orenburg State University, Orenburg, Russia
e-mail: kudelina.inna@mail.ru

T. Chaplina (ed.), *Processes in GeoMedia—Volume VIII*, Springer Geology,
https://doi.org/10.1007/978-981-97-6627-7_13

connection with the river flow regime under conditions of a changing climate and the prospects for increased technogenic load (Vasiliev 2018).

During the period from the discovery and development of the unique Orenburg oil and gas condensate field, a significant transformation of the geological environment, and especially the alluvial aquifer, has occurred (Gaev 2014). It is the most important source of household and drinking water supply for the population.

The Orenburg oil and gas complex improved the social-economic indicators of the region, but at the same time increased the technogenic load on the environment and water resources, causing their depletion, pollution and technogenic transformation of the chemical composition of waters. Waters of the sulfate-sodium and soda type in many areas have acquired the chloride-magnesium subtype and increased mineralization. This was established by V.S. Samarina, A.Y. Gaev, K.E. Drinking, O.M. and S.K. Sevostyanov and others in the 60–70 s. XX century (Gaev 2014). Currently, these trends have intensified, which was established in the process of searching for sources of water pollution within the Orenburg urban agglomeration and in the regions of Eastern Orenburg (Kudelina 2018a, b). This made it possible to identify the patterns of groundwater formation in the system water—host rocks—sources of pollution (Kudelina 2018a).

Due to the unevenness of river flow and its influence on the groundwater regime, a complex hydrogeological situation has arisen. Therefore, it is very important to consider the issues of groundwater formation in conditions of changing river flow and to develop scientific and methodological foundations for domestic and drinking water supply in urbanized areas with: (1) analysis of the current situation based on the developed methodology for studying the processes of pollution and depletion of groundwater of drinking quality; (2) studying the nature of the relationship between alluvial waters and river waters, and the processes of degradation of water quality under the influence of technogenesis; (3) development of recommendations for the protection of water intakes for domestic and drinking purposes from depletion and pollution (Kudelina 2018b; Leontieva 2021).

2 Source Materials

For hydrogeological assessment of the territory and analysis of technogenic transformation of groundwater were collected in the field and in the production funds of the Committee on Natural Resources, in the laboratories of the Orenburg State University, in the Department of Geoecology of the Russian Academy of Sciences, in the Orenburg Regional Committee of Statistics. The results of chemical analyzes of samples of natural waters, soils, snow, ground and silts were used. According to the method of V.S. Samarina (Gaev 2014) processed the results of water tests. According to A.Y. Gaev (1989–2018) (Gaev 2014), the results of ground-based and remote sensing studies were modeled.

3 Main Research Results

There are four stages in the history of research in the Ural basin. At the first stage (1800–1930), the first, but already significant data on the hydrogeological conditions of the basin were obtained. At the second stage (1931–1960), the mining industry actively developed, and virgin and fallow lands were developed. The third stage is the most intensive, associated with the end of the Soviet period (1961–1991), when targeted, systematic hydrogeoecological surveys and specialized studies were carried out, which were covered, in particular, in volume 43 "Hydrogeology of the USSR, Orenburg Region" (1972). Such great researchers worked as A.M. Chernyaev, O.I. I am. Anisimov, D.F. Zakharchenko, V.A. Gutsaki, V.V. Gudoshnikov, M.Ya. Tukmakova, A.P. Solodovnichenko, V.P. Vedenina, E.I. Tokmachev, T.A. Rudenko, A.A. Donetskova and others.

The fourth stage, modern, is associated with the transition to a market economy; it is characterized by optimally limited volumes of research only on those objects that are of interest to private initiative, namely, deposits and farms that are capable of providing a certain profit. Worked (Zektser et al. 2015; Hydrogeology of the USSR 1972; Borevsky 2014): Poladko M.F., Pavlova V.P. Geldt A.A., Gaev A.Y., Zakharova V.Y., Shevtsova L.F., Gatskov V.G., Taraborin D.G., Pampushka A.M., Lisov A.S., Alekseev G..A., Lyadsky, P.V., Kvasnyuk L.N., Zinchenko L.E., Lyasota I.A., Mafrakhova F.M., Igoshkina N.N., Tsvetkova N.V., Zatsepina A. A. and etc.

In the conditions of the central Orenburg region, a change in the chemical composition of water in alluvial water intakes occurs. Sodium sulfate waters are converted into soda water or magnesium chloride water, which increases their hardness and mineralization. During the flood period, the quality of water in the alluvial aquifer improves, and therefore, through replenishment technology, it is possible to minimize the transformation of water composition and overcome the shortage of drinking-quality water resources in low-water areas of the river basin. Ural (Kudelina 2018a, b; Leontieva 2021). It is proposed to overcome the shortage of water resources through a set of measures, accumulating flood waters in a cascade of small dams, each of which ensures a rise in the water level in the river by no more than 3 m.

In conditions of poorly dissected relief, dams will prevent flooding and inundation of large areas; their cascade will ensure the accumulation of sufficient water resources for the summer period. Replenishment of groundwater reserves will maintain the productivity of water intakes at a good level and the quality of their water will improve (Kudelina 2018a, b; Heviánková et al. 2016). With the replenishment of groundwater reserves and an increase in the productivity of water intakes, the transformation of the waters of the alluvial aquifer will be prevented.

The developed methodological approach to hydrogeological studies of water-scarce urbanized areas of the Orenburg region is intended to provide the population with drinking water by introducing technologies for replenishing groundwater reserves and protecting them from pollution and depletion using original barrier technologies (Blinov 2000; Perelman et al. 1990; Gutsaki 1963). The research results can also be used in complex water-scarce areas of Eastern Orenburg (Leontieva 2021).

In the conditions of the Eastern Orenburg region, the folded mountain spurs of the low-mountain Southern Urals are replaced to the east by the Trans-Ural peneplain. These spurs are composed of submeridional ridges of igneous, metamorphic and sedimentary Proterozoic-Paleozoic rocks. The relief is peneplain, having the appearance of a plateau, articulated in the west with the Irendyk-Krykty ridges. Their elevations of 700–900 m decrease to the east and south to 300–350 m. River valleys have a depth of up to 100 m. Hydrogeologically unstudied Mesozoic weathering crusts are developed from the surface. Hydrogeological patterns have been identified among them. Hydrogeological and hydrogeochemical maps have been constructed for a number of areas, from which the poor knowledge of fissure waters and weathering crusts is obvious. The collector of ancient weathering crusts, in our opinion, can be used to replenish groundwater reserves.

4 Climate, Relief and Hydrography of the Region

The climate is sharply continental, with evaporation 600–800 mm per year, the average annual precipitation is 400–150 mm (Vasiliev 2018). The hills are covered with screes. The region belongs to the Ural and Tobol basins and partly to the lake area of internal drainage. The density of the river and ravine network is 0.5 km/km^2. The main territory is occupied by the Ural basin. River valleys are confined to troughs. Up to 70–90% of the annual river flow occurs during floods. The water flow module decreases to the southeast of the region to 0.5 l/s km^2. In the east of the region there is the Zhetykol-Aiken closed lake basin. It consists of depressions of meridional orientation with shallow lacustrine flat basins. There are no permanent watercourses.

The predominant steppes are dark soils and southern chernozems, as well as wormwood-fescue steppes, plowed up to 30%. The soils are eroded, carbonate and solonetz-solonchak with a high content of exchangeable sodium. Less than 2% of the territory is covered by forest. It has been developed in the western part.

The region is composed of sedimentary and volcanogenic-sedimentary formations of the Paleozoic, which are complicated by uralides with intrusions of serpentinized hyperbasites, granitoids with gabbroid dikes. The oldest are secondary quartzites and metamorphic schists. Geographically, the main composition of Devonian volcanics (PR3 kl) of basic and acidic composition and volcanogenic-clastic rocks, as well as sandstones, gravelites, siltstones, mudstones, conglomerates and limestones of the Upper Famennian substage of the Devonian and Lower Carboniferous, predominate. In Mesozoic depressions, Triassic and Jurassic coal-bearing deposits with a thickness of 200–400 m with lenses of ferruginous sandstones and brown coal with marcasite and siderite were identified. The Middle Jurassic section is composed of kaolin clays, with layers of brown coal, siderite and sulfide nodules.

In young depressions and karst depressions, Paleogene and Pliocene sandstones, clayey sandstones and conglomerates, quartz sands, opoka, clays with sulfides and brown ironstone and gypsum are discovered. Well flow rates are low. Channel and floodplain alluvium and colluvial and aeolian clays and loams are of Quaternary age.

The Upper Mesozoic formation of weathering crusts transformed the Urals into a low-mountain, folded, blocky region with renewed ancient faults of meridional and sublatitudinal strike (Plotnikov 1978). The amplitude of uplifts in the Neogene-Quaternary period of the western part of the Eastern Orenburg region was up to 800 m. The extreme eastern part of the region was buried under the Meso-Cenozoic Turgai trough. Now the region is experiencing upward movements with an amplitude of up to 4–6 mm per year, which is reflected in the relief formed under the influence of water runoff and slope processes.

The waters in the mountainous part of the territory are fresh throughout. The rivers have a hydrocarbonate-calcium-magnesium (sodium) composition, and in the southeastern part they have a hydrocarbonate-sulfate, hydrocarbonate-chloride and chloride composition. Their mineralization increases from 0.3–0.5 to 3–5 g/l. The waters of lakes also change from fresh sulfate and soda types to chloride and salty ones in the Trans-Urals. Alluvium waters are of practical importance. In the area of internal runoff with an area of about 5 thousand km^2, dry lands are widespread, where runoff is observed only when snow melts and after rainstorms. The runoff module does not exceed 0.5 l/s per 1 km^2.

Lakes of the Trans-Ural region with a depth of 1–2 m are characterized by a rapid decline in level and drying out in the summer. Shelkar-Ega-Kara and Zhetykol, with an area of more than 9.6 and 5.0 thousand hectares, respectively, dry out every 10 years and freeze once every 3 years. The water in them is brackish. Lake Kupa in the Gai region is of karst origin.

To accumulate water resources in the region, the following reservoirs were created: Kumakskoye, Ushkotinskoye, etc. The largest is the Iriklinskoye reservoir with a volume of 3257 million m^3. Water is exchanged in it once a year.

The shortage of water resources slows down the socio-economic development of the region, despite the abundance of mineral resources in the Eastern Orenburg region. Water scarcity is caused by climatic conditions, weak relief, predominance of rocky species, and lack of forest plantations. All this limited underground flow and caused extreme poverty in groundwater. Technologies with the accumulation of water resources are required, since the region includes six administrative districts with a total area of 30,265 km^2. This is larger than the territory of a number of European countries. Despite its exceptional wealth in mineral resources, the region lags behind other areas in socio-economic development, and the main reason is the sharp shortage of water resources. Therefore, new technologies are proposed to replenish groundwater reserves and protect them from pollution using barriers (Blinov 2000; Perelman et al. 1990).

Most of the water intakes operated in the region are alluvial and are confined to the Ural River basin. River and alluvial waters are closely interconnected with each other and with the fissure waters of Paleozoic rocks. Zones of groundwater concentration gravitate toward river valleys. They are not protected from contamination.

The region has experienced marine transgressions in the past. Therefore, in the sedimentary rocks of the zone of active water exchange, there are relics of the marine salt complex that salinize both river and groundwater. This complicates the hydro-geochemical picture, for example, in the river basin. Suunduk, which is confined to the forest-steppe zone. Here, both the underground waters of the Suunduk water intake and the waters of the Suunduk River are saline.

In general, the region is characterized by extreme unevenness of water flow, up to 80–96% of it is confined to short-term periods of floods. Therefore, the proposed technology, focused on the accumulation of water during periods of floods, is designed to solve the problem of replenishing the water reserves of operated water intakes using groundwater.

Of the groundwater in the region, the predominant waters are fissure-type waters of the weathering zone (T6), fissure-karst waters in limestones (T8) and fissure-vein waters (T7) in tectonic disturbances. Weathering crusts are developed along them to depths of 600 m. The population is supplied with water mainly by alluvial waters of the sulfate and carbonate types. They are formed under the influence of many factors that are closely related to each other: landscape-climatic, lithological-stratigraphic, technogenic, etc.

The hydrogeological features of Mesozoic weathering crusts are practically not studied, but their reservoir capacity deserves attention. The formation of reservoirs and the distribution of water flow (underground and surface) in riverine and watershed zones are significantly influenced by neotectonic fracturing and technogenic factors. They are caused by the activities of the mining industry and agriculture, causing pollution processes. The distance of water management facilities from sources of pollution determines the intensity of water pollution processes and the depth of transformation of their chemical composition. Thus, in the villages of Kvarkeno and Krasnoyarsky, the water in the river and in water intakes does not meet sanitary standards for mineralization and hardness, since the host rocks contain relics of the sea salt complex. In addition, the technogenic transformation of the composition of waters in the region is significantly complicated by natural-historical and physical-geographical conditions, which together have led to a shortage of drinking-quality water. And this, in turn, complicates the processes of socio-economic development of the region. Further development of the region is impossible without solving water management and sanitary-epidemiological problems. These problems are associated with a high morbidity rate in the population. The regional government has set tasks to solve water management problems and improve the level of public health.

To solve these most important problems for the region, it is proposed to universally introduce technologies for replenishing groundwater reserves, accumulating flood runoff, using collectors located near reservoirs for this purpose.

When the water back up in the river is only 2–4 m, it is possible to increase the infiltration of flood waters into the aquifer and prevent the depletion of water resources at water intakes, reducing the level of water pollution, including the radiation level of radon and its decay products. Radon was mistakenly considered by some researchers to be a rare radioactive element, but the level of radiation on the planet depends on radon. It is super toxic and accumulates in granitoids. It is granitoids that play a leading role in the geological structure of Eastern Orenburg.

To implement these technologies, a set of maps and diagrams of water intakes and reservoirs was built, and the nature of the formation of alluvial waters was studied with an assessment of the influence of pollution sources on them. Typification schemes for the protection of the territory from pollution have also been completed. Using the example of the Upper Kumak reservoir area, the possibility and feasibility of introducing new technologies for water supply to the population in the region is shown.

5 Conclusions

The waters of the alluvial aquifer are of primary water economic importance for providing the population of the region with drinking water. In the case of an undisturbed regime, they, like river waters, belong to the sulfate-sodium type. When water flows from Permian deposits, they acquire the soda type. If polluted, these waters become hard, magnesium chloride with increased mineralization. During periods of floods, the alluvial aquifer is partially washed out of pollutants, and when replenishing reserves, the transformation of its composition is minimized.

Based on the results of surveillance observations, a new methodological approach to hydrogeological studies of the territory has been developed. It made it possible to establish patterns of changes in the dynamics and chemical composition of water at existing water intakes. The previously identified processes of depletion of water resources due to the uneven manifestation of seasonal and long-term changes in the elements of the water balance confirm the need for the use of new technologies. Their implementation will help overcome the shortage of water resources in the Orenburg region. When using water replenishment technology, the productivity of water intake wells increases and water quality improves, which can be supported by the introduction of barrier technologies. New water management technologies open up the possibility of providing the population of the Orenburg region with drinking water in the future.

References

Blinov SM (2000) Fundamentals of the use of geochemical barriers for environmental protection: author. dis. …c.g-m.s. 11.00.11 / S.M. Blinov. - Perm: Perm. un-t, p 23

Borevsky BV (2014) Typification of combined water intake systems in the joint use of underground and surface waters In: Borevsky BV, Kozak NS (eds) Exploration and protection of mineral resources, vol 5, pp 51–55

Gaev AY (2014) On the protection of water intakes for household and drinking purposes on the example of Orenburg In: Gaev AY, Kudelina IV, Malikov ON (eds) Region. prob. water use in a changing climate. Conditions: Mat. MNPK, 11–12.11.2014. Ed.: AM Gareev. Ufa: Aeterna, - S. 277–280

Gutsaki VA (1963) Weathering crust of the Orsk Trans-Urals. In: Gutsaki VA, Gudoshnikov VV (eds), Saratov: Research Institute of Geology at the University named after N.G. Chernyshevsky, p 281

Hydrogeology of the USSR. T. 43 (1972) Orenburg region. Tokmachev EI, Nedra M (eds), p 272

Heviánková S, Marschalko M, Chromíková J, Kyncl M, Korabík M (2016) Artificial ground water recharge with surface water. In: World multidisciplinary earth sciences symposium (WMESS 2016). IOP Publishing, pp 6–12

Kudelina IV (2018a) Ways to stabilize the regime of alluvial water intakes in a semi-arid climate. In: Kudelina IV (ed) Science, new technologies and innovations of Kyrgyzstan, vol 2. Bishkek, pp 82–86

Kudelina IV (2018b) On the possibility of stabilizing the operation mode of the water intakes of the Orenburg urban agglomeration. In: Kudelina IV (ed) Science, new technologies and innovations of Kyrgyzstan, vol 2. Bishkek, pp 68–72

Leontieva TV (2021) Hydrogeological aspects of the sustainable development of water-deficient areas of the Eastern Orenburg region. Abstract diss, Ph.D. Bishkek-Dushanbe, p 25

Perelman AI, Borisenko EN, Myrlyan NF et al (1990) Technogenic geochemical barriers. In: Nauka M (ed) Geochemistry of technogenic processes, pp 14–16

Plotnikov NI (1978) Hydrogeological foundations of artificial replenishment of groundwater reserves In: Plotnikov NI, Plotnikov NA, Sychev KI, Nedra M (eds), p 311

Sivohip JT (2021) Changes in river runoff during winter low water in the Ural River basin. In: Sivokhip ZT, Pavleychik VM, Chibilev AA (eds) Reports of the Russian academy of sciences, earth sciences, vol 499, no 2, pp 203–208

Sivokhip ZT (2019) Changes in the water regime of the rivers of the Ural river basin. In: Sivokhip ZT, Pavleichik VM (eds) Academician of the Russian academy of sciences, Chibilev AA reports of the academy of sciences, vol 488, no 5, pp 545–549

Sokolov AA (2020) Preservation of the tributaries of the Ural River in the Russian-Kazakhstan border. In: Sokolov AA, Bezuglov EV, Chibilev AA et al (eds) Orenburg: IS ural branch of the Russian academy of sciences, Russian geographical society, p 68

Vasiliev DY (2018) Long-term regime of air temperature and precipitation in the South Urals. In: Yu D, Vasiliev VM, Pavleychik VA (eds) Semenov and others reports of the academy of sciences, vol 478, no 5, pp 588–592

Zektser IS, Karimova OA, Chetverikova AV (eds) (2015) Resources of fresh groundwater in Russia and their use in emergency situations. Water Resour 42(4):351–366

Assessment of the Protection of Groundwater in the Syrtov Artesian Basin

I. V. Kudelina, T. V. Leontyeva, and M. V. Fatyunina

Abstract The territory of the Syrtov artesian basin is located in the eastern part of the East European Platform. The quality of groundwater in the territory of the Syrtov artesian basin is formed under the influence of a number of natural and man-made factors. The likelihood of contamination of the geological environment increases in the locations of oil refining and chemical industry enterprises, and energy complex enterprises. The protection of groundwater from the ingress of polluting components plays a huge role in preserving its natural state. The purpose of the study is to assess the protection of groundwater in the Syrtov artesian basin. As a result, it was concluded that almost all major aquifers and complexes containing fresh water and used for household and drinking water supply to the population have a low degree of protection, especially from the penetration of pollution from the surface. It is possible to prevent negative processes of technogenic water pollution through the introduction of new technologies and, in particular, the use of complex barriers.

Keywords Groundwater protection · Polluting components · Complex barriers

1 Introduction

The territory of the Syrtov artesian basin is located in the eastern part of the East European Platform. Administratively the research area includes the western part of the Orenburg, southern Samara and southeastern Saratov regions, the southwestern end of the basin area covers a small territory of the Volgograd region. The total research area is 121.21 thousand km^2 (Iceberg 2002; Korotkov 1997).

Tectonic maps of transboundary territories of the south of the East European Platform were taken as the tectonic basis of the territory of the Syrtov artesian basin (Iceberg 2002; Korotkov 1997). The features of the tectonic structure of the territory under consideration are determined by its location at the junction of the Volga-Ural anticline, the Cis-Ural trough and the Caspian depression. The basin's foundation is

I. V. Kudelina (✉) · T. V. Leontyeva · M. V. Fatyunina
Orenburg State University, Orenburg, Russia
e-mail: kudelina.inna@mail.ru

T. Chaplina (ed.), *Processes in GeoMedia—Volume VIII*, Springer Geology,
https://doi.org/10.1007/978-981-97-6627-7_14

composed of dislocated crystalline and metamorphic sediments and is characterized by a clearly defined block structure caused by faults of various amplitudes and spaces. The relief of the foundation contains large arches and deflections, many of which are reflected (manifested) in the sedimentary cover. They appear along the fault—in the form of uplifts, flexures, faults, zones of increased fracturing, which determine the conditions for the distribution and formation of objects of hydrogeological units.

In the sedimentary cover, the raised foundation blocks mainly correspond to arched uplifts, and the lowered ones correspond to depressions. The exception is the eastern half of the basin, where the lowered basement block corresponds to an arched uplift (Orenburg arch), and in the relief this positive structure corresponds to a synclinal depression. This is explained by the peculiarity of the tectonics of this region, in which the dominant role, in the zone of transition from the platform to the trough, is played by salt-dome tectonics.

In the west, the Syrtovsky artesian basin borders on the Volga-Khopersky, in the northwest on the Volga-Sursky, in the northeast on the Kama-Vyatsky, in the south on the North Caspian artesian basins of the second order. In the east, the adjacent second-order hydrogeological structure is the Belsky foothill artesian basin. The western border of the Syrt artesian basin runs along the top of the Zhigulevsk-Pugachevsky arch and the extreme southwestern part covers the Pachelma aulacogen (Zektser et al. 2015; Kudelina 2018a).

In modern relief, the western border of the Syrtov artesian basin is confined to the right bank of the river. Volga, which is the main drainage of the basin. The southern border of the Syrtov artesian basin is confined to the southern foreland arch of the Caspian depression. The northwestern border with the Volga-Sura basin runs along the contact of the Buzuluk depression and the Kuznets graben. The northeastern and eastern border runs along the axial part of the Orenburg arched uplift. The regional aquitard on the main area of the Syrtov basin is the aquiferous Kungurian (carbonate-sulfate-halogen) horizon.

It is the presence of this horizon in the section of the sedimentary cover that determines the number of hydrogeological levels. The first hydrogeological floor includes hydrogeological units of the sedimentary cover from the Holocene to the Kungurian inclusive. The second hydrogeological level covers part of the geological section below the Kungur and includes hydrogeological units of Permian, Carboniferous, Devonian and crystalline basement deposits. The absence of the aquiferous Kurgur horizon in the section of the sedimentary cover combines the mapped hydrogeological units into one hydrogeological level. The presence of local water-resistant, relatively water-resistant horizons in the hydrogeological stage determines a different number and combination of hydrodynamic zones. Based on the accepted principles of hydrogeological zoning, within the artesian structure of the second order of the Syrtov artesian basin, four hydrogeological structures of the third order and fourteen structures of the fourth order are distinguished. Hydrogeological structures of the third order include the Zhigulevsko-Pugachevsky and Orenburg arch structures, the Zavolzhskaya and Buzulukskaya hydrogeological structures. In the Zhigulevsko-Pugachevskaya arch structure, three structures of the IV order are distinguished: East-Zhigulevskaya, Pugachevskaya and Klintsovskaya, the Orenburg arch structure is also subdivided into three smaller structures: Sharlykskaya, Nezhinskaya and Sol-Iletskaya. The Zavolzhskaya hydrogeological structure has the fewest hydrogeological structures of the fourth order; there are only two of them—Saratov and Nizhnezavolzhskaya. The Buzuluk hydrogeological structure of the third order has five hydrogeological structures of the fourth order: Kinelskaya, Sorochinskaya, Irgizskaya, Central Syrtovskaya, Rubezhinskaya (Hydrogeology of the USSR 1972; Pityeva 1999).

The quality of groundwater in the territory of the Syrtov artesian basin is formed under the influence of a number of natural and man-made factors. It is often difficult to separate them from each other, since intensive economic activity often activates the action of natural factors that provoke a deterioration in the quality of groundwater, their medical and biological properties: general mineralization, hardness, chemical type of water, as well as the content of biologically necessary and toxic microcomponents (Gutsaki 1963; Kovalevsky 2001).

There is a whole series of objects in the basin that negatively affect the state of the geological environment. The main load on the geological environment comes from large industrial enterprises located in large cities, the development of oil and gas fields, construction materials, intensive extraction of groundwater from the subsoil for water supply and other purposes. The probability of contamination of the geological environment increases in the locations of oil refining and chemical industry enterprises, and energy complex enterprises (Goldberg 1984; Kudelina 2018b). The degree of their influence is manifested in the disruption of natural water exchange, and, as a consequence, contamination of ground and underground waters, and flooding of the territory.

In this regard, a complex hydrogeological situation arose in the study area. Therefore, it is very relevant to consider the issues of groundwater protection depending on natural and anthropogenic factors (Gutsaki 1963; Kovalevsky 2001).

Source materials for hydrogeological assessment of the territory and analysis of groundwater protection were collected in the field and in the production funds of the Committee on Natural Resources, in the laboratories of the Orenburg State University, in the Department of Geo-ecology of the Scientific Center of the Ural Branch of the Russian Academy of Sciences, in the Orenburg Regional Committee of Statistics. The results of chemical analyzes of samples of natural waters, soils, snow, soils and silts were used. According to the method of V.S. Samarina (Hydrogeology of the USSR 1972) processed the results of water tests. According to A.Ya. Gaev (1989–2018) (Hydrogeology of the USSR 1972), the results of ground-based and remote sensing studies were modeled.

2 Results and Discussion

The protection of groundwater from the ingress of polluting components plays a huge role in preserving its natural state. This becomes especially relevant in connection with the economic development of the territory, the construction of industrial, agricultural and other facilities on it that affect groundwater, and, in connection with this, the justification of environmental measures to protect groundwater and water intake structures from pollution.

The natural protection of aquifers depends on many factors: soil types, humus content in them, the content of the clay fraction in soils and rocks of the aeration zone, the thickness of the aeration zone, its lithological composition, filtration properties and facies variability of rocks, the depth of the aquifer in question, its hydraulic relationships with adjacent hydrogeological units, water exchange rates (Blinov 2000; Plotnikov 1978). When determining the degree or category of groundwater protection, VSEGINGEO methodological recommendations were used, which provide for the assessment of groundwater protection according to several indicators (Table 1).

The thickness of the aeration zone is one of the main and often determining criteria for the protection of groundwater from the penetration of polluting waters. For the same aquifer, depending on the thickness of the aeration zone, the degree of protection will be different. When the same hydrogeological units occur on watershed spaces, the degree of protection of groundwater is much higher than on slopes. It increases from the slopes to the central part of the watersheds. The degree of protection of groundwater is usually minimal when aquifers occur in river valleys and in erosional incisions.

Of great importance when assessing the protection of groundwater are the lithological composition and facies variability of rocks in the aeration zone, the presence of filtration windows through which pollutants can penetrate into groundwater. The more low-permeability water-resistant rocks and clay deposits in the section, the more protected the aquifer is. In the absence of clayey low-permeability barriers in the section, even with a significant thickness of the aeration zone, the protection of groundwater from the penetration of polluting components "from above" will be insignificant (Marschalko et al. 2016).

Groundwater pollution is largely determined by the intensity of physicochemical processes that control the migration of polluting components into groundwater. These are the speed of groundwater flow and the sorption capacity of rocks. The flow rate or lateral filtration indicates the intensity of water exchange. The more intense the water exchange, the less the possibility of long-term accumulation of polluting components.

In conditions of layered strata, when sands are interlayered with clays and loams in the aeration zone, opportunities are created for the accumulation of polluting components on sorption geochemical barriers, which is typical for most aquifers of the Tatar and Kazan ages.

In fractured, gravel-pebble rocks there are practically no conditions for the sorption of pollutants.

The protection of non-confined aquifers also depends on the possibility of the entry of polluting components from underlying pressure or subpressure horizons, the presence of impermeable or low-permeability layers in the base, their thickness, permeability, and the presence of facies windows in them.

Humus contained in soils has a high absorption capacity of individual components. Its average content, judging by soil maps of the constituent entities of the Federation,

Table 1 Criteria for assessing the protection of groundwater from pollution

Factors determining the degree security	Degree of protection			
	protected	Relatively protected	Weakly protected	Unprotected
1. *Protection of groundwater from the penetration of pollutants through the rocks of the aeration zone*				
Types of soils and humus content in them	Chernozems Humus > 7% (1 point)	Gray forest Humus 4–7% (4 points)	Podzolic Humus 2–4% (7 points)	Light chestnut-new, gray earth Humus < 2% (10 points)
Clay fraction content, % in soils of the aeration zone	> 30 (1 point)	20–30 (4 points)	8–20 (7 points)	< 8 (10 points)
Thickness of the aeration zone, m	> 20 (1 point)	10–20 (3 points)	5–10 (5 points)	< 5 (7 points)
Lithological composition of rocks in the aeration zone, facies variability of rocks in plan and section	Predominance of water-resistant rocks: filtration windows are practically absent (1 point)	Interlayering of low-permeability and water-resistant rocks: filtration windows on an area < 5% (3 points)	Interlayering of permeable, weakly permeable and water-resistant rocks: filtration windows on an area from 5 to 30% (5 points)	Interlayering of permeable and weakly permeable rocks: filtration windows on an area of over 30% (7 points)
2. *Protection of groundwater from the lateral flow of regulated substances into an unconfined aquifer*				
Groundwater outflow rate m/year	Above 2000 (1 point)	500–2000 (2 point)	50–500 (3 point)	Less 50 (4 point)
Sorptive capacity of rocks	High (alternation of clays with loams, sandy loams, fine-grained sands) (1 point)	Medium (interlayering of loams, sands, sandy loams) (2 points)	Sands with separate layers of loams and clays (3 points)	Gravel and pebbles, coarse sands, lithified fractured rocks (4 points)

(continued)

Table 1 (continued)

Factors determining the degree security	Degree of protection			
	protected	Relatively protected	Weakly protected	Unprotected
3. *Protection of the unconfined aquifer from penetration pollutants with pressurized water*				
Interaction of free-flow and sub-pressure horizons (in case the levels of the sub-pressure horizon are higher than those of the free-pressure horizon)	Power aquitard more than 20 m, facies window None (1 point)	The thickness of the aquifer is 10–20 m, facies windows are less than 5% of the area Aquitards (2 points)	The thickness of the aquifer is 5–10 m, facies windows are on 5–30% of the area Aquitard (3 points)	The thickness of the first aquitard from the surface is less than 5 m, facies windows On the square over 30% (4 points)

rarely exceeds 4%. The soils of the Syrtov basin are low in humus and have a low protective capacity for the underlying groundwater.

All of the above criteria relate mainly to the first aquifers from the surface (usually free-flowing), which are the most vulnerable from the point of view of pollution in all directions. Pressure horizons are in more favorable conditions. With a significant area of distribution of a hydrogeological unit, the conditions of its occurrence may be different, both in areas of occurrence first from the surface, and in a situation where they are covered by other hydrogeological units. In any case, an important role in determining the protection category of the main aquifers and complexes is assigned to the overlying hydrogeological units, composed predominantly of clayey-loamy varieties with very low filtration properties and high thickness. In this case, it is more correct to use not the aeration zone, but the thickness of the overlying sediments (Voudouris 2011).

All of the above factors were taken into account when determining the protection of groundwater, which was ranked in four categories: unprotected, weakly protected, relatively protected and protected. Each protection factor was assessed using a point system, the integral assessment of groundwater protection was made according to the average score: protected—less than 2 points; relatively protected—2.1–4.4 points; weakly protected—4.5–6.8 points; unprotected—over 6.8 points.

The natural features of the Syrtov artesian basin made it possible to identify within its boundaries territories with all four categories of protection. The most protected from pollution are the underground waters of pressure horizons and complexes, the roof of which contains weakly and impermeable sediments. Almost the groundwater of all aquifers, lying under Neogene and Eopleistocene deposits, as well as their coeval aquifers, lying at the base or under a layer of clay, are classified as protected.

Groundwater is protected when it occurs at considerable depths on the surfaces of large watersheds, covered on top by deluvial loams and when there are rocks of varying permeability in the aeration zone (under conditions of significantly heterogeneous section).

Weakly protected underground waters are those of free-flowing or low-pressure hydrogeological units with a small thickness of the aeration zone (up to 10 m) and the absence of sustained water-resistant sediments in their roof. Usually this is groundwater from the first floodplain terraces of rivers and fractured limestones, overlain by thin sandy-loamy material.

The least protected are groundwaters located in river floodplains. Their vulnerability here is due to the close occurrence of the roof of water-saturated rocks, the presence of filtration windows in the aeration zone, that is, layers and lenses of sand among sandy loams and loams, and sometimes the predominance of sandy varieties in the section of the aeration zone.

The rest of the territory, partly within the watershed, almost entirely slope space, terraces of the highest rank, is characterized by the presence of groundwater, relatively protected from pollution. Below we will briefly discuss the definition of the category of natural protection from pollution of the main aquifers and complexes in various conditions of occurrence—regardless of age, but only on the typical factors that determine it.

The aquiferous Quaternary alluvial horizon is confined to river valleys, floodplains and above-floodplain terraces. Within the distribution of this horizon, unprotected, weakly protected and conditionally protected groundwater are distinguished.

Unprotected groundwater is common in river floodplains. The aeration zone here usually does not exceed 2–5 m, composed mainly of sands with sandy loams, sometimes sands, or layers and lenses of sand among loams, causing the presence of filtration windows, which favors the filtration of the polluting component. Here, due to the close occurrence of groundwater, regardless of the humus content in the soils and weak sorption activity, the likelihood of groundwater contamination is greatest, both from the surface and from underlying and adjacent hydrogeological units, which are characterized by discharge into river valleys. These factors significantly increase the possibility of pollutant complexes entering the horizon.

Groundwater within the first floodplain terraces of large rivers, such as the Volga, Ural, Ilek, Irgiz and others, is classified as weakly protected from pollution. The largest territories with weakly protected groundwater of alluvial deposits can be traced near the Samarskaya Luka and stretch along the left bank of the river. Volga to the southern border of the basin. Such conditions for the protection of groundwater here are due to the predominance of sandy varieties in the section of the aeration zone and the insignificant thickness of the aeration zone, usually not exceeding 10–15 m.

The remaining areas of distribution of the alluvial aquifer are characterized by the presence of groundwater that is conditionally protected from pollution. The conditions of protection here are determined by the predominance of loams and clays in the section of the aeration zone. And the thickness of the aeration zone on all terraces, except the first, is very significant. The thickness of the aeration zone on the terraces reaches 15–20, in rare cases 30 m. Akchagyl clays often occur at the base of the horizon. From the surface, the horizon is covered with loess-like loamy deposits, which increases the thickness of the overlying deposits and increases the degree of protection of groundwater.

Aquifers Tithonian and Bajocian-Bathonian horizons. On the left bank of the Volga River they are distributed under Akchagyl and alluvial deposits. The main outcrops are located within the Buzuluk depression in the Samara-Buzuluk and Samara-Chagan interfluves. When located on watersheds, the Bajocian-Bathonian complex is covered by water-resistant Callovian-Kimerid clays, creating reliable conditions for their protection from pollution. On the slopes of watersheds, groundwater of Bajocian-Bathonian deposits is covered by a thin cover of eluvial-deluvial loams and is relatively protected. The same pattern can be traced for the groundwater of the Tithonian aquifer complex. Where the groundwater of this complex lies under Cretaceous low-permeable sediments and Quaternary loams, they have the highest category of protection. In places where groundwater comes to the surface, its protection category is reduced to relative.

Aquiferous Lower Triassic complex. The complex has its main outcrops in the central and southeastern parts of the Syrtov basin. In the southern and southwestern parts of the basin it underlies younger sediments. In this case, the thickness of the overlying rocks reaches 80–120 m or more, which makes it possible to consider groundwater in such conditions to be protected from pollution.

On the slopes, the Lower Triassic groundwater is covered by deluvial-proluvial low-permeable sediments. The thickness of the aeration zone on the slopes rarely exceeds 30 m. The relative protection of the groundwater of the complex in these conditions of occurrence, in addition to the overlying loams, is determined by the predominance in the upper part of the complex of low-permeable clayey-siltstone varieties.

Aquifer Tatar and Urzhum complexes. In the northern and eastern parts of the Syrtov artesian basin, both complexes lie mostly first from the surface or close to it, only in river valleys and paleodolias they submerge under younger sediments. The complexes have their greatest exposures to the surface in the eastern part of the basin, composing the main watersheds of the Volga and Ural rivers (Samara—Bol. Kinel, Bol. Kinel—Kon, Samara—Sakmara and Sakmara—Ural). In the rest of the territory, the complexes are almost completely covered by a thick layer of Neogene-Quaternary sediments and contain brackish waters. Water in such conditions is pressurized and protected from pollution.

On the slopes of watersheds, along the left banks of rivers, they are often, in narrow areas, covered by Eopleistocene and Akchagyl deposits containing water-resistant clays. Under such conditions, the groundwater of both complexes is reliably protected, since the overlying low-permeability and water-resistant clay deposits are a good barrier that prevents the penetration of polluting components from the surface. Everywhere the complex is represented by a complex layered strata of weakly permeable, impermeable and permeable sediments. This combination of sediments creates favorable conditions for the protection of groundwater associated with them. The thickness of the aeration zone of these complexes increases from rivers to watersheds from 15–20 to 40–60 m or more. In this direction, the conditions for the protection of groundwater are also becoming more reliable. Moreover, almost all watershed spaces where the above-mentioned aquifer complexes are common are covered with colluvial loams with a thickness of 5 to 10–15 m, which also improve the conditions for the protection of groundwater. It should be noted that the protection of groundwater on watersheds is generally higher than on slopes, which allows them to be classified as protected. The groundwater of complexes located on slope areas is classified as relatively protected, since the thickness of the aeration zone here is much less.

Aquiferous Upper Kazan complex. It comes to the surface or lies close to it in the Samarskaya Luka area on the left bank of the Volga River, in small areas near the Bol River. Kinel and in the eastern part of the basin, where it is exposed along the right tributaries of the Sakmara River. A distinctive feature of the Kazan deposits is the predominance of mudstone-like clays, siltstones and marls in the upper part of the section, which, in combination with overlying weakly permeable Quaternary deposits, determines the relative protection of groundwater. In the main territory of the Syrtov artesian basin, the aquiferous Upper Kazan complex lies at a considerable depth under younger sediments (often Upper Permian or Neogene), which makes them reliably protected from pollution. The groundwater of the Kazan complex is almost universally underlain by coeval sulfate-halogen practically impermeable sediments.

In local areas, the thickness of the overlying loams does not exceed 0.5–1.0 m, the depth of water is 10–15 m. These are areas with poorly protected groundwater. Typically, such areas have small areas and are found in the Samarskaya Luka area. Weakly protected waters are confined to places where karst limestones come to the surface. There is usually no screening layer here, the thickness of deluvial weakly permeable loams is insignificant and does not exceed 3–5 m. The depth of the groundwater level in limestones is mainly 15–20, less often 30 m. The rocks in the upper part are fractured, have low sorption activity, which sharply reduces protective properties of groundwater associated with them.

Such areas are found only in the western part of the basin, have small areas and are confined to the aquiferous Lower Moscow-Artinsky complex.

The groundwater of Cretaceous and Paleogene deposits, which lie first from the surface or close to it, are relatively protected from pollution. They are usually covered by a cover of deluvial-proluvial loams up to 8–10 m thick. The thickness of the aeration zone in such hydrogeological units is usually 10–20 m. The aeration zone is represented by clays, marls, opoka, chalk, siltstone and sandstone. There is no sustained aquifer in the roof, which does not allow them to be classified as protected.

The scale of the research does not allow dividing the territory of the Syrtov artesian basin into more detailed areas with varying degrees of groundwater protection. The information reflects the prevalence of groundwater within certain territories with varying degrees of protection. All this does not exclude the possibility of the presence of local areas in these territories with a different degree of protection.

In general, it can be noted that the main part of the territory of the Syrtov artesian basin is characterized by the presence of relatively protected and sheltered groundwater (Kovalevsky 2001). But this is an areal characteristic of an artesian basin.

The waters of the alluvial aquifer are of primary water economic importance for providing the population with drinking water. In the case of an undisturbed regime, they, like river waters, belong to the sulfate-sodium type. When water flows from Permian deposits, they acquire the soda type. If polluted, these waters become hard, magnesium chloride with increased mineralization. Almost all major aquifers and complexes containing fresh water and used for domestic and drinking water supply to the population have a low degree of protection, especially from the penetration of pollution from the surface.

As a result of the work carried out, a methodological approach to hydrogeological studies of the territory was developed. It revealed that the main resources of drinking water are concentrated under river floodplains in unprotected areas, and the main sources of pollution are confined to drainage areas. Streams and areas of pollution from them have already penetrated into river valleys, and in some cases into open water bodies. Large industrial enterprises have been operating since the 30–70 s. twentieth century, which explains the unsatisfactory state of water resources. It is possible to prevent negative processes only by introducing new technologies and, in particular, complex barrier technology.

References

Blinov SM (2000) Fundamentals of the use of geochemical barriers for environmental protection: abstract. Ph.D. Dissertation. 11.00.11. In: Blinov SM (eds) Perm State University, Perm, 23 pp

Goldberg VM (1984) Hydrogeological basis for the protection of groundwater from pollution. In: Goldberg VM, Gazda S (eds) M.: Nauka, p 262

Gutsaki VA (1963) Weathering crust of the Orsk Trans-Urals, VA 12. In: Gutsaki, Gudoshnikov VV (eds). Research Institute of Geology at the University named after N.G. Chernyshevsky, Saratov, 281 pp

Hydrogeology of the USSR (1972) T. 43. In: Tokmacheva EI (ed) Orenburg region. M.: Nedra, p 272

Iceberg PE (2002) Precambrian of the East European Platform: geology and oil and gas potential. In: Iceberg RE, Aksamentova NV, Belonin MD (eds) VNIGRI, St. Petersburg, p 391

Korotkov AI (1997) Formation of underground brines of the South Tatar arch. In: Korotkov AI, Kharkhordin IL (eds) VI Tolstikha Readings, St. Petersburg, pp 46–49

Kovalevsky VS (2001) Combined use of surface and groundwater resources. M.: Scientific World, 332 pp

Kudelina IV (2018) Ways to stabilize the regime of alluvial water intakes in a semi-arid climate. In: Kudelina IV (eds) Science, new technologies and innovations of Kyrgyzstan, Bishkek, vol 2, pp 82–86

Kudelina IV (2018) On the possibility of stabilizing the operating mode of water intakes in the Orenburg urban agglomeration. In: Kudelina IV (eds) Science, new technologies and innovations of Kyrgyzstan, Bishkek, vol 2, pp 68–72

Heviánková S, Marschalko M, Chromíková J, Kyncl M, Korabík M (2016) Artificial ground water recharge with surface water. In: World Multidisciplinary Earth Sciences Symposium (WMESS 2016). IOP Publishing, pp 6–12

Pityeva KE (1999) Hydrogeoecological studies in the areas of oil and gas fields. In: Pityeva KE (eds) M.: Nedra, p 199

Plotnikov NI (1978) Hydrogeological principles of artificial replenishment of groundwater reserves. In: Plotnikov NI, Plotnikov NA, Sychev KI (eds) M.: Nedra, 311 pp
Voudouris K (2011) Artificial recharge via boreholes using treated wastewater: possibilities and prospects. Water 3(4):964–975
Zektser IS, Karimova OA, Chetverikova AV (2015) Resources of fresh groundwater in Russia and their use in emergency situations. Water Resour 42(4):351–366

Regularities of Noble Metals Distribution in Halogenous Formations of the Pre-Ural Marginal Trough

G. A. Ponomareva

Abstract The results of the determination of noble metals in rock salts are shown: oil and gas fields of Sakmaro-Ileksky, Yuzhno-Buzuluksky, Sol-Iletsky, Teplovsky-Derkulsky oil and gas and geological regions and Iletsky salt deposit. Platinum, palladium, gold and silver were determined in the samples of the Kungur stage of the Irenian horizon by atomic absorption spectrometry using their own patented method. Elevated levels of all noble metals have been established. In addition, higher concentrations of platinum, palladium, gold and silver occur in rock salts associated with hydrocarbon fields. Perhaps this indicates the presence of a geochemical barrier at the interface between hydrocarbon and halogen media. The above facts of the detection of noble metals in the strata of rock salts of the Orenburg Pre-Ural marginal trough are also of practical importance, demonstrating the relationship of elevated metal concentrations with halogenesis and the effects of salt tectonics.

Keywords Halogen formations · Noble metals · Distribution · Platinoids · Geochemical barrier

1 Introduction

The first information about noble metals (NM) in waste salts of the Verkhnekamsk potassium-magnesium salt deposit in the Perm region was obtained by Dytnersky in 1984 (Smetannikov and Kudryashov 1995). Later, Smetannikov et al. revealed the increased content of gold, platinum, palladium and silver both in the salts themselves and in the insoluble residue, which represents production waste and the content of insoluble residue can reach one fourth of the total volume of ores (Smetannikov and Kudryashov 1995; Smetannikov et al. 2013). In continuation of the works the forms of metals were determined: for platinoids it is, as a rule, complexes with participation of organic matter and chlorides; for gold—the nugget form, also complexes with participation of organic matter and, probably, carbonyl complex compounds

G. A. Ponomareva (✉)
Orenburg State University, 13 Pobeda Ave., Orenburg 460018, Russia
e-mail: galy.ponomareva@mail.ru

T. Chaplina (ed.), *Processes in GeoMedia—Volume VIII*, Springer Geology,
https://doi.org/10.1007/978-981-97-6627-7_15

and chlorides. Silver is associated, according to the authors, with salts of thioacids, is present in the nugget form and presumably in the form of chlorides. Silver and platinoids are observed in the non-salt part of the rocks, the latter being found in the organic matter. In halite Au, Ag and platinum group metals (PGM) are associated with the organic phase of insoluble residue (Smetannikov and Kudryashov 1995; Smetannikov et al. 2013, 2006; Sedikh et al. 2017). Prognostic calculations indicate that the content of gold, platinum and palladium in the salt rocks of the Verkhnekamskoe deposit is comparable to the resources of the largest ore provinces (Smetannikov et al. 2006). These data allowed Zoloev et al. (2001) to refer the Verkhnekamskoe potassium-magnesium salt deposit in the Pre-Ural trough to a non-traditional type of platinometallic mineralization—PGM in salt sediments.

The Department of Geology, Geodesy and Cadastre at the Orenburg State University has been conducting long-term studies of the distribution patterns of platinoids (together with gold and silver) in the platform and folded parts of the Orenburg region. The purpose of the present work is to study the distribution patterns of PGM in halogen formations of the Southern Ural marginal trough. The object of the study is associations of PGM (platinum, palladium, gold and silver) in halites. The subject of the study is the regularities of PM distribution in halite salts.

On the territory of the Orenburg region halide formations of the Early Permian period were widely developed. The thickness of saline strata increases to the south and southeast, reaching maximum values in the Pre-Ural trough (Baranov et al. 1997). The detection of NM in halogenic salts of the Southern Urals was reported in Ponomareva and Ponomarev (2021).

2 Materials and Methods of Research

Geochemical samples of the studied halogen objects of the Orenburg region were formed from samples of the Iletsk salt deposit, as well as from well cores from oil and gas fields of Sakmaro-Ilekskoye, Yuzhno-Buzulukskoye, Sol-Iletskoye, Teplovsko-Derkulskoye (northern side of the Pre-Caspian Syneclise) oil and gas geological areas and were accompanied by geological documentation presented in Ponomareva and Ponomarev (2021). Salt structures of the Orenburg Pre-Ural marginal trough are associated with a network of disjunctive tectonic faults. The sampling characterizes the confinement of the sampled saline strata to the boundaries of structural-tectonic elements of the 1st order, as well as superorder structures. In the Orenburg part of the Volga-Urals oil and gas province, rock salts are usually fluid supports in natural hydrocarbon reservoirs. The structure of sedimentary strata of the territory is characterized by the presence of gypsums, anhydrites, and rock salts of Early Permian age. The strata have a rhythmic structure, alternation of halogen and sulfate-halogen salts is observed, which are repeated many times in the section. The thickness of saline formations tends to increase in the southern direction of the Volga-Ural Anticline (VUA). Halokinesis (ramparts, including those of meridional strike, diapirs, etc.) has been widely developed at the junction of the supra-row structures of the East

European Platform—VUA, PCP and PS. Individual salt formations reach more than 5000 m (Baranov et al. 1997; Blizeev 1971; Petrishchev and Noreika 2018; Petrishchev 2017). Deposits of the Irenian horizon overlie the riffogenic Artinsk limestones and underlie marine sediments of the Ufa Stage and continental sediments of the Neogene and Quaternary systems (pebbles, clays, sands) (Baranov et al. 1997). Salt samples were taken from strata that serve as covers (regional fluid supports) for reservoir rocks of oil and gas fields.

The Iletskoye halite deposit is located on the Sol-Iletskiy vault of the Volga-Ural anteclise. The deposit is a salt dome more than 1200 m thick (Petrishchev 2017; Antipova 1977). Samples of Iletskii rock salt were taken from mine workings.

In the samples of rock salt from the above mentioned deposits and areas the content of the following PMs was determined—Au, Ag, Pt, Pd, as these metals are industrially significant.

Geochemical rock salt samples were crushed (0.07 mm), mixed and reduced to an average of 50 g. Then, a 10 g sample weight was taken according to the quantitative chemical analysis procedure.

Instrumental determination of metals was carried out by atomic absorption spectrometry with electrothermal atomizer on the device of "Lumex" company (St. Petersburg). The samples were subjected to preliminary chemical preparation according to the method of quantitative chemical analysis, developed in VSEGEI with application of own proprietary method of preliminary additive oxidation of carbon-containing components of salt samples (Russian Federation Patent No. 2409810). It should be noted that the author's data, as well as the results presented in the works of Smetannikov and his colleagues (Smetannikov and Kudryashov 1995; Smetannikov et al. 2013, 2006; Sedikh et al. 2017), note the presence of carbon-containing compounds in the studied salt samples (Ponomareva and Ponomarev 2021) and in their production wastes. In the final acid solutions or finely dispersed systems, the concentrations of noble metals were measured, taking into account the conditions of their determination, the spectrometer operation manual, and the chemical composition of the samples (Ponomareva and Ponomarev 2021). The data on the content of noble metals in samples of halogen salts from different areas and Iletskiy rock salt deposit are given in Table 1.

Studies of the statistical distribution of NM were performed using the STATISTICA 8 software package.

Table 1 Noble metals concentrations in saline formations of the Southern Urals

Objects	Pt	Pd	Au	Ag	ΣPt, Pd	ΣБМ
Iletskoye field (9)	$\frac{43-311}{164}$	$\frac{3-98}{39}$	$\frac{20-1465}{416}$	$\frac{62-1545}{623}$	203	1242
Salt of oil areas and fields (19)	$\frac{52-3128}{880}$	$\frac{3-248}{54}$	$\frac{5-2064}{537}$	$\frac{53-3079}{766}$	934	2237

The numerator shows the range of contents, the denominator shows the average values, and the number of samples is given in parentheses. Metals contents are given in mg/t

3 Results and Their Discussion

The results obtained by the author during the determination of NM in rock salt indicate their uneven distribution. The highest concentrations of platinum (3128 mg/t) and gold (2064 mg/t) are confined to halites of Peschanoye oil field. Maximum silver content (3079 mg/t) is observed in rock salt of Vershinovskoye oil field. Palladium is highest in salts of Akobinskoye gas field (248 mg/t). The lowest concentrations of platinum (43 mg/t) and palladium (3 mg/t) are observed in samples from Iletskoye rock salt deposit.

The range of values of concentrations of gold, platinum, palladium is greater in salts of oil and gas deposits. And in samples of Iletskiy rock salt deposit the range of platinum and palladium contents is lower by an order of magnitude. This may indicate the presence of a geochemical barrier at the salt-oil boundary.

Distribution of platinum metals in halogen salts of Orenburg region by absolute values and platinum to palladium ratio is of ultramafic character: average platinum and palladium contents are 100n mg/t and 10n mg/t respectively, platinum to palladium ratio is greater than one. Arranging metals in the order of decreasing average values of their concentrations in salts of Iletskoye deposit, we obtain a series of Ag, Au, Pt, Pd, demonstrating specialization of this deposit in noble metals. According to PGM—palladium-platinum, with the ratio of platinum to palladium much more than one.

It should be noted that maximum concentrations of noble metals have been detected in halogen salts conjugated with oil and gas deposits. These deposits are characterized by peculiarities of their position in geological and structural zones. Thus, the Akobinskoye gas field is confined to the Pre-Ural marginal trough (western side). The Peschanoye field is confined to the junction zone of the Volga-Ural Anteclise and the Pre-Caspian Syneclise. The Vershinovskoye oil field is located at the junction of overlying structures of the VEP: VUA, northern side of the PS and Pre-Ural trough. The fault-block structure and stepwise dipping of the crystalline basement of the Caspian syneclise in the southwestern direction and of the Pre-Ural trough in the southeastern direction have been established (Baranov et al. 1997). In the sedimentary cover of the described territory the presence of sublatitudinal and submeridional strike-slip discontinuities has been revealed (Baranov et al. 1997). These are areas of increased permeability. Salt domes of more than 4.0 km are observed in the Kungur sediments. The Pre-Ural trough rock salts are characterized by confinement to zones of juxtaposition of two types of structures: areas of drifting and feeding by clastic matter and areas of accumulation and burial of these sediments.

Salts of hydrocarbon deposits demonstrate specialization by metals—Pt, Ag, Au, Pd (metals are arranged in the order of decreasing average values of their concentrations). For PGM, as well as in the salts of Iletskoye deposit—palladium-platinum specialization, with the ratio of platinum to palladium much more than one.

The use of rank correlation method allowed to reveal interrelations of metals. In Iletsk salt and in halogen screens of oil and gas fields platinum is directly related to gold and silver (correlation coefficients 0.42 and 0.39 respectively for Iletsk field and for hydrocarbon fields—0.3 and 0.44). The lack of correlation between platinoids suggests that in the halogen formations of the Southern Urals the distribution of palladium is independent of the distribution of platinum.

The distribution of NM in halogen salts and in oil of oil and gas deposits has common features (Table 2). Specialization of salts of oil deposits and oil itself (Ponomareva 2015) have similar distribution of PM. For PGM, as well as in salts—palladium-platinum specialization with platinum to palladium ratio also more than one, however it is less contrasting.

According to the results of correlation analysis, the similarity of the relationship of Pt with Au 0.7 in oil deposits with oil field salts with Ag 0.44 and with Au 0.3 and with Iletsky halite with Ag 0.42 and Au 0.42 is also observed.

The interrelationships of noble metals in oil and gas deposits and halite salts are also similar in many respects: in oil there is a direct significant relationship of metals with the exception of platinum with silver (Ponomareva 2015). In salts confined to hydrocarbon deposits, Au, Ag and Pd have direct close relationship. And in Iletskoye deposit Au, Ag and Pt are directly related.

In oil of Orenburg fields in general (in oil of Buzuluksky and Sol-Iletsky oil and gas geological regions including) and in halogen screens of hydrocarbon fields the same specialization on platinoids is observed.

Application of an analog of dispersion analysis—nonparametric Friedman's rank criterion in the study of BM geochemistry in halogen salts of the Southern Urals allowed us to obtain a "box" diagram, which shows some peculiarity of palladium distribution (Fig. 1). The diagram makes it possible to compare the values of the Friedman parameter (mean values of contents for PM, range, dispersion).

The nonparametric Friedman rank criterion is suitable in case of distribution of values different from normal (which is observed in this case).

According to the results of statistical analysis of PM contents in halogen formations, the comparative homogeneity of rock salts of the platform part of the Orenburg region (meaning the statistical distribution of metals) was found, which can be explained by the relatively small sample size. Nevertheless, higher contents of both individual metals and their total amount in the salts of hydrocarbon deposits have been established, which possibly proves the existence of a geochemical barrier at the salt-oil boundary. However, confirmation of this idea needs additional research.

Halogen formations within evaporites are widely spread not only in the Orenburg region and neighboring territories, but also in the whole world. Interest in

Table 2 Noble-metal specialization of halogen formations and hydrocarbons of Western Orenburg region

No. (n/a)	Deposit	Noble-metal specialization	Pt/Pd
1	Iletskoe	Ag, Au, Pt, Pd	$\gg 1$
2	Halogen screens of hydrocarbon fields	Pt, Ag, Au, Pd	$\gg 1$
3	Hydrocarbon fields (39)	Pt $\geq$ Ag, Au, Pd	2

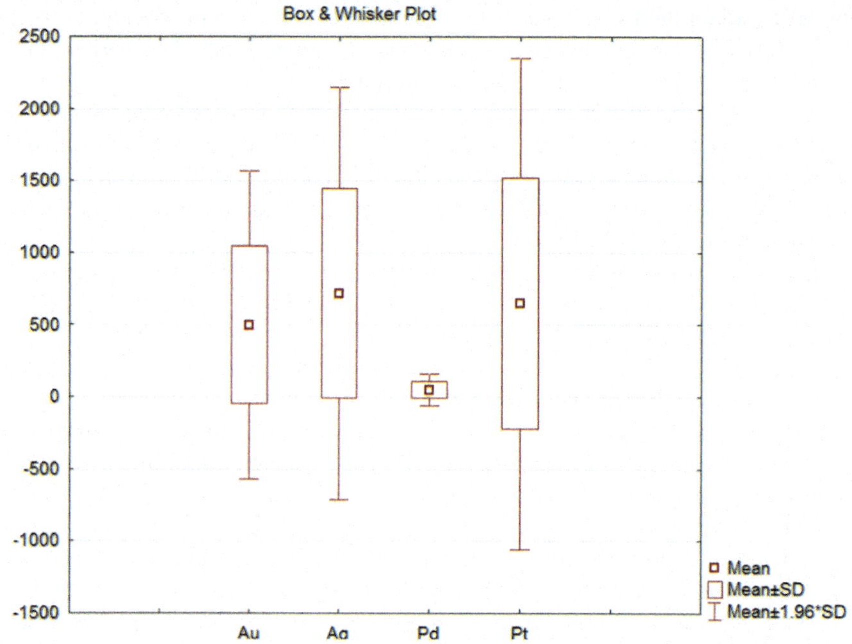

Fig. 1 Friedman test. Diagram of distribution parameters of platinum, palladium, gold and silver in halogen formations of the Southern Urals (average values of contents, dispersion, range)

minerals associated with saline strata has not weakened to this day. Thus, the relationship between salt diapirs and stratiform deposits of zinc and lead sulfides (galena, sphalerite, pyrite) in the Basque-Cantabrian basin is studied (Perona et al. 2018). According to the authors (Rddad et al. 2019) Ba–Pb–(± Zn) mineralization associated with the salt diapir in Tunisia should be attributed to the Mississippian type. The fourth largest zinc deposit in Asia Jinding (Jinding Zn–Pb) in Yun-Nan Province of China is confined to the evaporite dome, the genesis of the ores has not been fully identified (Leach et al. 2017). Researchers' views on the history of development of salt-dome structures are based on the stage-by-stage development, with a significant role played by the ratio of the rate of their formation and exogenous activity. It has been established that salts also influence oil and gas systems (Jackson and Hudec 2016). It is quite likely that BMs can be associated with lead–zinc mineralization. It is well known that sulfide minerals are concentrator minerals for gold and silver.

The above facts of detection of noble metals in saline strata convincingly indicate that ore-bearing structures are associated with the effects of salt tectonics. This increases the probability of ore deposits detection in similar geological conditions by the principle of analogy.

4 Conclusion

1. Elevated contents of Pt and Pd, Au, Ag are revealed in halogen formations of the Kungurian age of the Southern Urals.
2. The range of Pt, Pd contents is greater in halogen formations of oil and gas deposits in comparison with Iletsk salt deposit.
3. Higher contents of both individual PM and their total amount in halogen salts of hydrocarbon deposits were found, which may indicate the existence of a geochemical barrier at the salt-oil boundary.
4. Detection of noble metals in saline strata of the Orenburg Urals is also of applied importance, demonstrating the relationship of elevated metal concentrations with the processes of halogenesis and the effects of salt tectonics.

References

Antipova AS (1977) Features of salt karst development of the Iletskoye rock salt deposit. In: Problems of salt accumulation: collected works, vol 2. Nauka, Novosibirsk, pp 79–83

Baranov VK, Galimov AG, Dontskevich IA et al (1997) Geological structure and oil and gas content of the Orenburg region. Orenburg Book Publishing House, Orenburg, p 272

Blizeev BI (1971) Structure of the Kungur saline strata of the Southern Urals. In: Geology and Genesis of deposits of mining and chemical raw materials: collection of works, Kazan, pp 165–171

Jackson MPA, Hudec MR (2016) Salt tectonics: principles and practice. Cambridge University Press, Cambridge, 498 pp. https://doi.org/10.1017/9781139003988

Leach DL, Song Y-C, Hou Z-Q (2017) The world-class Jinding Zn–Pb deposit: ore formation in an evaporite dome, Lanping Basin, Yunnan, China. Miner Deposita 52:281–296

Perona J, Canals À, Cardellach E (2018) Zn-Pb mineralization associated with salt diapirs in the basque-cantabrian Basin, Northern Spain: geology, geochemistry, and genetic model. Econ Geol 113(5):1133–1159

Petrishchev VP (2017) Structural-tectonic and geomorphologic contiguity of solanic-dome morphostructures of the Orenburg Urals. Prob Reg Ecol 5:82–87

Petrishchev VP, Noreika SY (2018) Regional peculiarities of the formation of solanodome morphostructures. Geomorphology 1:66–74

Ponomareva GA (2015) Geochemical peculiarities of noble metal distribution in oil and gas fields of the Orenburg region. Bull Orenburg State Univ 7(182):167–172

Ponomareva GA, Ponomarev AA (2021) Distribution patterns of platinoids in halogen rock masses in the Southern Urals. Mining J 12, 10–14. https://doi.org/10.17580/gzh.2021.12.02

Rddad L, Jemmali N, Sośnicka M, Cousens B (2019) The genesis of the salt diapir-related Mississippi Valley-type Ba–Pb–(± Zn) ore of the Slata district, Tunisia: the role of halokinesis, hydrocarbon migration, and alpine orogenesis. Econ Geol 114:1599–1620. https://doi.org/10.5382/econgeo.4687

Sedikh EM, Smetannikov AF, Bannykh NL, Shanina SN (2017) Features of Au, Pt and Pd determination in salt rocks of Verkhnekamskoye deposit. Geochemistry 4:363–367

Smetannikov AF, Kudryashov AI (1995) Content and distribution of Au, Ag and Pt-metals in sylvinite of Verkhnekamskoe deposit. Geochemistry 9:1348–1351

Smetannikov AF, Shanina SN, Sinegribov VNN, Sinegribov VA, Yudina TB, Sedykh EM (2006) Noble metals of the Verkhnekamskoye salt deposit. Min J 6:62–66

Smetannikov AF, Onosov DV, Sinegribov VA, Novikov P, Yu N (2013) Noble metals in the salts of the Verkhnekamskoe deposit and the technology of their extraction. Min J 6:55–58
Zoloev KK, Volchenko YA, Koroteev VA, Malakhov IA, Mardirosyan AN, Khrypov VN (2001) Platinum-metal mineralization in the geological complexes of the Urals. Yekaterinburg, 199 pp (in Russian)

Origin of Solar System Comets

Azariy Barenbaum

Abstract The author's "cometoasteroid" concept of Solar System comets origin, based on Galactocentric paradigm ideas, is discussed. It is argued that long- and short-period comets we observe are daughter formations, products of collisions of galactic comets of Orion-Cygnus jet stream with asteroid belt bodies, captured by Sun attraction into orbits more diverse and distant than at asteroid belt bodies. These daughter objects are composed of substance of galactic origin comets frozen together with debris and vaporized material of asteroids. Long-period and short-period comets were formed in time interval from 5 to 0.7 Ma, and differ from each other in that the former have aphelion of its orbits is ~ 10^4–10^5 AU. Therefore, today they are returning back to the Sun for the first time. While, the second have never left Solar System limits, they gravitate towards the ecliptic plane and today have survived only in outer space between giant planets. At random convergences with these planets, they are thrown towards the Sun, and then we are observed them as members of the comets family of corresponding giant planet. It is shown that the stated concept explains well all main properties of observed comets in Solar System.

Keywords Galactic comets · Origin of long- and short-period comets · Galactocentric paradigm

1 Introduction

In connection with discovery of phenomenon of jet outflow of matter in spiral galaxies (Barenbaum 2002) and creation of Galactocentric paradigm (Barenbaum 2010), the author has no doubt that in the very near future this paradigm will become a unified worldview basis for all space sciences. In recent works (Barenbaum 2021, 2022, 2023), the author substantiated this conclusion using the example of solving a number of problematic issues in cosmology, astronomy and geology. This article discusses

A. Barenbaum (✉)
Oil and Gas Research Institute RAS, Moscow, Russia
e-mail: azary@mail.ru

T. Chaplina (ed.), *Processes in GeoMedia—Volume VIII*, Springer Geology,
https://doi.org/10.1007/978-981-97-6627-7_16

the solution to another important problem—the origin of Solar System comets, posed to modern science at the turn of the eighteenth and nineteenth centuries.

At the "2-th Vsekhsvyatskikh Readings on Physics and Dynamics of Comets" in 1990 in Kyiv the author for the first time put forward a new hypothesis of Solar system comets origin (Barenbaum 1990), based on Galactocentric paradigm representations. Later, this hypothesis in the form of a more general physical concept was substantiated in books (Barenbaum 2002, 2010).

The new concept states that Solar System comets are collision products of comets galactic origin with bodies of asteroid belt, captured by Sun's gravitational pull into elliptical orbits that are more diverse and distant than ones of the belt bodies. These "daughter" comets were born in asteroid belt of 5.0–0.7 Ma during period of Sun's stay in the Orion-Cygnus jet stream. They consist of icy substance of galactic comets (primordial comets) frozen together with solid (debris) and diffuse (evaporated) material of asteroids.

Thus, the comets we observe cannot be considered objects of only galactic or only circumsolar origin and they should be considered as a category of bodies of dual genesis specific to Solar System—"cometoasteroids". Below we offer a rationale for this "cometoasteroidic concept" that explains the origin and formation mechanism of observed comets. The focus is on the following issues:

- General information about observed comets,
- Review of origin hypotheses of observed comets,
- Brief description of Galactocentric paradigm representations,
- Information about galactic comets and their properties,
- Issues of galactic comets collision with asteroid belt bodies,
- Mechanism of long-period and short-period comets formation,
- Estimate of the number of both group comets in Solar system.

2 Observed Comets and Their Properties

Comets have been observed by people since ancient times. These are huge space objects, with a diameter of up to 10–20 km and a mass of up to 10^{18} g, consisting of agglomerates of ice, dust and gases mixed with rocky material. They move around Sun in more varied and distant orbits than the orbits of planets and asteroids. Depending on the nature of their orbits, comets are divided into two main categories. The first category includes comets with a long period of revolution around Sun $T > 200$ years. The second category includes comets with short periods $T < 200$ years.

Although short-period and long-period comets are composed of essentially the same material, their orbits differ significantly. Short-period comets overwhelmingly move in the same direction as planets and asteroids. They have less eccentric orbits, lying closer to the ecliptic plane. While long-period comets have highly elongated orbits with eccentricity $e = \sim 1$, randomly located relative to Sun.

For most of its trajectory, the comet is inaccessible to observation from Earth. However, when the comet approaches Sun at a distance of less than 3–5 AU, the process of evaporation of the comet's icy core begins under the influence of solar radiation. This process is accompanied by a bright glow around the comet's nucleus, called a coma, and a powerful ejection of gas and dust from the nucleus. From this moment on, the brightness of the comet increases sharply, making it accessible to observation.

Over time, after several passages by Sun, comets lose some of their matter, as a result of which their glow first decreases and then disappears completely. Wherein, in the place of "extinct" comets, there remain streams of meteors and small asteroids, the characteristics of which are almost no different from ordinary objects in the asteroid belt. There are known cases when comets fell apart or were completely destroyed, and also died when they collided with planets or Sun. Every year, astronomers record the appearance of about five comets, half of which are previously unknown objects.

Currently, most of the properties of comets observed in the Solar System have been physically explained. However, the fundamental question about the origin of the comets themselves remains open, replete with a wide range of contradictory, often mutually exclusive judgments (Dobrovolsky 1966; Vsekhsvyatsky 1967; Kazimirchak-Polonskaya 1978a, b; Churyumov 1980; Marochnik 1985; Shulman 1987; Tomanov 1989; Tsitsin 2009, etc.).

3 The Problem of Comets Origin

Hypotheses for the origin of comets, in accordance with their two main types, are usually divided into two groups: "capture", which assume the capture of interstellar comets by Sun from the Galaxy, and "eruptive", which assume the comets birth in the Solar System itself.

The founder of the first hypothesis was Laplace (1796). He believed that comets are interstellar objects that can be captured by the Sun gravitational pull. Empirical evidence in favor of this point of view was first presented by Svedstrup (1884), who discovered the connection between the perihelia of cometary orbits and the plane of the Galaxy. This connection was later confirmed by Oppenheim (1924), Littleton (1948), Witkovsky (1971), Radzievsky and Tomanov (1973), Shorma and Khonna (1988) and other astronomers. Galactic idea of comets origin was also developed by Bobrovnikoff (1929), McCrea (1953), Yabusbita and Hasegawa (1978) and Kazimirchak-Polonskaya (1978a, b).

The first to suggest that comets could arise as a result of the ejection of matter from the bowels of planets was Lagrange (1812). Proctor (1881) was the first to support this eruptive hypothesis. He explained the comets origin by strong volcanic activity on Jupiter that occurred 10 million years ago. In our time, the eruptive idea was also shared by Drobyshevsky (1978), Guliev and Vsekhsvyatsky (1981), and Guliev (1992). A consistent supporter of the hypothesis of the ejection of comets during volcanic eruptions not only on large planets, but also on their satellites was Vsekhsvyatsky (1967).

A number of supporters of the eruptive hypothesis associate the formation of comets in one way or another with asteroids. Thus, for example, Koval (1976) believes that comets are formed when asteroids bombard the icy surfaces of the satellites of giant planets. On the contrary Orlov (1936, 1960) argued that comets arise in the asteroid belt itself as a result of collisions of its bodies. This point of view was supported by Fesenkov (1951) and later Davydov (1981), Radzievsky (1995).

Of greatest interest to us are two hypotheses of Oort (1950, 1951), put forward by him in the middle of the last century. In the first hypothesis, Oort (1950) associate the comets origin with the explosion of the planet Phaethon, which, according to Olbers' assumption, previously existed on the site of modern asteroid belt. The explosion of Phaethon was accompanied by the formation and ejection of many comets, which, according to Oort hypothesis, created a huge "comet cloud" on the outskirts of Solar System. From this cometary reservoir, comets are currently coming to Sun under the influence of various disturbances.

A year later, Oort (1951) abandoned the first version of the comets formation in Solar System and in the second his hypothesis, following Schiaparelli (1871) and Schmidt (1945), proposed to consider the comet cloud as a remnant of the protosolar nebula, preserved after the Solar System formation. This idea was then supported by Kuiper (1951), and later by Safronov (1969), Öpik (1975), Tsitsin et al. (1985) and many others. For some time this Oort hypothesis was considered generally accepted.

However, Van Flandern (1976, 1992) has a different opinion on this matter. Based on a detailed analysis of the orbital elements of 60 long-period comets whose perihelia are located in the asteroid belt, Van Flandern came to the conclusion that the explosion of a planet between Mars and Jupiter was real, and occurred only 3 million years ago.

The listed hypotheses of the comets origin, as well as other, not mentioned ideas (Vsekhsvyatsky 1967; Churyumov 1980; Tomanov 1989; Marov 1994), which more or less felicitously explain the formation of comets of one type, invariably encounter difficulties when explaining comets of another type. For hypotheses of the capture of interstellar comets, these difficulties are associated with the complete absence of comets moving relative to Sun along strongly hyperbolic trajectories, as well as with the very low probability of such comets being captured by Sun and their subsequent transformation into short-period comets. For eruptive hypotheses the difficulties are different. They consist in the lack of a plausible explanation for the long-period comets origin and their connection with Galaxy plane.

The Oort comet cloud hypothesis is also not without fundamental difficulties. Calculations have shown (Van den Bergh 1982; Bailey 1986; Dolgopolova and Marochnik 1987; Chepurova and Shershkina 1989) that when Sun passed through dense clouds of gas and dust and as a result of the passage of other stars in Galaxy near it, Solar System would certainly have lost a significant part of its comets. Attempts made by a number of authors (Whipple 1964; Hills 1982; Tsitsin 1999, etc.) to move the cometary reservoir closer to Sun cannot save the situation.

4 Galactic Model

We begin the substantiation of the comet-asteroid hypothesis with brief information about the galactic model that describes the jet outflow phenomenon (Barenbaum 2002, 2010, 2018a; Barenbaum and Titorenko 2020).

According to Galactocentric paradigm, in the core of our Galaxy there is a continuous destruction of stars with dust and gas formation. This dust-gas matter first accumulates in the rapidly rotating gas-dust nuclear disk of Galaxy, from where they then flows out in the form of two jet streams that fan-like spread in galactic plane. Moving at a speed of $\approx$ 300 km/s from Galaxy center to its periphery, the substance of jet streams condenses into dense gas and dust clouds, comets and stars. These processes are sharply intensified at the places intersection of two jet streams with four spiral galactic arms of electromagnetic nature, rotating around Galaxy center at a constant angular velocity and accumulating gas-dusty matter.

However, objects emerging from galactic arms and jet streams behave differently. Those of them that were formed from gas-dust matter of jet streams inherit their high radial velocity and leave visible limits of Galaxy forever. Others, formed from gas and dust of galactic arms, remain in the arms and with time move into independent orbits around Galaxy center.

Our Sun (Solar System) belongs to the second type stars. Having arisen earlier in one of spiral arms, at now Sun is currently moving in galactic plane around Galaxy center in a slowly rotating elliptical orbit, quasiperiodically crossing the jet streams and galactic arms. During periods of such crossings, all objects in Solar System are exposed to intense bombardments by galactic comets. In Earth's geological history, the times of such bombardments are marked as epochs of global geological events (tectonic, climatic and biotic), which act as boundaries of different ranks of the Phanerozoic and Precambrian stratigraphic scale.

In Fig. 1 in projection onto galactic plane shows Sun's position of the relative to 4 arms and 2 jet streams of Galaxy. Here, we also showed the Sun orbit and its current position on it (Barenbaum 2018a).

Due to uneven Sun's motion in orbit, rotation of the line apsides of orbit itself, as well as rotation of nuclear disk and Galaxy arms with their angular velocities, the picture (Fig. 1) changes rapidly over time. The model tracks this changing picture and also makes it possible to calculate with high accuracy the times of Sun crossing jet streams and spiral arms of Galaxy (Barenbaum 2010; Barenbaum and Titorenko 2020).

Figure 1 shows that Sun is currently on the inner side of Orion-Cygnus branch, which is the jet stream closest to Sun. Geological evidence suggests that Sun may have entered this jet stream about 5.0 Ma and left it about 0.7 Ma, so that for about 4 million years Solar System was bombarded by galactic comets (Barenbaum 2002, 2010).

According to our comet concept, it was the bombardment of Solar System by comets of Orion-Cygnus jet stream that played a major role in formation of comets

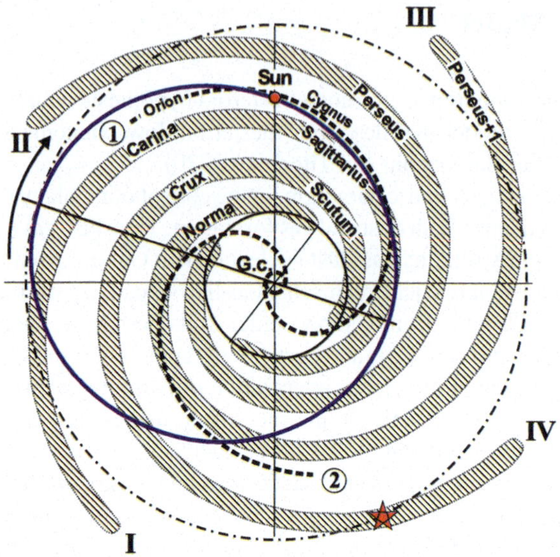

Fig. 1 The current position of the Sun in orbit (ellipse) relative to four galactic arms (Roman numerals) and two jet streams (Arabic numerals in circles) in projection onto Galaxy plane. The solar orbit is shown for the last revolution of Sun around Galaxy center (G.c.). The small circle in the center, where the jet streams flow out, is Galaxy nuclear disk. The larger circle where galactic arms begin determines the size of Galaxy isothermal core. The outer dotted ring corresponds to Galaxy corotation radius R*. The arrow indicates the direction of Sun moving on orbit, which coincides with rotation of Galaxy and rotation of the apses line (straight line) of solar orbit. The asterisk shows the place in arm Crux-Scutum (IV) where Solar System was formed earlier (Barenbaum 2018a)

observed today. Note that the connection of Solar system comets with the Orion-Cygnus branch has been repeatedly suggested by specialists (Napier 1985; Tomanov 1989).

In our comet concept this cometary bombardment also explains other “extraordinary” phenomena in Solar System (Barenbaum 2002): the “peculiar” state of Sun (Glushneva 1994); increased Earth’s geological activity (Artyushkov 1994); strong turbulence of giant planets atmosphere (Marov 1981); excited state of bodies in asteroid belt (Ruskol 1986); significant dustiness of interplanetary space (Brownlee 1984); as well as the presence in it of a large number of comets, meteorites and asteroids with a dynamically short lifetime (Simonenko 1985).

5 Galactic Comets

It was previously noted that galactic comets, like young stars, arise in places where galactic arms interact with jet streams. Therefore, one should distinguish between comets of jet streams and galactic arms. The former move relative to Sun at a very high speed, the latter with speed is one or two orders of magnitude less. Both groups of galactic comets are currently inaccessible to detection from Earth, since they can be observed when Sun is respectively in jet streams and galactic arms. Today this is not the case. Therefore, everything that we know about galactic comets was obtained on the basis of an analysis of consequences of their falls on Earth and other planets (Barenbaum 2010).

The most studied comets of Orion-Cygnus jet stream, in which Sun was in period from 5.0 to 0.7 Ma. These comets moved relative to Sun at a speed of 450 km/s, had a core diameter of 0.1 – 3.5 km, a mass of $10^{12} - 10^{17}$ g, and an energy of $10^{20} - 10^{25}$ J. They consisted mainly of water ice with an admixture of carbon-containing components and had density $\approx$ 1.0 g/cm^3. The flow density of comets in Orion-Cygnus jet stream was 10^{-9} comets per year per km^2 (Barenbaum and Shpekin 2021).

In addition to their icy composition and very high rate falls onto planets, galactic comets differ from comets observed today in Solar System in other ways. Let's list them briefly:

1. The nuclei of galactic comets are exponentially distributed in diameter and not according to the inverse square law as in asteroids and comets of the Solar System. Accordingly the geological structures created by their falls, namely: seamounts (Earth); craters, seas and mascons (Moon, Mars and Mercury); montes, coronas, diatremes, and shield volcanoes (Venus) are also characterized an exponential size distributions (Barenbaum 2015; Barenbaum and Shpekin 2019).
2. Falls of galactic comets have character of comet "showers" lasting 2–5 million years, which are repeated in 20–37 million years. During one shower ~ 10^4–10^7 galactic comets can fall to planets. As a result, most of craters with diameter D > 10 km on terrestrial planets (except for Earth) are created by galactic comets. Due to the high frequency of comets falling, the surfaces of Moon, Mars and Mercury bombarded by them are 100% saturated with craters with D > 10 km. This saturation limit by craters is the same for Moon, Mars and Mercury and amounts to 150 craters per million km^2 (Barenbaum and Shpekin 2021).
3. Unlike asteroids and comets of Solar System, which move in ecliptic plane, galactic comets move in Galaxy plane inclined to the ecliptic at an angle of 62°. As a result, at Sun's orbital movement, galactic comets alternately bombard the southern and northern hemispheres of planets (Barenbaum 2022). The last time they bombarding southern hemisphere of all planets.
4. The specificity of galactic comets also lies in the different physical mechanisms of their interaction with planets with and without an atmosphere. If for asteroids with D $\approx$ 100 m or more, Earth's atmosphere has little effect on results of their fall (Melosh 1989), then galactic comets of the same size in Earth's atmosphere are completely destroyed, and their enormous kinetic energy causes heating asthenosphere rocks. On continents, this leads to rise of their surface, and in oceans, to the outpouring of lava to bottom and formation of seamounts (Barenbaum 2010, 2013, 2016).

6 Collisions of Comets and Asteroids

The issue of interaction of galactic comets with bodies of asteroid belt deserves special attention. It is known that 98% of all asteroids discovered today in Solar System are concentrated between orbits of Mars and Jupiter in form of a wide toroidal ring (Fig. 2).

In this region of interplanetary space, they can exist for more than 4.6 billion years from moment of their origin, only slightly changing their orbital parameters over time. The distribution of asteroids by diameter in this ring is power law and with exception of largest of them is well obeys the inverse quadratic relationship (Hartman 1978): $N(D_\alpha) = 10^5\, D_\alpha{}^{-2}$, where: $N(D_\alpha)$ is number of asteroids with diameters more than D_α (in km). This expression gives distribution of spectral density of asteroids over sizes $n(D_\alpha)$ in following form:

$$n(D_\alpha) = \frac{dN}{dD_\alpha} = -2 \times 10^5 D_\alpha^{-3} \qquad (1)$$

Obviously, collisions of galactic comets with asteroids are in the nature of an explosion, in which energy is released, which is close in magnitude to kinetic energy of colliding bodies. Depending on size ratio of comet and asteroid, the processes initiated by their collision should be different.

When hit by a comet in large asteroid with mass ma significantly exceeding comet mass m_k, energy of explosion mainly goes on formation of crater with accompanying crushing and evaporation of asteroid substance and removal of ejection products from crater as well as on formation of a powerful shock wave. The comet body completely collapses and its substance evaporates and dissipates, mechanically and chemically mixing with asteroid material. The total mass of matter ejected at formation of the funnel can exceed 10–1000 times mass of comet itself. The rate of output of the main mass of rocks lies in limits from meters to several kilometers per second, and

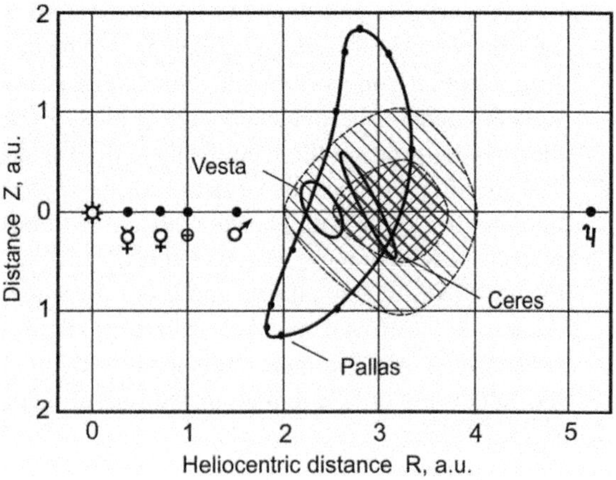

Fig. 2 The cross-section of asteroid belt by a plane perpendicular to the ecliptic plane and passing through Sun. The regions with different spatial density of asteroids are shaded. Shown the loops that asteroids Pallas, Ceres and West are making in this plane (according to L. Kresak)

of evaporated matter can reach ~ 10 km/s (Bazilevsky et al. 1983). At this for planets and very large asteroids most of ejected material falls back to their surface, while for small ones—is lost (Simonenko 1979).

For smaller asteroid bodies (m_α ~ m_k) the impact energy may be enough to cause, together with cumulative effect, the disintegration of asteroid and also break it into parts. In the case comet collision with very small stones ($m_\alpha \ll m_k$), both bodies go fully to ruin. This process is inevitably accompanied by release into interplanetary space of huge quantities of cometary and meteoritic dust.

Of great interest is a question about minimal size of asteroids D_α*, collisions with which are disastrous for galactic comets and about total number of such bodies in asteroid belt.

We find the value of critical diameter of asteroids D_α* from considerations of equality of binding energy E_k of cometary nucleus to kinetic energy of asteroid E_α in the comet rest frame. Assuming that comet nucleus is held by gravitational forces, we write its binding energy as $E_k = 6m_k{}^2G/5D_k = \pi^2\rho_k{}^2D_k{}^5G/30$, where G is gravitational constant, D_k is diameter of comet nucleus. The expression for the kinetic energy of asteroid has the form $E_a = m_aV^2/2 = \pi\rho_aD_a{}^3V^2/12$, where V is mutual velocity of galactic comet and asteroid. Taking densities of cometary nuclei and asteroid equal, respectively, $\rho_k = 1$ g/cm^3 and $\rho_\alpha = 4$ g/cm^3 and assuming velocity $V = 450$ km/s, dependence of value D* km on diameter of cometary nucleus, we obtain in form $D_\alpha{}^* = 4.7 \times 10^{-7}\ D_k{}^{5/3}$. It follows that for comet with core diameter of $D_k \approx 3.5$ km, the cause of destruction is its collision with a half-meter asteroid, and at diameter D_k ~ 200 m, a comet can be destroyed by centimeter stone.

The total number of asteroids in belt $N(D_\alpha{}^*)$ collisions with which are disastrous for galactic comets can be estimated with formula (1) under assumption that given law is satisfied not only for great asteroids but also for more small unobserved objects. In this case, we have $N^* = N(D_\alpha{}^*) = 4.5 \times 10^{25}\ D_k{}^{-10/3}$. Whence we find that for galactic comets with a core diameter $D_k = 3.5$ km the number of "dangerous" asteroids in belt will be ~ 7×10^{13}, and for comets with D_k ~ 100 m it will increase to ~ 10^{19}.

Since asteroid belt occupies space volume of ~ 1.6×10^{26} km^3 (Simonenko 1979), in the first case, their concentration in belt will be ~ 4×10^{-13} km^{-3}, and in the second ~ 6×10^{-8} km^{-3}. For comparison, let us indicate volume of Earth ~ 10^{12} km^3.

Thus, the concentration of "dangerous" asteroids in belt is negligible. At the same time, probability of colliding with such bodies for galactic comets is not small. As a measure of this probability, we introduce cross section comet destruction in collisions with asteroids S*. Assuming that destruction causes any passage of comet at distance from asteroid less than the half-sum of diameters of both bodies, we have

$$S*(D_k) = \frac{\pi}{4}\int_{D^*}^{D_{\alpha m}} (D_\alpha + D_k)^2 n(D_\alpha)\cdot dD_\alpha$$
$$= N*\left[D_k^2 + 2D_kD_\alpha^* + D_\alpha^{*2}\ln\left(D_{\alpha m}/D_\alpha^*\right)\right], \quad (2)$$

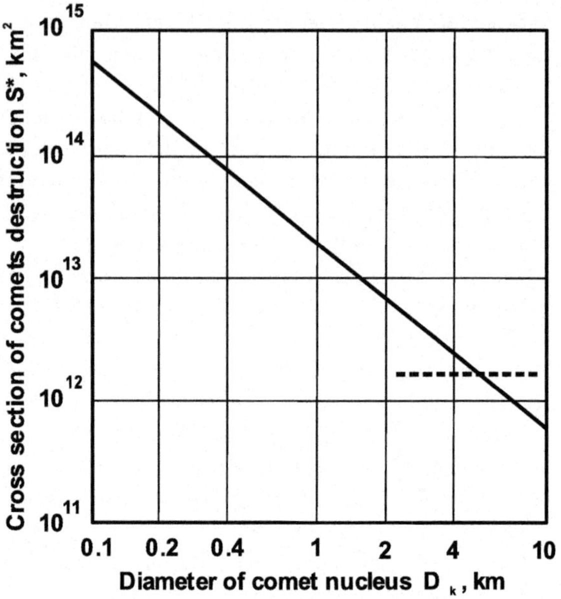

Fig. 3 Cross sections of galactic comet destruction in asteroid belt, depending on diameter of cometary nucleus. Dotted line is Sun's cross sectional area

where $D_{\alpha m} = 1000$ km is maximum diameter of asteroids.

The graph of cross section $S^*(D_k)$ is shown in Fig. 3 (Barenbaum 2002).

It follows from our calculation that for galactic comets with core diameter $D_k =$ 0.1–3.5 km, their destruction cross section in asteroid belt is $S^* \sim 10^{12}$–10^{15} km^2. This value is more than Sun cross-sectional area $S_{\odot} = 1.6 \times 10^{12}$ km^2 and even more so of all planets. In particular, for Jupiter $S_j = 1.6 \times 10^9$ km^2, and for Earth is 1.3×10^8 km^2.

Thus, the probability of galactic comets destruction in asteroid belt is two orders of magnitude greater than probability of their fall on all objects of Solar System. Hence, it is clear that asteroid belt is the main place where interaction of galactic comets with Solar System takes place.

From structure of formula (2) also follows that for $D_k \approx 0.1$ km a decisive contribution to the S^* value is due to collisions of comets with small-sized asteroids, while on share of large asteroids accounts for only ~ 10^{-4} part of all collisions. Therefore in asteroid belt, processes of evaporation and crushing of galactic comets must be significantly dominated, accompanied by ejection in interplanetary space of large amount of cometary gas and dust. But this conclusion does not mean that collisions of comets with large asteroids do not play a significant role, and they can be ignored. Because of significant cumulative effects, such collisions lead to intense disintegration of asteroids and ejection of their fragments from belt (Barenbaum 2010).

Thus, in collisions with galactic comets in asteroid belt, albeit with various probability, appear all known types of small bodies of Solar system: (1) large blocks of rocks—asteroids; (2) their small debris—meteorites; (3) destroyed nuclei of galactic comets; and also (4) finely dispersed substance, represented by cometary and meteoritic dust.

Let us be interested about the further fate of these neoplasms.

6.1 Interplanetary Dust

According to existing ideas (Mironova 2015), the suppliers of dust particles to interplanetary space can be: Sun's crown, asteroid belt, Kuiper belt, Oort cloud comets, giant planets, and Galaxy.

About interplanetary dust we, first of all, judge by micrometeorites falling to Earth, a smallest particles of a cosmic substance with dimensions from a micron to several millimeters. In its basic bulk, they are very porous brittle aggregates of dark color, similar in chemical composition to carbonaceous chondrites or, less frequently, consisting of associations of crystalline grains of minerals characteristic of other known types of meteorites (Brownlee 1984).

In our concept, dust particles appear in asteroid belt, the evaporation of its bodies in collisions with galactic comets. At low temperatures, vapors and gases, as well as the smallest droplets of the substance melted by the impact, quickly cool and condense into dust aggregates. At the same time, a part of dust grains easily freezes with the debris of asteroids, settling on their surface, or are captured and included in nuclei of secondary comets being formed. Most of dust grains remain in a free state.

The behavior of such free dust particles in Solar System, in the main, is determined by their mass and sizes. Dust grains of very small diameters (d < 10^{-5} cm) and masses (10^{-15}–10^{-13} g) are quickly swept out of interplanetary space under influence of Sun's light pressure. Larger particles by action of Pointing-Robertson effect (Physics of Space: Little encyclopedia 1986), on the contrary, move in spiral orbits towards Sun. The time of their fall on Sun (in years) is: $\tau = 3.5 \times 10^6 d\rho\ R^2$, where: ρ is density of dust particles mater, R is radius (AU) of initial orbits of dust particles, from which they begin move towards Sun.

For dust particles with a diameter of d ~ 10^{-4} cm and density of $\rho = 2.5$ g/cm^3, which begin to move from asteroid belt, the value $\tau = 5 \times 10^5$ years, and for the same dust particles thrown behind orbit Saturn $\tau = 50$ million years. All other things being equal, larger and/or denser particles fall on Sun later.

When moving towards Sun, some of dust particles are intercepted by planets, their satellites and asteroids. So, for example, in Iapetus—satellite of planet Saturn, which rotates being constantly turned to Saturn by one side, the albedo of surface of its front hemisphere in direction of motion is an order of magnitude darker than the back hemisphere.

However, the most intense deposition of dust particles occurs in asteroid belt (Fig. 4), which is reflected in the peculiar arrangement of its bodies, covered with dust particles "heavily" (C-asteroids) and "weakly" (S-asteroids).

Figure 4a, borrowed from (Simonenko 1979), shows the dependence of the spatial density of C- and S-asteroids in belt on the Sun distance. And in Fig. 4b, the

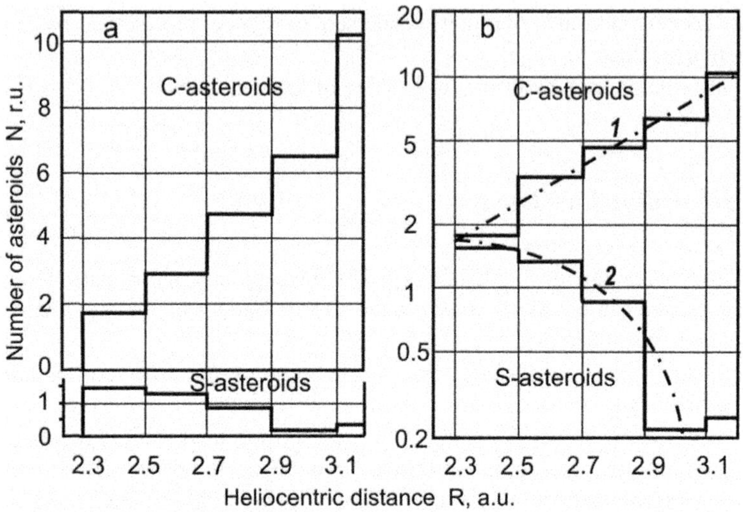

Fig. 4 Spatial density in asteroid belt of "dark" (C) and "light" (S) asteroids with diameter of more than 50 km as function of heliocentric distance in ordinary (**a**) and semi-logarithmic (**b**) coordinate system. The dotted dashed lines (1) and (2) correspond to formulas (3) and (4)

same dependences are presented in semi logarithmic coordinate system (Barenbaum 1990a). We see that the distributions in belt of C- and S-asteroids are well approximated by formulas:

$$N_C(R) = 1.75 \exp[(R - R_0)/L], \tag{3}$$

$$N_S(R) = 1.75\big[1 - \exp(-(R - R_0)/L\big], \tag{4}$$

where $R_0 = 2.3$ AU and $L = 0.42 \pm 0.02$ AU are constants.

Formulas (3) and (4) indicate the diffusive mechanism of dust penetration into asteroids belt. In this case, constant L has physical meaning of diffusion length $L = (n\sigma)^{-1}$, where n is spatial density of asteroids in belt, and σ is cross section for capture by asteroids of dust particles.

Analysis shows (Barenbaum 2002) that value of constant $L = 0.42$ AU can be explained by absorption of dust by asteroids with $D > 10$ cm. Other things being equal, dust deposition rate on asteroids surface grows as ~ $D^{8/5}$. Therefore, it is no accident that among bodies with diameter $D > 80$ km about 83% is of type C, that is they are covered with rather "crust" of dust and only 17% relate to type S (Simonenko 1979).

There is greatest probability of dust deposition on biggest asteroids: Ceres, Pallas and Vesta. These asteroids often experience of collisions with galactic comets and capture fragments of rocks and dust that are formed at collisions. All three asteroids have close satellites and rich asteroid families, and their surface is characterized by a sharp heterogeneity of reflective properties. The latter circumstance does not allow us to classify these asteroids as S and C type, and forces them to pick out into separate class U (Simonenko 1985).

Asteroids of smaller sizes are destroyed in collisions with galactic comets and may lose dust crust. Some of their fragments are lost and go beyond asteroids belt, while others remain (Eos, Koronis, Nyca, etc.). Some asteroids acquire satellites (Hebe, Herculina, etc.). Satellite systems are extremely unstable. All these asteroids as well as their family members have S-type, i.e. are "young". They had not yet covered themselves by dark dusty crust and, in contrast to "old" C-asteroids they do not avoid forbidden hatches (Simonenko 1985).

Smaller asteroids, when colliding with galactic comets, are crushed and can lose their dust crust. Part of their debris is lost and leaves asteroid belt (for example, the groups of Apollo and Cupid), others remain and form families (Eos, Koronis, Nyca, etc.). Some asteroids acquire satellites (Hebe, Herculina, etc.). Satellite systems are extremely fragile.

All the named asteroids and members of their families are S-type, i.e. "young". They have not yet managed to become covered with a dark dusty crust and, unlike "old" C-asteroids, do not avoid forbidden hatches (Simonenko 1985). Over time, dust layer on asteroids surface increases and their color changes from light to dark.

A vivid example of C → S asteroids transformation are families Eos ($R = 3.02$ AU) and Koronis (2.87 AU). Time destruction of parental bodies these systems is ~ 1 Ma (Simonenko 1985). We believe that destruction parent asteroids in families Eos and Koronis occurred during last cometary bombardment. Therefore now their fragments of S-type sharply differ from all other bodies of outer zone belt.

Thus, the main reason for differences between C- and S-asteroids is varying degree of dustiness their surface. Character of distribution C- and S-asteroids in belt can be explained (Barenbaum 1990a) as a result of simultaneous existence two competing processes: on the one hand, deposition of dust particles on asteroids surface, and on the other, destruction of asteroids during their collisions with galactic comets.

6.2 Meteoric and Larger Bodies

For all small bodies of Solar System with masses greater than those of dust grains, the main factor determining their further fate is value of the initial velocity V, which they receive at birth. If the parent asteroid is great and does not fully destroying by the impact of galactic comet and velocity of secondary products is small and does not exceed $V_1 = D_\alpha\sqrt{\pi\rho_\alpha G} \approx 10^{-3}\, D_\alpha$ km/s, then material ejected by explosion is not able to overcome attraction of parent asteroid and falls back onto its surface.

If the velocity V of the ejected debris is in the range from V_1 to $V' = \sqrt{2}\, V_1$, then under certain conditions they can begin to rotate around parent body with mass m_α at a distance from its center $r < R\,(m_\alpha/M_\odot)^{2/5}$. The presence of one or even several close satellites has been found in many large asteroids. The lifetime of such "satellite" systems is short. Under the influence of disturbances from Jupiter and Sun, as well as other asteroids, the satellites must quickly fall on the central body, or leave it. Consequently, the very fact of existence of such systems testifies to their very recent formation.

The situation $V > V'$ is of the greatest interest to us. In this case, as in the case of complete destruction of asteroid, the newly formed bodies lose their connection with the parent asteroid. At the same time, a sufficiently large impulse can be transferred to some of them, which will allow them to leave not only asteroids belt, but also Solar System. The condition of "knocking out" asteroids from belt in the first approximation can be written in form: $\Delta V > V_x$, where ΔV is increment in velocity of asteroid (or its breakaway part) after collision with comet, V_x is the average chaotic velocity of asteroids in belt. In accordance with the momentum conservation law, the maximum ΔV obtained by asteroid in elastic collision with comet will be $\Delta V = Vm_\alpha/(m_\alpha + m_k)$, where V is comet velocity relative to asteroid. Hence, the condition for knocking bodies out of asteroid belt is reduced to the restriction $m_a \leq m_k/(V/V_x - 1)$.

Assuming $V = 450$ km/s, $V_x = 5$ km/s, and taking mass of large galactic comets $m_k \sim 10^{16}$ g, we obtain $m_\alpha \leq 10^{18}$ g. At a density of asteroid matter $\rho_\alpha = 4$ g/cm^3, specified mass corresponds to bodies with diameter $D_\alpha \leq 8$ km. Thus, a chance to go beyond belt limits have mainly bodies of small and medium-sized, and larger ones remain in belt. By this circumstance we explain the fact (Simonenko 1985) that almost all asteroids outside belt, in particular, of the Apollo, Amur and Aton families, are mainly represented by bodies with a diameter of less than ten kilometers. This conclusion is fully valid also for daughter comets leaving belt.

Having become "free", daughter bodies begin to move along independent Keplerian orbits, determined by magnitude and direction of velocity V, which they acquire relative to Sun at birth moment. The influence of this speed on their further fate can be taken into account by representing V in center of mass system of colliding parent bodies:

$$V = |W + U| = \sqrt{W^2 + U^2 + 2WU \cdot \cos(\vartheta)}, \tag{5}$$

where W is bodies velocity in the center mass system of "parents", U is speed of the mass center relative to Sun, ϑ is angle between vectors W and U.

For secondary bodies captured by Sun, their velocity V at heliocentric distance R equal to the aphelion distance is related to eccentricity e and magnitude of orbit semi-major axis α by the expressions:

$$V = \sqrt{GM_\odot(2/R - 1/\alpha)} = V* \sqrt{1 - R/2\alpha} = V* \sqrt{\frac{R}{2\alpha}\left(\frac{1-e}{1+e}\right)}, \tag{6}$$

where $V* = \sqrt{2GM_\odot/R}$ is critical speed of the secondary bodies ejection from belt, at which Sun is still able to hold them by its attraction. Assuming $R = 2.8$ a.e. that is equal to average radius of asteroid ring, for critical speed we will have $V^* = 25$ km/s.

Formula (6) indicates that only bodies remain in Solar System, whose initial velocity relative to Sun's lies within $V^*/\sqrt{2} \leq V < V^*$. If body is ejected from belt with speed $V > V^*$, it leaves Solar System. If body speed is less than $V^*/\sqrt{2}$, then it will quickly fall on Sun. Therefore, for all newly formed bodies captured by the Sun gravity, orbital radii must satisfy the inequality $0 < R/\alpha \leq 1$, where upper limit corresponds to bodies remaining in belt, and lower ones to cometoasteroids whose orbits go far of belt.

Combining expressions (5) and (6), for capture of daughter bodies by Sun, we obtain the following condition

$$W^2 + U^2 + 2WU\cos(\vartheta) + \left(\frac{R}{2} - 1\right)V* < 0 \tag{7}$$

Formula (7) indicates that probability of secondary bodies to be captured by Sun is anisotropic and strongly depends on ratio of the masses of colliding comets and asteroids. To take into account the effects of anisotropy, we introduce velocity distribution function of daughter bodies F (W, U), defining it as follows:

$$F(W, U) \cdot dW \cdot d\Omega = \frac{F_0}{4\pi}\left(\frac{W}{V*}\right)^{-\gamma} \sin(\vartheta) \cdot dW \cdot d\vartheta, \tag{8}$$

where γ is constant, F_0 is constant corresponding to the normalization condition

$$4\pi \int_{V'}^{\infty} dW \int_{0}^{\pi} F(W, \vartheta)\sin(\vartheta) \cdot d\vartheta = 1, \tag{9}$$

where V' is velocity of «running away» for daughter bodies.

In formulas (8) and (9), it is assumed that the distribution function F (W, U) is power-law in the velocities of daughter objects W, and isotropic in the angles ϑ of their ejection from belt in system of center mass of the parental bodies.

Let us analyze the consequences of formulas (7–9) for two limiting, but important cases. In the first case, galactic comets collide with very large asteroids ($m_\alpha \gg m_k$), and in the second, with very small ones ($m_\alpha \ll m_k$).

6.2.1 Collisions of Galactic Comets with Large Asteroids

When comet collides with a very large asteroid, the speed of mass center U of colliding bodies practically coincides with speed of asteroid motion around the Sun $U = \sqrt{GM_\odot/R} = V*/\sqrt{2}$, and ejection of daughter bodies occurs symmetrically to ecliptic plane. Substituting the value of velocity U in units of V^* into formula (7)

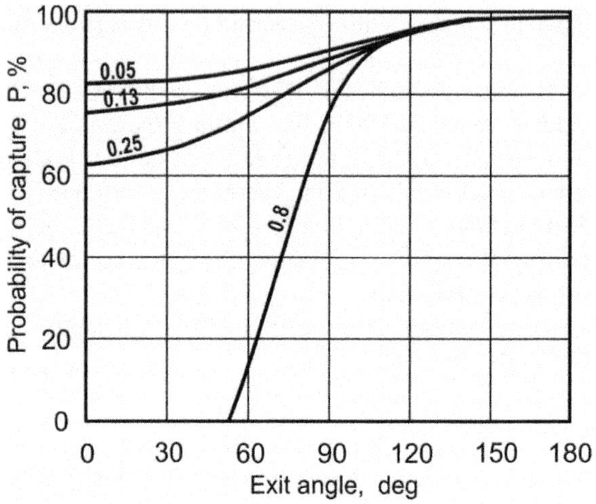

Fig. 5 Probability of capture by Sun of daughter bodies had arisen owing to collisions of galactic comets with "large" asteroids as a function of angle their ejection from belt and orbit radius. Curves are correspond parameter *R/α* of daughter bodies which reach orbits of Uranus (0.05), Saturn (0.13), Jupiter (0.25) and outward border of asteroid belt (0.8)

and solving quadratic equation, the capture condition is obtained in the form:

$$W\left(\vartheta, R/\alpha\right) \le W* = \frac{V*}{\sqrt{2}}\left[\sqrt{1+\cos^2(\vartheta) - \frac{R}{\alpha}} - \cos(\vartheta)\right] \tag{10}$$

The probability of capture by Sun of daughter bodies $P(\vartheta, R/\alpha)$, leaving belt at different azimuthally angles, but characterized by the same values of semi-major axis of orbit, for the case $m_\alpha \gg m_k$ is shown on Fig. 5.

The calculation was carried out according to formula

$$\mathrm{P}(\vartheta, \mathrm{R}/\alpha) = \int_{V'}^{\mathrm{W}*} \mathrm{W\,dW} \Big/ \int_{V'}^{\infty} \mathrm{W\,dW} \tag{11}$$

for isotropic distribution of daughter bodies on angle ϑ and power-law mass distribution in formula (8) with constant $\gamma = 2$.

Figure 5 shows that bodies ejected in the opposite direction to motion parent asteroid have the greatest chance to remain in Solar System. Moreover, for bodies in asteroid belt, this case is the only possible one. With an increase in radius of orbit, the capture probability of daughter bodies by Sun $P(\vartheta, R/\alpha)$ begins to depend weakly on angle ϑ of their exit from belt.

Let us also pay attention to the fact that rotation periods of daughter bodies around Sun turn out to be more than those of parent asteroids, therefore, as a result of many acts of crushing, asteroids are displaced to outer edge of belt. Whereas very large asteroids, having experienced a small number of crushing, remain in the belt to their former places (see Fig. 2).

6.2.2 Collisions of Galactic Comets with Small Asteroids

In the case of $m_\alpha \ll m_k$, the situation changes radically. If previously, the predominant type of daughter bodies were debris of asteroids, which freezing with more or less amount of icy cometary material, but now secondary bodies, mainly are represented by substance of destroyed galactic comets.

In addition, the center mass of colliding bodies, it is no longer connected with asteroid, but with galactic comet. In this case, the speed is $U \approx 18$ V*, and the solution to Eq. (6) takes the form:

$$W* = 18V * \left[\sqrt{3.1 \cdot 10^{-3}\left(1 - \frac{R}{2\alpha}\right) - \sin^2(\vartheta)} - \cos(\vartheta) \right] \qquad (12)$$

Function (12) has a physical meaning under more stringent restrictions on speed and direction of ejection bodies from asteroid belt. Now only bodies can remain in Solar System, which leave the belt with velocity $U - V^* < W < U + V^*$ and in a cone of angles $\Delta\vartheta \leq \pm 3°$, and in direction opposite to motion of galactic comet (Fig. 6).

Lets we note that angles ϑ in Fig. 6a, b are postponed of sundry directions. The dependence (Fig. 6a) is plotted for ecliptic plane, and angle ϑ is measured from direction of Sun motion in Galaxy. While the calculation (Fig. 6b) refers to galactic plane making an angle of 62° with the ecliptic, and angle ϑ is measured from direction to Galaxy anticenter.

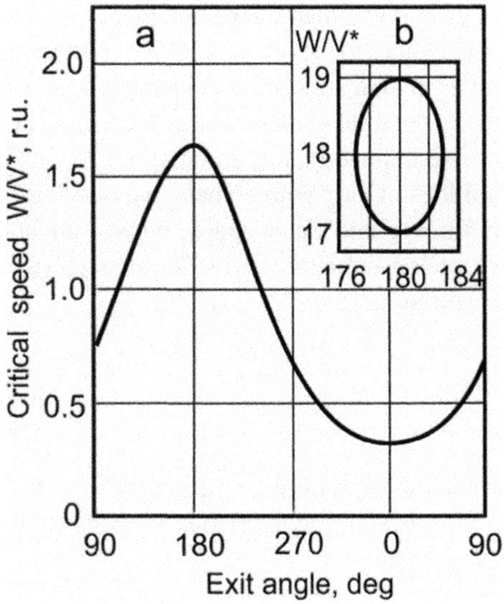

Fig. 6 Dependence of critical velocity of bodies on their ejection angle from belt in collisions of galactic comets with large (**a**) and small (**b**) asteroids: calculation using formulas (10) and (12) for $R/\alpha = 0$

The probability of Sun capture of bodies formed in collisions with small asteroids is of the order of magnitude $(V^*\Delta\vartheta/U)^2 \sim 3 \times 10^{-5}$, which is much lower than in the case when galactic comets are scattered by large asteroids.

With an increase in the asteroid mass, which comet collided with, the vector of preferred orientation of line apsid of daughter bodies orbits captured by Sun changes direction. In the case $m_\alpha \gg m_k$, it lies in the ecliptic plane and coincides with direction of Sun's motion in Galaxy.

The connection of comets with galactic plane, first established by Svedstrup (1884), was later confirmed by increasingly representative statistical observational data. Therefore, supporters of galactic origin of comets in Solar System (Shulman 1987; Tomanov et al. 2013), in a dispute with adherents of the eruptive idea, usually put forward this connection, as well as the obvious similarity in chemical composition of comets and interstellar gas, as very important, if not decisive argument.

7 Mechanism of Daughter Comets Formation

It has been noticed that young craters on planets are often bordered by radiation ejections of matter. These emissions have form "rays" of light color tapering towards the end. The rays are usually on order of several kilometers wide and ~ 100 km long. In very large craters, the emissions reach of ~ 1000 km in length and hundreds of kilometers in width. Often along the rays there are located clusters of small secondary craters (Bazilevsky et al. 1983; Melosh 1989).

It is believed that the rays are formed as a result of cumulative ejection from crater of evaporated material of the impactor mixed with crushed material of surface rocks. The light color of the rays is given by presence of frozen water vapor and ice of other gases in their substance. Assuming that most of rays consist of water ice layer ~ 1 cm thick, the total matter mass of individual emissions will be at least $m = 10^{12} - 10^{15}$ g.

We believe that cumulative emissions of this kind occur when galactic comets collide not only with planets, but also with asteroids. In the latter case, the substance is thrown into outer space, where, under the action of self-gravity, it can form a secondary cometary body. The characteristic time of body formation can be estimated by formula

$$t \approx \left(\frac{2r^3}{Gm}\right)^{\frac{1}{2}}, \tag{13}$$

where m is the mass of ejected substance, r is the ejection length.

Since Sun gravity also acts on ejected matter, the initial size of ejection cannot exceed value

$$r \leq R\left(\frac{m}{M_\odot}\right)^{\frac{2}{5}}, \tag{14}$$

where R is heliocentric distance of nucleus "assembly" of daughter cometary.

From formulas (13) and (14) it follows that if the length of ejection is ~ 5–500 km, and velocity dispersion in it does not exceed the first meters per second, then daughter comets with mass of ~ 10^{12}–10^{19} g can be formed already in asteroid belt itself.

Thus, ideas of supporters of eruptive formation of comets receive, at least on a qualitative level, quite acceptable theoretical justification. This mechanism in principle explains the composition and structure nucleus of cometary bodies in the model of Whipple (1950, 1951) as well as allows one to understand dimensions of observed comets themselves.

At $m \leq 10^{10}$ g, agglomeration of daughter comets becomes ineffective due to decrease in volume of space for matter collection, and at $m \geq 10^{20}$ g, the requirements for dispersion of ejection velocities are sharply tightened due to the short time (13). Of course, this mechanism does not exclude the possibility of daughter comets growth also due to additional addition to them of scattered gas and dust matter enveloping Solar System during period of bombardment by galactic comets (Barenbaum 2010).

Thus, our consideration shows that all secondary bodies, whether they remain in the belt or are thrown out of its limits and even outside the Solar System, are objects associated with their birth with asteroid belt. In this case, with the exception of the smallest dust grains, the rest bodies, be they fragments of asteroids, meteorites or daughter comets, move in the Solar System, obeying to formula (6).

Therefore, comets of Solar System and asteroids should be considered as extreme members of single series of secondary bodies "cometoasteroids" (Shestaka 1995), which, according to our concept, are formed during collisions of galactic comets with asteroid belt bodies. Depending on radius of orbits, that cometoasteroids have acquired, ratio in their substance of "rocky" and "ice" components can vary: from the complete absence of ice in asteroid belt, where ice sublimes under the action of solar radiation, to its significant presence in comets, which spend most of their life far from Sun.

Let us now turn to a discussion of range of questions relating directly to observed comets, and also explain the reasons leading to appearance of long-period and short-period cometoasteroids.

8 Long-Period Comets

It has already been noted that comets with orbital period $T > 200$ years and semi-major axis of orbit $\alpha = T^{2/3} > 34$ AU are referred to as long-period comets. However, historically, long-period comets are often understood (Marochnik 1985) more homogeneous group of bodies with periods of ~ 10^6 years, which is more than half of all comets of this type. The aphelion of such comets are at distance of more than 10^4 AU from the Sun, which causes the appearance of narrow maximum in their distribution in the $1/\alpha$ parameter at coordinates origin of (Fig. 7).

The frequent occurrence of such cometary orbits led Oort (1950, 1951) to assumption of the presence at distance of ~ 10^4–10^5 AU from Sun of quasi-spherical relict

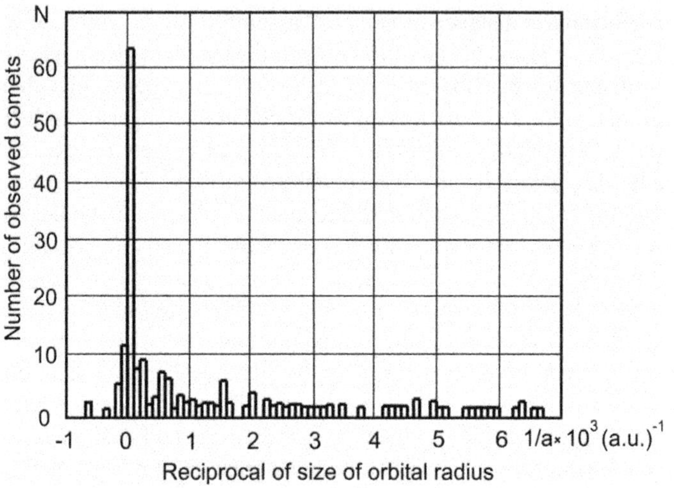

Fig. 7 Distribution of long-period comets by reverse large semi-axis of their orbit (Marochnik 1985)

cometary cloud, numbering 10^{12}–10^{13} comets, from where they penetrate into the inner region of Solar System. In view of the need to explain origin of other groups of long-period comets, the Oort hypothesis was subsequently transformed into ideas of existence of a whole system of "cometary reservoirs" located closer to Sun: of Whipple, Kuiper, Hills et al. (Tsitsin 2009).

The Galactocentric paradigm in a different way solves the problem of observed comets origin. In Solar System there are no "external" or "internal" reservoirs of relict comets preserved from the time of its formation, but there are daughter cometoasteroids. These objects formed and were ejected from asteroid belt in the time between 5.0 and 0.7 Ma during Sun's stay in Orion-Cygnus jet stream, and now rotation around Sun in orbits of greater radius than asteroids.

Therefore, long-period comets with periods of ~ 10^6 years are nothing more than the highest-speed group of cometoasteroids that left the belt with velocities close to $V*$. They have an orbital semi-major axis $\alpha = t^{2/3}$ ~ 10^4 AU and now are returning back to Sun. Since Sun is currently moving in Galaxy with an acceleration of + 1.8 km/s per 1 million years (Barenbaum 2023), for cometoasteroids with periods of ~ 10^6 years, the reference frame associated with the Sun is not strictly inertial. As a result, their initially elliptical orbits are transformed into weakly hyperbolic and weakly parabolic.

Let's discuss this issue in more detail. For this purpose, among entire multitude of cometoasteroids captured by Sun with orbital parameters $\alpha/R \gg 1$, we will single out three of their main groups. In the first group of "New" we included those that return to Sun for first time after their ejection from asteroid belt, and to the other two groups of "Repeated" and "Tertiary" we include cometoasteroids completing their not first revolution around Sun.

Let's consider the features each of these cometoasteroids groups.

8.1 "New" Long-Period Cometoasteroids

According to given earlier definition, orbital period T of New cometoasteroids must be identically equal to the time t, that has elapsed since time of their formation in asteroid belt. Therefore, in accordance with formula (6), between initial velocity V, aphelia distance $R_{\alpha f} = (1 + e)\alpha \approx 2\alpha$, and time of formation such comets t, there exists a rigid functional connection:

$$V = V * \sqrt{1 - \frac{T_0^{2/3}}{2} t^{-2/3}} \approx V * \sqrt{1 - t^{-2/3}}, \tag{15}$$

where T_0—period of orbital motion bodies in asteroid belt.

Believing that the formation of New cometoasteroids occurred during the last bombardment of Solar system by galactic comets, i.e. in the time interval $0.6 \leq t \leq 5.0$ Ma, it should be concluded:

(1) New cometoasteroids observed today should be ejected from the belt in a narrow range of initial velocities in vicinity of $V*$.
(2) Aphelion of orbits New cometoasteroids should be in interval distances from Sun $1.4 \times 10^4 \leq R_{\alpha f} \leq 5 \times 10^4$ AU.
(3) Ejection of New cometoasteroids from asteroid belt should be almost isotropic in all directions;
(4) Eccentricity of New orbits cometoasteroids should be practically equal to 1.

These parameters of New cometoasteroids orbits almost exactly determine the shape and dimensions of "cometary cloud", which according to the Oort hypothesis surrounds Solar System since its formation. However, as noted earlier, the long existence of such a cometary cloud near the Sun is physically impossible.

Therefore, it should be recognized that there is no excess of comets on the periphery of Solar system, and the so-called "Oort cloud" represents the geometric position of aphelion of cometoasteroids orbits that first time are returning back to Sun after their ejection from asteroid belt.

In other words, "Oort cloud" is not a place of real comets concentration, but only a spatial area of concentration of aphelion points of the most high-speed cometoasteroids that are returning to Solar System today. Since such objects return to Sun only for the first time after their ejection from the belt and have not yet lost high concentrations of gas and dust, they are perceived as "young", i.e. large and bright comets, which, in accordance with ideology of the Oort hypothesis, are called "New".

Let $n(V, t)$ is cometoasteroids number flying from the belt in unit velocity interval per unit time. Suppose also that all these bodies were formed during the last Sun stay in Orion-Cygnus jet stream, at that with equal probability throughout this time range.

$$n(V,t) = \begin{cases} 0 & t_1 > t > t_2 \\ n_0(V) & t_1 \le t \le t_2 \end{cases} \tag{16}$$

Then, the number of New comets returning to Sun every year we will found as

$$n = \int_{V_1}^{V_2} n(V) \cdot dV, \tag{17}$$

where velocities V_1 and V_2 are calculated by formula (15) for Sun's entry time (t_1 = 5.0 Ma) and exit time (t_2 = 0.7 Ma) from jet stream Orion-Cygnus.

Bearing in mind that velocities V_1 and V_2 differ little from the value of V^*, at integrate (18) we obtain $n = n_0(V^*)(V_2 - V_1)$. Assuming that unknown cometary objects are detected annually in an amount of $n \approx 5$, of which New long-period comets are about half (Churyumov 1980), we find the value of $n_0(V^*) \sim 10^4$ comets per year in interval of velocities 1 km/s. As a first approximation, we can also estimate number N of New long-period cometoasteroids currently existing in the solar system. Taking into account also those of them that have never appeared near Sun, their number in order of magnitude is $N \sim 10^6 - 10^7$ (Barenbaum 2002).

8.2 "Repeated" Long-Period Cometoasteroids

This group of bodies is represented by cometoasteroids, which were ejected from the belt with lower initial velocities and, accordingly, have smaller periods and radiuses of orbits. On the histogram of Fig. 5 they form a long "tail" of distribution.

Each of observed "Repeated" cometoasteroids can in principle to finish its not first turnover around the Sun. Therefore, one would expect that number cometoasteroids of this type in Solar System should significantly exceed the number of "New" and grow linearly with a decrease in their period.

In reality however it does not happen that way. Due to motion Sun on galactic orbit with acceleration the reference system associated with Sun for the long-period cometoasteroids is not inertial. Therefore, when they return to Sun their originally elliptical trajectories are transformed becoming weakly hyperbolic or weakly parabolic.

But even small deformations of orbits long-period cometoasteroids usually lead to their loss by Solar System. The systematization of cometary trajectories performed by Marsden (1986) shows that of 545 observed long-period comets, 162 comets had an elliptical orbit, and 285 and 98 comets, respectively, were weakly parabolic and hyperbolic. This means that of all cometoasteroids returning to Sun, only $p = 162/545 \approx 30\%$ has a chance to complete its subsequent turnover around Sun, and 70% will be irretrievably lost.

Hence it is clear that the region of orbits with distances of aphelion $R_{\alpha f} \leq 1.4 \times 10^4$ AU should not only must be deprived of New cometoasteroids, but also will very much depleted by Repeated.

Let's consider this issue in more detail. Let $n_1(T) = n_0$ is number of cometoasteroids with a period (T) are fly out from asteroid belt per unit time, and p—probability of their "survival" for each return to Sun. Then after k turnovers around Sun, number of such bodies in Solar System will be equal to $n_k(T) = n_1(\mathrm{T})p^{k-1}$.

Since we only observe cometoasteroids whose orbital period is $T = t/k$, where t is the time elapsed since their fly out from asteroid belt, we can write

$$n(T) = n(T/k) \cdot p^{\frac{t}{T}-1}, \tag{18}$$

where, as before, we will count.

$$\mathrm{n(t/k)} = \begin{cases} n_0 \; t_2 \leq t \leq t_1, \\ 0 \;\; t_2 > t > t_1. \end{cases} \tag{19}$$

Passing from the intensity cometoasteroids ejection $n(T)$ to the frequency of their detection, total number of cometoasteroids with periods within $t_1 \leq T \leq t_2$, returning annually to Solar System, we find by summation

$$\mathrm{N(t_1 \leq T \leq t_2) = n_0(t_2 - t_1)}\left\{1 + \sum_2^5 \left(\frac{1}{\mathrm{k}} - \frac{1}{\mathrm{k}+1}\right) \cdot \mathrm{p}^{\mathrm{k}-1}\right\} \tag{20}$$

Formula (20) characterizes the frequency of cometoasteroids arrival in the main maximum in Fig. 7 and includes both New, as also part of Repeated cometoasteroids with period $t_1 \leq T \leq \mathrm{t}_2/2$. If we set $p = 0.3$ (Barenbaum 2002), then according to (20) we get $N = 1.0794\, n_0$. Since generation rate of New cometoasteroids is assumed equal to $\mathrm{n_0}$, the proportion of repeated cometoasteroids in maximum (Fig. 7) is 7.94%. And at this 84.6% of repeated cometoasteroids have committed two, 12.7%—three, 2.3%—four and 0.4%—five complete turnovers around Sun.

8.3 "Tertiary" Long-Period Cometoasteroids

The character of function $N(1/\alpha)$ at the tail of distribution (see Fig. 7) gives we the suggests that some of these cometoasteroids could have formed after Sun left Orion-Cygnus jet stream. The reason for their appearance could be interactions in Solar system of cometoasteroids themselves. And, consequently, such objects in relation to New and Repeated cometoasteroids are, as it were, "Tertiary".

Tertiary cometoasteroids are characterized by a more complex and late originate. Due to this circumstance, in Fig. 7 they form a statistically extremely nonregular long "tail". It is not excluded, of course, that some cometoasteroids of tail are also a repeated cometoasteroids that lost mass in results of close spans near Sun or changed initial trajectory due to perturbations from planets.

Distinctive feature of this objects group is that as their periods decrease, they are less dependent on orbital Sun acceleration, and therefore their trajectories remain predominantly elliptical, and probability of returning to Sun approaches $p = 1$. The time formation of such cometoasteroids is $t \ll t_1$.

It follows that at least some "Tertiary" cometoasteroids made many revolutions around Sun. Therefore, the number of such cometoasteroids in Solar System should be small, of the same order as Repeat ones.

9 Short-Period Cometoasteroids

Short-period cometoasteroids are objects ejected from asteroid belt at a speed that prevents them flying out far from Sun. This group includes cometoasteroids with periods of $T < 200$ years around Sun. Such periods correspond to semi-major axis of orbits less than 36 AU, which determines localization of these objects in Solar System within the area of existence of known planets.

Short-period cometoasteroids are characterized by direct direction of motion around Sun, gravitate towards the ecliptic plane and, on average, have not very elongated orbits. Compared to long-period ones, they, on average, have lower masses and contain in their composition a larger percentage of solid clastic material of asteroids. Due to the short period T, these cometoasteroids practically "do not feel" the Sun orbital acceleration, therefore, from the moment of Sun exit from Orion-Cygnus jet stream, they were able to complete many, many thousands of revolutions around Sun.

It is obvious that in the epoch of Sun stay in jet stream there were much more such cometoasteroids than now. Therefore, today we see as short-period ones, only those of them that (1) escaped evaporation by solar radiation, (2) did not die in collisions in asteroid belt, (3) did not fall on surface of Sun and planets, and, finally, (4) were not captured by planet-giants and thrown out of Solar System by them.

The first two conditions for the "survival" of short-period cometoasteroids actually come down to the same requirement—perihelion of their orbits exceeding the outer radius of asteroid belt. We justify this conclusion by following simple considerations. Assuming that cometoasteroids reemit the incident solar radiation as absolutely black bodies, then their surface temperature T, depending on heliocentric distance R, in the first approximation must obey the law (Physics of Space: Little encyclopedia 1986): $T = 216/R^{1/2}$, where R is determined in AU, and T in degrees Kelvin.

Taking into account that significant part of cometoasteroids matter represented by water ice and frozen gas hydrates (Churyumov 1980; Shulman 1987) the temperature $T \approx 135$ °K is critical. At this temperature, water ice under low pressure conditions turns into vapor, bypassing the liquid phase. Therefore, at higher temperatures, ice either does not form or quickly sublimates. In the process of sublimation, a refractory layer of dust particles is formed on an open surface, which has high thermal insulation properties, which prevents further sublimation (Ibadinov 1999). As a result of the formation of such a layer, cometoasteroid ceases to be observed as comet and in its manifestations qualifies as an asteroid.

The critical temperature T $\approx$ 135 °K is reached at distance of $R' = 2.6$ AU from Sun, that actually coincides with inner boundary of the asteroid belt. From this it is clear that long-term existence of cometoasteroids, which are either directly in the belt, or often cross it, is physically impossible. During long-term orbital motion with perihelion $R_p = \alpha(1 - e) \leq R'$, cometoasteroids turn into swarms of meteoric bodies, which differ little from the same bodies of asteroid belt. Long-term orbital motion of cometoasteroids with perihelion $R_p \leq R'$ is also impossible for another reason. As a result of multiple passages through asteroid belt area, sooner or later they must experience a collision with its bodies and cease to exist as comets.

Thus, the place of survival of short-period cometoasteroids in Solar System, where they can persist as comets for relatively long time, is interplanetary space behind asteroid belt. However, this area cannot be considered completely safe for them, because it is here that cometoasteroids experience random encounters with giant planets.

Calculations by Kazimirchak-Polonskaya (1978a, b,) show that the behavior of short-period cometoasteroids in interplanetary space can be characterized as their rather long motion along fairly stable elliptical orbits, sometimes disrupted by their approach to planets. As a result of rapprochement, the transition of cometoasteroids from one quasi-stable orbit to another most often occurs. However, with a very close rapprochement, the comet ejection from Solar System and its "capture" by a giant planet are also possible. In the latter case, there is a high probability that cometoasteroid will be thrown away towards Sun at a distance of $R \leq 3$–4 AU, and then we will be able to detect it as object belonging to the cometary family of this giant planet.

The indicated character of cometoasteroids motion in interplanetary space is confirmed by the statistics of observations of short-period comets in inner region of Solar System (Fig. 8).

It can be seen from the figure that the distribution function N (k) of short-period cometoasteroids by the number of their appearances at small values of k is universal for all giant planets. In comet families of different giant planets the number of cometoasteroids that have completed $k = 2-3$ revolutions around Sun is related to number objects that have made only one revolution, by the ratio

$$N(k) = N(1) \cdot p^{k-1}, \tag{21}$$

where constant $p = 0.606$.

Formula (21) indicates that transition of cometoasteroids from one quasi-stable orbit to another is a random short-term process, which is probabilistic in nature and clearly differs from cometoasteroids transition mechanism to orbits in the region k > 3.

The probability that cometoasteroids acquire "transitional" orbits is much higher than translating them on long-term "sustainable" ones. So, for example, the total number of random captures obeying rule (21) in Jupiter family is $N = \sum N(k) = 55$,

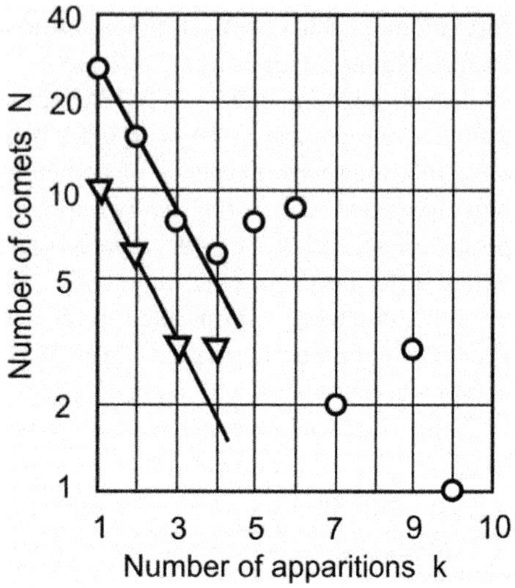

Fig. 8 Distribution of short-period cometoasteroids by the number of detections (k < 10) for Jupiter family (circles) and for families of Saturn, Uranus and Neptune combined (triangles)

and in combined Saturn-Uranus-Neptune family $N = 22$. Consequently, in Jupiter only for 37% of cases, the capture carried more or less prolonged character, and in 63% cases it was reduced only to the transfer of cometoasteroids from one unobservable state to another with a short-term entry into the glow zone. For planets farther from Sun, the proportion of such captures is higher and reaches 80%.

To explain the formation mechanism of cometary families and the distribution over them of the number of short-period cometoasteroids, let us turn to the following simplest model. Suppose that at heliocentric distance $R > 3$–4 AU in Solar System to the present time, a certain number of cometoasteroids have survived, whose orbits lie in ecliptic plane, and they revolve around Sun in forward direction.

Consider a cometoasteroid that accidentally flew distance (d) from some planet with mass M_i. Let, due to planet attraction, trajectory of this cometoasteroid deviates by angle χ. Then, in the center-of-mass system coinciding with planet, the value of angle χ will be determined by the relation (Putilin 1953):

$$\sin\left(\frac{\chi}{2}\right) = \left[\sqrt{1 + \left(\frac{V^2 d}{GM_i}\right)^2}\right]^{-1}, \tag{22}$$

where $V = |V_C - V_i|$ is cometoasteroid speed relative to the i-planet.

We use expression (22) to estimate magnitude of cometoasteroid capture radius by planets. In this case, under "capture" we will understand such cometoasteroid interaction with planets, when it is thrown in direction of Sun and becomes accessible to detection as a comet. In the first approximation, such capture is accompanied by change in initial direction of cometoasteroid motion through an angle $\chi = \pi/2$. In accordance with formula (22), this is equivalent to its flew at distance from planet $d_i \approx GM_i/V$. The probability w_i of such flew will be $W_i \approx d_i/(2\pi R_i)$, where R_i is the heliocentric distance of the i-planet. Substituting the expression for d_i into this formula, we obtain

$$W_i \approx \frac{1}{2\pi}\frac{M_i}{M_\odot}\left[\frac{V_i}{V_x}\right]^2 \tag{23}$$

where V_x denotes chaotic component of cometoasteroid speed relative to the average speed $V_i = \sqrt{GM_\odot / R_i}$

The ratio V_i/V_x in formula (23) is a weak function of heliocentric distance. Therefore, to within a factor of order of unity, probability W_i of cometoasteroid ejection toward Sun is proportional to value of $M_i/M_\odot$. The calculation by formula (23) and actually observed number k_i of short-period comets in families of giant planet are shown in Fig. 9 (Barenbaum 2002). The calculated values were found as $k_i = KM_i/\sum M_i$, where $K = 114$ is total number of comets in Jupiter, Saturn, Uranus and Neptune families (Churyumov 1980).

Despite the simplified nature of our model, agreement between the results of calculations and observations is quite satisfactory. So, according to our calculations, the ratio of cometoasteroids number in Jupiter and Saturn-Uranus-Neptune families $W_U/(W_S + W_U + W_N) = 2.47$, whereas according to actual data it is 3.22. However, if this ratio is determined not for all comets, but only for the "transitional" group

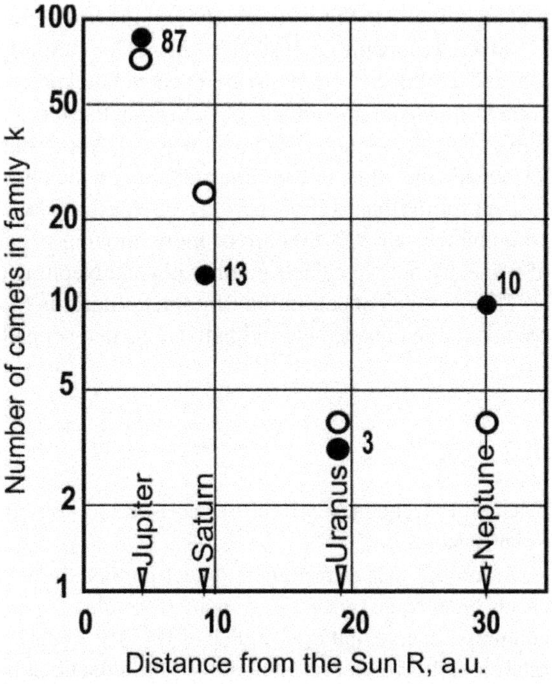

Fig. 9 Distribution of comets number by families of giant planets: black circles—observations (Churyumov 1980), open circles—theoretical calculations (Barenbaum 2002); Numbers—number of observed comets in families of individual giant planets

comets, for which, strictly speaking, the formula (23) is applicable, then the indicated ratio will be equal to 55/22 = 2.5 and practically coincide with our calculation.

The explanation for this fact, of course, should be sought not so much in the theoretical model itself, as in the correctness of initial assumptions laid down in it. There are two of them: (1) the presence of many unobservable cometoasteroids behind asteroid belt, quite evenly distributed in interplanetary space and strongly gravitating towards the ecliptic plane, and (2) the random nature of cometoasteroids interaction with giant planets.

It also follows from structure of formula (23) that most probable interaction of cometoasteroids with planets occurs at aphelion portion of their trajectories, where cometoasteroids speed in magnitude and direction is as close as possible to planet orbital speed V_i. Therefore, for cometoasteroids of one family, aphelion distances turn out to be approximately equal to orbit radius of planet that captured them. This circumstance leads to grouping of known short-period comets into families of individual giant planets.

Taking the average speed of chaotic motion of cometoasteroids $V_x = 5$ km/s, we get that as a result of capture by giant planets, approximately one of ~ 10^3–10^4 cometoasteroids that accidentally find themselves near the planet can be thrown towards Sun and discovered as a short-period comet by astronomers. It is clear that such formation mechanism of cometary families is much more effective than transferring long-period cometoasteroids with orbital speed of ≈ 25 km/s to short-period orbit and even more so capture by giant planets of galactic comets, which moving relative to Sun at a speed of ≈ 450 km/s.

Interactions with giant planets throw low-velocity cometoasteroids both toward Sun and in other directions. With repeated bombardments of Solar system by galactic comets, this led to movement of such cometoasteroids beyond the outer boundary of Sun's planetary system (Barenbaum 2010). As a result, they acquired trajectories that lay outside the orbits of Neptune and Pluto, which provided them with a significantly longer lifetime than at short-period cometoasteroids. Today they are very large objects with diameter of 200–800 km or more, moving in ecliptic plane and showing signs of cometary activity. Their number beyond Neptune orbit is ≈ 3000.

The average concentration of cometoasteroids (n) in interplanetary space in the ecliptic plane outside asteroid belt can be determined from the relation:

$$\nu = \pi n V_x \sum_{1}^{4} w_i d_i^2, \tag{24}$$

where ν is frequency arrival of short-period cometoasteroids into inner region of Solar System.

Assuming that astronomers detect previously unknown short-period cometoasteroids on average twice a year, using formula (24), we find $n \sim 3 \times 10^{-22}$ км$^{-3}$. For comparison, note that this value is ~ 10^5–10^8 times less than bodies concentration in asteroid belt, which are "dangerous" for galactic comets.

The total amount of cometoasteroids N in outer region of planetary system we determine by multiplying their concentration n in space volume occupied by these bodies $V = \pi h (R_N^2 - R_0^2)$, where R_N is Neptune orbit radius, h is "thickness" of localization zone of cometoasteroids. Assuming $h \sim 1$ AU, we arrive at value $N \sim 2.4 \times 10^7$. Knowing N we get estimate of lifetime τ short-period cometoasteroids in interplanetary space. Normalizing N by quantity $\nu \sim 2$ year^{-1}, we find $\tau = N/\nu \sim 10^7$ years.

This result means that unlike long-period cometoasteroids, which almost all originated during the last Sun's stay in Orion-Cygnus jet stream some part of short-period cometoasteroids observed today could have been preserved since previous bombardment of Solar System by galactic comets.

10 Main Results and Conclusions

The author's concept of the formation of comets in Solar System, based on ideas of Galactocentric paradigm, is presented (Barenbaum 2010).

The creation of this concept became possible, firstly, as a result of discovery the phenomenon of jet outflow of gas and dust matter from the core of our and other spiral galaxies. And, secondly, due to the rejection of Kant-Laplace cosmogony hypothesis, which allows for the one-moment formation of Solar System and its planets, as well as their further evolution without taking into account physical processes in Galaxy.

It is shown that the short-period and long-period comets observed today are the collision products of galactic comets of Orion-Cygnus jet stream with bodies of the asteroid belt, captured by Sun attraction into orbits more diverse and distant than at asteroids. Such objects—"cometoasteroids" consist of comets matter of Galaxy origin, which is frozen together with debris and evaporated material of asteroids.

Short-period and long-period cometoasteroids were formed from 5 to 0.7 Ma, and are distinguished differ from each other in the fact that the first, never left limits system of Sun's planets, gravitate towards the ecliptic plane and today are preserved in outer space between giant planets. In the event of an accidental approach to these planets, they are thrown towards Sun. At a heliocentric distance of less than 3–5 AU such cometoasteroids become detectable, and astronomers interpret them as short-period comets of the family of corresponding planet.

In contrast, long-period cometoasteroids have spent almost their entire life far beyond planetary system of Sun. They have very elongated orbits, and therefore many of them today return to Sun only for the first time after birth in asteroid belt. A high concentration of aphelion points of their orbits at a distance of 10^4–10^5 AU from Sun is a direct consequence of these bodies ejection from belt in the time interval 5–0.7 Ma, when Sun was in Orion-Cygnus jet stream.

The new concept not contradictory explains all the main properties of long-period and short-period comets in Solar System, combining the best aspects of known hypotheses of comets origin and at the same time eliminating their shortcomings. It favorably differs from “interstellar” hypotheses in that it does not encounter difficulties in explaining the formation of families of short-period comets, and from “eruptive” hypotheses in that it easily explains the peculiarities of orbits of various groups of long-period comets.

The author’s concept also preserves the idea of existence near Sun of cometary bodies cloud. However, unlike the hypotheses of Oort, Kuiper, Hills, Whipple, Tsitsin and others, this cloud consists not only of comets, but also of gas and dust and is an extremely non-stationary formation. This comet-gas-dust cloud arises during periods of Solar System bombardments by galactic comets and exists after that for ~ 10^7 years. It currently such cloud contains ~ 10^7 cometoasteroids of all types.

Thus, the new concept postulates a close connection between cometoasteroids and other small bodies in Solar System, primarily with asteroids and cosmic dust. There is no fundamental difference between cometoasteroids and bodies in asteroid belt. The differences between these two groups of bodies are determined by parameters of their orbits. Most asteroids move in belt zone, where ice is evaporated by solar radiation, while cometoasteroids located outside this zone retain ice for a longer time.

And, finally, no less important dignity of cometoasteroidic concept is that it is a necessary component part of a more general system of representations—Galactocentric paradigm, which connects phenomena in Solar System and on its planets with physical processes in Galaxy (Barenbaum 2002, 2010).

References

Artyushkov EV (1994) Proc Acad Sci 336(5):680–683

Bailey ME (1986) Mon Not Roy Astron Soc 218(1):1–30

Barenbaum AA (1990) Kometnyy Tsirkulyar 418:11–12

Barenbaum AA (1990a) Kometnyy Tsirkulyar 419:9–10

Barenbaum AA, Shpekin MI (2019) J Phys Conf Ser 1147:012057

Barenbaum AA (2002) Galaxy, solar system, earth. In: Subordinate processes and evolution, PH: GEOS, Moscow, p 393

Barenbaum AA (2010) Galaxycentric paradigm in geology and astronomy. PH: LIBROKOM, Moscow, p 544

Barenbaum AA (2013) Physical-chemical and petrophysical studies in earth sciences. In: Proceedings of XIV international conference IGEM RAN, Moscow, pp 31–42

Barenbaum AA (2015) J Phys Conf Ser 653:012073

Barenbaum AA (2016) Tectonophysics and topical issues of earth sciences. In: Conference on proceedings of institute for earth physics RAN, Moscow, vol 2, no (5), pp 430–438

Barenbaum AA (2018) Eng Phys (9):30–47

Barenbaum AA (2018a) Proceedings of all-Russian annual seminar on experimental mineralogy, petrology and geochemistry (VESEMPG-2018), GEOKHI RAN, Moscow, pp 294–298

Barenbaum AA (2021) Hubble’s law and the cosmic microwave background in the absence of the big bang. https://doi.org/10.9734/bpi/rtcps/v4/13528D

Barenbaum AA (2022) Solution of problem questions of astronomy and geology using the optimized galactic model. https://doi.org/10.1007/978-3-031-25962-3

Barenbaum AA (2023) On the issue of "True polar wander" phenomenon and its alternative physical interpretation based on galactic model. https://doi.org/10.1007/978-981-99-6575-5
Barenbaum AA, Shpekin MI (2021) J Phys Conf Ser 1787, 012021
Barenbaum AA, Titorenko AS (2020) Proceedings of all-russian annual seminar on experimental mineralogy, petrology and geochemistry (VESEMPG-2020). GEOKHI RAN, Moscow, pp 210–215
Bazilevsky AT, Ivanov BA, Florensky KP et al (1983) Impact craters on the moon and planets, PH: Nauka, Moscow, p 200
Bobrovnikoff NT (1929) Lick Obs Bull 408:28–38
Brownlee DE (1984) Comets and origin of life. PH: Mir, Moscow, pp 69–77
Chepurova VM, Shershkina SL (1989) Kinematika i Fizika Nebesnykh Tel 5(4):82–87
Churyumov KI (1980) Comets and their observation. PH: Nauka, Moscow, p 160
Davydov VD (1981) Kosmicheskiye Issledovaniya 19(5):749–762
Dobrovolsky OV (1966) Comets. PH: Nauka, Moscow, p 288
Dolgopolova EN, Marochnik LS (1987) Astronomicheskii Vestnik 21(4):331–334
Drobyshevsky EM (1978) The moon and the planets. Holland 18:145–194
Fesenkov VG (1951) Sov Astron A J 28(2):98–111
Glushneva IN (1994) Astron Rep 71(4):652–656
Guliev AS (1992) Kinematics Phys Celestial Bodies 8(4):51–60
Guliev AS, Vsekhsvyatsky SK (1981) Sov Astron 59(3):630–635
Hartman WK (1978) Solar System, PH: Mir, Moscow, pp 167–186
Hills JG (1982) Astron J 87(6):906–910
Ibadinov KhI (1999) Astrono Vestn 33(4):363–368
Kazimirchak-Polonskaya EI (1978a) Astrometriya i nebesnaya mekhanika. Moscow-Leningrad, no 7, pp 340–383
Kazimirchak-Polonskaya EI (1978b) Astrometriya i nebesnaya mekhanika. Moscow-Leningrad, no 7, pp 384–418
Koval VI (1976) Astron Tsirkulyar 927:4–6
Kuiper GP (1951) Astrophysics. In: Hynek JA (ed) McCraw-Hill. Co. Inc., New York, pp 357–424
Lagrange JL (1812) Sur l' origine oles cometes. Mem. VII. Paris, pp 381–395
Laplace PS (1796) Exposition du systeme du Monde. Paris, Ed 1–5
Lyttleton RA (1948) Month Not Roy Astron Soc 108(1):465–475
Marochnik LS (1985) Svidaniye s kometoy. PH: Nauka, Moscow, p 208
Marov MYa (1981) Planets of solar system, PH: Nauka, Moscow, p 256
Marov MYa (1994) Astron vestn 28(4):5–85
Marsden BG (1986) Catalogue of cometary orbits. In: Smithsonian Astrophys. Obs., Cambridge
McCrea WN (1953) La Physique des cometes. Liege, pp 337–350
Melosh HJ (1989) Impact cratering. A geological process, Oxford University Press Inc., New York, p 336
Mironova EN (2015) Dust in the solar system: Preprint FIAN, No. 10
Napier WM (1985) Dynamics comets. Origin and evolution: Pros. 83-rd Collog. Rome. 1984. Dordrecht., pp 31–41
Oort JH (1950) Bull Astron Inst Netherlands 11:91–110
Oort JH (1951) The Observatory 71:120–147
Öpik EJ (1975) Irish Astron J 12(31):1–48
Oppenheim S (1924) Probleme der Astronomie. Festschrift fur H. Seeliger, Berlin, pp 131–143
Orlov SV (1936) Sov Astron A J 13(3):232–238
Orlov SV (1960) On the nature of comets, PH: AN SSSR, Moscow, p 191
Physics of space: little encyclopedia (1986) PH: Sov. Entsiklopediya, Moscow, p 783
Proctor RA (1881) The poetry of astronomy. London, p 156
Putilin II (1953) Minor planets, PH: Gostekhteoretizdat, Moscow, p 412
Radzievsky VV (1995) Astron Vestn 29(2):185–190
Radzievsky VV, Tomanov VP (1973) Astron Vestn 7(2):73–82

Ruskol EL (1986) Results of science and technology. Astronomy, PH: VINITI, Moscow, vol 28, pp 3–116
Safronov VS (1969) Evolution of preplanetary cloud and formation of Earth and planets. PH: Nauka, Moscow, p 244
Schiaparelli GV (1871) Siebente Note. 1871. Verlag von Th. von der Nahmer, Stettin, pp 261–267
Schmidt OYu (1945) Doklady AN SSSR 49(6):413–416
Shestaka IS (1995) Astron Vestn 29(5):468–471
Shorma SD, Khonna M (1988) Month Not Roy Astron Soc 235(4):1467–1471
Shulman LM (1987) Comet nuclei. PH: Nauka, Moscow, p 230
Simonenko AN (1979) Meteorites—asteroid fragments. PH: Nauka, Moscow, p 224
Simonenko AN (1985) Asteroids. PH: Nauka, Moscow, p 208
Svedstrup A (1884) Astron Nachr 107:113–126
Tomanov VP (1989) Comet cosmogony, PH: VGPI, Vologda, p 96
Tomanov VP, Rodin DA, Shilyaev AS (2013) The origin and evolution of comets. PH: VGPI, Vologda, p 166
Tsitsin FA (1999) Zemlya i Vselennaya, 1:60–69
Tsitsin FA, Rastorguev AS, Chepurova VM (1985) Astron Tsirkulyar 1408:5–8
Tsitsin FA (2009) Essays on the modern cosmogony of the solar system: origins problems Horizons, PH: Feniks+, Dubna, p 356
Van den Bergh S (1982) J Roy Astron Soc Can 76(5):303–308
Van Flandern TC (1976) Proceedings of the 41st colloquium of the international astronomical union held in Cambridge, England, 17–19 Aug, pp 89–90
Van Flandern TC (1992) Bul Amer Aston Soc 24(3):1062
Vitkovsky II (1971) Astronomicheskii Vestnik 5(2):82–88
Vsekhsvyatsky SK (1967) The nature and origin of comets and meteoric matter, PH: Prosveshcheniye, Moscow, p 184
Whipple FL (1950) Astrophys J 111(2):375–394
Whipple FL (1951) Astrophys J 113(3):464–474
Whipple FL (1964) Astron J 69:563
Yabusbita S, Hasegawa I (1978) Month Not Roy Astron Soc 185(2):549–553

Content of Hydrocarbons in the Sea Bottom Sediments of the Kara Sea North-Eastern Part

Yu. S. Tkachenko, E. A. Tikhonova, and O. V. Soloveva

Abstract The study of the sea bottom sediments (0–5 cm) of the northeastern part of the Kara Sea was carried out during of the voyage of the SRV "Academician Boris Petrov". Data on the content and composition of hydrocarbons in the sea bottom sediments of the studied water area were obtained. The purpose of this work was the qualitative and quantitative content assessment of hydrocarbons in the sea bottom sediments of the northeastern part of the Kara Sea, as well as the identification of their probable origin. The material for the study were samples of sea bottom sediments upper layer collected by means of a coring tube in the various water areas of the north-eastern part of the Kara Sea in the summer of 2022. The determination of qualitative and quantitative hydrocarbon's composition was carried out by means of gas chromatography ("Crystal 5000.2" chromatograph with flame ionization detector). Diagnostic indices of the origin of hydrocarbons were used to identify the likely sources of organic matter. It was found that the total amount of hydrocarbons in sea bottom sediments of the northeastern part of the Kara Sea varied from 25.8 to 152.4 mg/kg, on average—68.21 ± 7.76 mg/kg. According to the results of chromatograms analysis, the predominance of high-molecular *n*-alkanes was noted. The most abundant *n*-alkanes were homologues of *n*-C_{25}–C_{29}, which are markers of allochthonous organic matter. When considered the diagnostic indices (CPI_1, CPI_2, ACL, LWH/HWH, TAR, C_{31}/C_{29}, Paq, TMD, $\Sigma C_{25-35}35/\Sigma C_{15-21}$, Pr/Ph), it was found that the increased concentrations of hydrocarbons were mainly due to the natural processes and the predominance of allochthonous organic substances in bottom sediments.

Keywords Hydrocarbons · Diagnostic indices · Bottom sediments · Kara Sea

Yu. S. Tkachenko (✉) · E. A. Tikhonova · O. V. Soloveva
A. O. Kovalevsky Institute of Biology of the Southern Seas of RAS, Sevastopol, Russia
e-mail: yulechkatkachenko.90@mail.ru

T. Chaplina (ed.), *Processes in GeoMedia—Volume VIII*, Springer Geology,
https://doi.org/10.1007/978-981-97-6627-7_17

1 Introduction

The interest in the study of hydrocarbons (HC) in the bottom sediments of the Kara Sea is due to the high oil and gas content of the high-latitude region (Kontorovich 2015). The water area of this sea is one of the most oil and gas promising on the Arctic shelf, accounting for about 25% of all hydrocarbon resources of the northern seas (Kontorovich 2015). Unlike the southern part of the Kara Sea, the northern side of this region has not been studied by deep drilling, the prospect of oil and gas potential of the studied water area can be based only on geological and geophysical data (Kontorovich and Kontorovich 2019). In this regard, there is increased interest in the waters of the northern part of the Kara Sea, so there is a need to regularly conduct studies of the ecological state of seawater and sea bottom sediments.

An important and integral part of monitoring of marine ecosystems is the assessment of pollution of waters and sediments with petroleum products, since oil and petroleum products are one of the most common sources of pollution of the seas and oceans (Tikhonova 2021; Tikhonova et al. 2021; Mironov and Mironov 2020). Moreover, in high-latitude areas, the decomposition of petroleum hydrocarbons is slowed down due to a wide range of natural and climatic conditions (Nemirovskaya 2013).

In recent years, there observed a serious transformation of the Arctic ice cover due to climatic changes (Flint et al. 2018; Kazennov et al. 2018; Aksenov and Ivanov 2018; Alekseeva et al. 2018; Kravchishina et al. 2019), which affected the natural hydrocarbon background, therefore, the study of hydrocarbons in this region is becoming particularly relevant. To assess the anthropogenic load on the water area, it is necessary to determine the natural hydrocarbon background resulting from the natural vital activity of marine organisms and oil seepage from the seabed (Nemirovskaya 2013).

The study of the composition, distribution and genesis of hydrocarbons in sea bottom sediments makes it possible to predict the oil and gas potential of the subsurface, and to assess the ecological state of the water area (Tikhonova et al. 2020). Petroleum hydrocarbons accumulate at the bottom of reservoirs due to sedimentation (Kotova 2020). The processes of biochemical oxidation in the bottom sediments are much slower than in the water. So, the sediments often concentrate substances of anthropogenic origin, thus benthic organisms can accumulate toxic substances, in turn, being harmful to other living organisms (Kotova 2020; Soloveva et al. 2022).

Therefore, the study of anthropogenic pollution of sea bottom sediments is an important and integral part of environmental studies of aquatic ecosystems.

The purpose of this work was the qualitative and quantitative assessment of hydrocarbons content in the sea bottom sediments of the northeastern part of the Kara Sea and the determination of probable sources of their origin.

The following diagnostic indices were used to identify the genesis of HC: the ratio of allochthonous substance to autochthonous compounds (terrigenous/aquatic ratio, TAR), the average length of the hydrocarbon chain (average chain length, ACL) and the ratio of low-molecular weight to high-molecular *n*-alkanes (Low-molecular

Weight to High-molecular Weight Homologies ratio, LWH/HWH) (Silliman and Schelske 2003; Zhang 2014; Duan 2000; Huang et al. 2011; Blumer et al. 1971). Some markers make it possible to determine which type of vegetation prevails in the ecosystem and to identify the main source of HC input, namely ACL (Average chain length) and C_{31}/C_{29} (the ratio of the contribution of herbaceous vegetation to woody vegetation) (Huang et al. 2011). To identify the petroleum and biogenic origin of hydrocarbons, indices such as the CPI_1 odd index calculated for lighter *n*-alkanes and CPI_2 calculated for the high molecular weight homologues, as well as ACL, LWH/HWH (Silliman and Schelske 2003; Huang 2011; L(u) and Zhai 2008) were used.

2 Materials and Methods

The material for the study was samples of the upper layer (0–5 cm) of sea bottom sediments collected by a coring tube in the summer of 2022 in various parts of the water area of the northeastern part of the Kara Sea. A total of 16 samples were taken (Fig. 1). The sampling depth varied from 54 to 151 m. The temperature of the upper layer of bottom sediments ranged from + 0.3 to + 1.8 °C, the pH value was from 7 to 8.

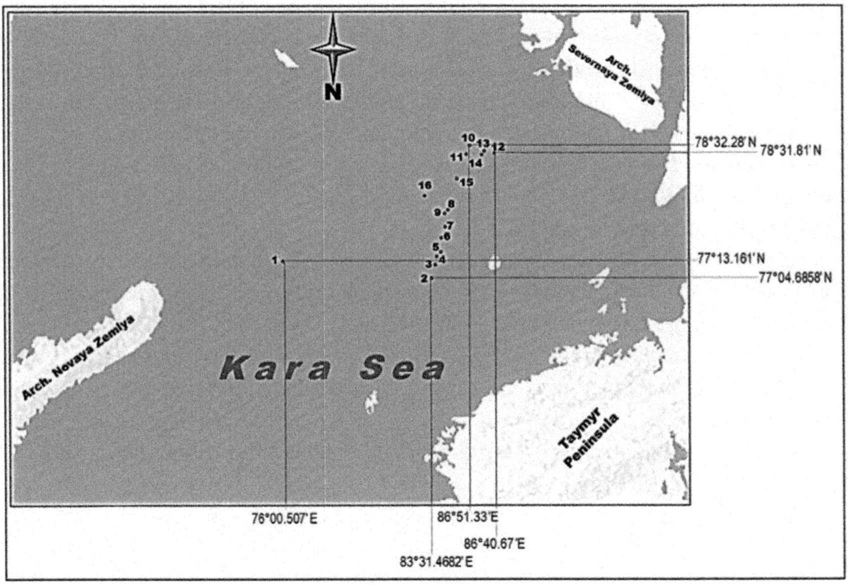

Fig. 1 The scheme of sea bottom sediment sampling stations in the northeastern part of the Kara Sea

The samples collected were cleaned of stones, algae, etc., and then dried to an air-dry state in natural conditions on the ship. Then the samples were ground in an inert mortar with a pestle and sifted through a sieve with a mesh diameter of 0.25 mm. After that, the samples were placed in zip packages for further transportation to the A. O. Kovalevsky Institute of Biology of the Southern Seas laboratory.

Determination of HC in sea bottom sediments was carried out according to method.[1] A sample (5–7 g) was placed in a conical flask, 20 ml of *n*-hexane was added and shaken for 30 min. After the extract had settled (10 min), it was transferred to a clean conical flask. The extraction was repeated 2 more times, and 60–70 ml of the extract was obtained at the output. To separate the polar compounds, the extract was passed through a glass column with a retracted lower end filled with aluminum oxide. The resulting sample volume was concentrated to 1 ml in a fume hood at room temperature.

The determination of the qualitative and quantitative composition of HC and *n*-alkanes in sea bottom sediments was carried out in the collective use center "Spectrometry and Chromatography" by means of gas chromatography ("Crystal 5000.2") chromatograph with flame ionization detector (FID).

An aliquot (1 mcl) of the obtained extract was injected using a micro-syringe into a gas chromatograph evaporator heated to 250 °C. For HC separation, a TR–1MS capillary column of 30 m × 0.32 mm with a fixed phase thickness of 0.25 microns (Termo Scientic) was used. HC separation was ensured by the following conditions of the chromatographic system: detector temperature—320 °C; column temperature was programmed from 70 to 280 °C (temperature rise rate of 8 °C min^{-1}); carrier gas flow (nitrogen) in the column—2.5 ml min^{-1} without flow division (see Footnote 1).

The quantitative determination of the total HC content was determined by the method of absolute calibration of the FID using a mixture of HC prepared by the gravimetric method. A operational standard solution with a concentration of 1 mg/ml was prepared from the initial standard solution by diluting it with *n*-hexane. The standard sample ASTM D2887 Reference Gas Oil (SUPELCO, USA) was used to separate the hydrocarbon mixture.

The content of *n*-alkanes was determined using a standard sample of paraffin hydrocarbons in hexane with a mass concentration of each component of 200 mcg min^{-1}, pristan + phytane—100 mcg min^{-1} in hexane (SUPELCO, USA).

Chromatek Analyst 3.0 software was used to process the results. Concentrations of HC and *n*-alkanes were calculated by absolute calibration and percentage normalization.

The following diagnostic markers were used to identify the origin of HC (Table 1).

---

[1] Drugov and Rodin (2020).

Table 1 Main diagnostic indexes for determining the genesis of hydrocarbons

Index	Formula	References
TAR	$\sum C_{27} + C_{29} + C_{31}/\sum C_{15} + C_{17} + C_{19}$	Silliman and Schelske (2003)
LWH/HWH	$\sum(C_{13}-C_{21})/\sum(C_{22}-C_{37})$	Blumer et al. (1971)
ACL	$(27{\cdot}C_{27} + 29{\cdot}C_{29} + 31{\cdot}C_{31} + 33{\cdot}C_{33} + 35{\cdot}C_{35} + 37{\cdot}C_{37})/(C_{27} + C_{29} + C_{31} + C_{33} + C_{35} + C_{37})$	Zhang (2014)
C_{31}/C_{29}	C_{31}/C_{29}	Huang et al. (2011)
Pr/Ph	Pr/Ph	Nemirovskaya (2013)
CPI_1	$(1/2)\{(C_{15} + C_{17} + C_{19} + C_{21})/(C_{14} + C_{16} + C_{18} + C_{20}) + (C_{15} + C_{17} + C_{19} + C_{21})/(C_{16} + C_{18} + C_{20} + C_{22})\}$	L(u) and Zhai (2008)
CPI_2	$CPI_2 = (1/2)\{(C_{25} + C_{27} + C_{29} + C_{31} + C_{33} + C_{35})/(C_{24} + C_{26} + C_{28} + C_{30} + C_{32} + C_{34}) + (C_{25} + C_{27} + C_{29} + C_{31} + C_{33} + C_{35})/(C_{26} + C_{28} + C_{30} + C_{32} + C_{34} + C_{36})\}$	L(u) and Zhai (2008)
Paq	$(C_{23} + C_{25})/(C_{23} + C_{25} + C_{29} + C_{31})$	Duan (2000)
TMD	$(C_{25} + C_{27} + C_{29} + C_{31} + C_{33})/(C_{15} + C_{17} + C_{19} + C_{21} + C_{23})$	Nemirovskaya (2013)
$\Sigma C_{25-35}/\Sigma C_{15-21}$	$(C_{25} + C_{26} + C_{27} + C_{28} + C_{29} + C_{30} + C_{31} + C_{32} + C_{33} + C_{34} + C_{35})/(C_{15} + C_{16} + C_{17} + C_{18} + C_{19} + C_{20} + C_{21})$	Blumer et al. (1971)

3 Results and Discussion

The sea bottom sediments of the northeastern part of the Kara Sea were represented by dense sandy-clay siltstones, mostly dark gray and brown in color. The total HC content ranged from 25.8 to 152.4 mg/kg, with an average of 68.21 ± 7.76 mg/kg. The maximum concentration was noted at station 14, and the minimum at station 3 (Fig. 2). It is known that the maximum of petroleum hydrocarbons is concentrated in bottom sediments in areas with a slow flow. The erosion of contaminated sediments causes secondary water pollution throughout the water area. Increased concentrations of HC lead to deterioration of the vital activity of hydrobionts, physiological disorders, fertility changes, mortality may increase, and the biomass and number of organisms also change (Nemirovskaya 2013; Sanitary and biological studies of coastal waters of the southwestern Crimea at the beginning of the XXI century 2018).

According to the data (Nemirovskaya 2015) in 2011, in different water areas of the Kara Sea, the average HC content in the surface layer of sea bottom sediments ranged from 11.3 to 15.6 mg/kg. In this study, the average value of HC concentrations exceeded by 4 times the maximum, recorded in 2011 (Nemirovskaya 2015). In 2016, at the mouth of the Ob River, the concentration of these substances in sea bottom sediments was in the range of 3.5–42.6 mg/kg (Nemirovskaya 2021). Thus, the maximum HC content (42.6 mg/kg) was 1.6 times less than the average value recorded by us in this study.

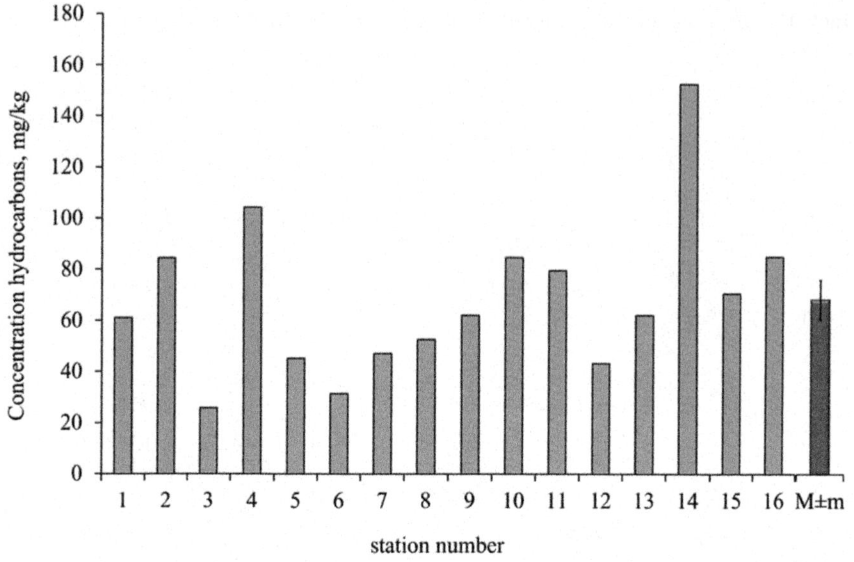

Fig. 2 Concentrations of hydrocarbons in bottom sediments of the northeastern part of the Kara Sea

For comparison, in the Barents Sea in 2016, the HC content in the surface layer of sea bottom sediments varied in the range of 3–44 mg/kg, in 2017 it amounted to 4–57 mg/kg (Nemirovskaya et al. 2021a). In 2020, the HC content in the sea bottom sediments of the Barents Sea increased and was in the range of 3–186 mg/kg (Nemirovskaya et al. 2021a). In general, there was a tendency to increase HC concentrations in the Barents Sea. The values we obtained for HC in the northeastern part of the Kara Sea were higher than in 2016–2017 in the Barents Sea, and in 2020 the HC content in the Barents Sea increased and became higher than in our studies.

According to the authors (Nemirovskaya et al. 2021a, b, 2019), from 2016 to 2022, there was an increase in HC concentrations both in water and in sea bottom sediments of the Arctic seas, but in general, there were no excess of their concentrations and there was no chronic oil pollution. In addition, the increase in HC content in the northern seas of Russia may be due to natural processes that should not be neglected (Nemirovskaya 2013). Since the climate has changed significantly in recent years, subsequently the area of long-term ice is reduced and the natural hydrocarbon background is changing (Flint et al. 2019, 2020). Earlier (Flint et al. 2020) in 2020 an abnormally early release of seasonal ice from its marine basin had been discovered in the Kara Sea. And in September of the same year, in some areas of the sea, foci of high abundance of planktonic algae were observed, which are in the spore-forming phase of development characteristic of the summer season (Flint et al. 2020). These phenomena could lead to an increase in the HC content in the bottom sediments.

The concentration of HC in marine sediments is not regulated by Russian regulatory documents, therefore, the norms set out in the so-called "Dutch lists" are often used to assess the level of contamination of sea bottom sediments with HC.[2] In accordance to these standards, the sea bottom sediments of the northeastern part of the Kara Sea exceed the permissible level (50 mg/kg) in 68% of samples.

Currently, there are other domestic classifications for assessing the level of pollution of bottom sediments of hydrocarbons. One of these is the classification of Uvarova (1989). According to it, bottom sediments can be divided into.

- pure—0 to 5.5;
- moderately polluted—25.6 to 55.5;
- polluted—55.6 to 205.5;
- dirty—205.6 to 500.0;
- very dirty—over 500.0 mg/kg.

According to this classification, the bottom sediments of the northeastern part of the Kara Sea were moderately polluted—37% and polluted—62%.

Pollution of sea bottom sediments by petroleum hydrocarbons leads to the restructuring of benthic communities. Despite the fact that benthic organisms are more resistant to oil pollution than planktonic ones, there is a poverty of species composition with a high abundance and biomass of pollutio*n*-resistant forms. Increased HC levels can cause changes in the qualitative and quantitative characteristics of bottom communities, an increase in the relative and sometimes absolute number of detritophages, phytophages and carnivores (Nemirovskaya 2013; Sanitary and biological studies of coastal waters of the southwestern Crimea at the beginning of the XXI century 2018).

There is an opinion, that exists a clear "concentration limit" (40–60 mg/kg) of oil pollution determination, which is detected by bacterial communities (Kuznetsova and Dzyuban 2006). Below this boundary, the microbial cenoses of water-soil boundary still cope with the organic substances entering the sediments and stabilize the situation. With an excessive amount of petroleum products in sea bottom sediments, the number and biomass of bacteria increases and, due to the active transformation and degradation of petroleum hydrocarbons, their structure changes and substances more resistant to biological processing or toxic products are obtained (Kuznetsova and Dzyuban 2006).

According to these data, 87% of the samples exceeded the "concentration limit" of oil pollution. It is noted (Nemirovskaya 2015) that background concentrations of HC in marine sediments are usually below 50 mg/kg in silty and 10 mg/kg in sandy sediments. If this rule is followed, then in 68% of the samples, HC concentrations exceed background levels. It is worth noting that all of the approaches mentioned above are not decisive, they are relative in nature and do not allow us to determine the nature of petroleum hydrocarbons accurately.

The origin of HC reflects the composition of *n*-alkanes (Nemirovskaya 2013). *N*-alkanes in the range C_{16}–C_{33} have been identified in the sea bottom sediments of the northeastern part of the Kara Sea. The homologue of C_{16} was determined at

[2] URL: esdat.net/Environmental%20Standards/Dutch/annexS_I2000Dutch%20Environmental%20 Standards.pdf (date of application: 03.05.2023).

two stations—stations 1 and 5, n-alkane C_{33} was identified in all the samples except stations 6 and 7. The remaining homologues were ubiquitous.

In the composition of the n-alkanes in sea bottom sediments (Fig. 3) predominated the odd most stable homologues of n-C_{25}–C_{29} which are associated with terrestrial vegetation (Nemirovskaya 2021). The number of autochthonous n-alkanes decreases during sedimentation, as they are more susceptible to biodegradation (Belyaev and Ponyaev 2022). In addition, the LWH/HWH index varied in the range from 0.063 to 0.48 (on average 0.27) (Table 2), which demonstrated the terrigenous input of organic matter. The C_{17}/C_{25} ratio ranged from 0.05 to 0.5 (on average 0.22), which indicated the advantage of allochthonous homologues (Nemirovskaya 2021).

The distribution of n-alkanes of sea bottom sediments at individual stations is shown in Fig. 4. The predominance of high-molecular homologues was noted. The maximum of n-alkanes distribution occured to homologues C_{25}–C_{29} (Fig. 4). When compared with the calculated markers (CPI, $\Sigma C_{25-35}/\Sigma C_{15-21}$), it was seen that the terrigenous origin of n-alkanes prevailed.

The sum of homologues C_{15}–C_{21} (ΣC_{15-21}) gives an idea of autochthonous organic compounds (Blumer et al. 1971). The sum of n-alkanes C_{25}–C_{35} (ΣC_{25-35}) is mainly terrigenous substances. The ratio of these groups makes it possible to identify the origin of organic matter most accurately and at the same time exclude the influence of other factors, such as the granulometric composition of bottom sediments and sedimentation (Duan 2000).

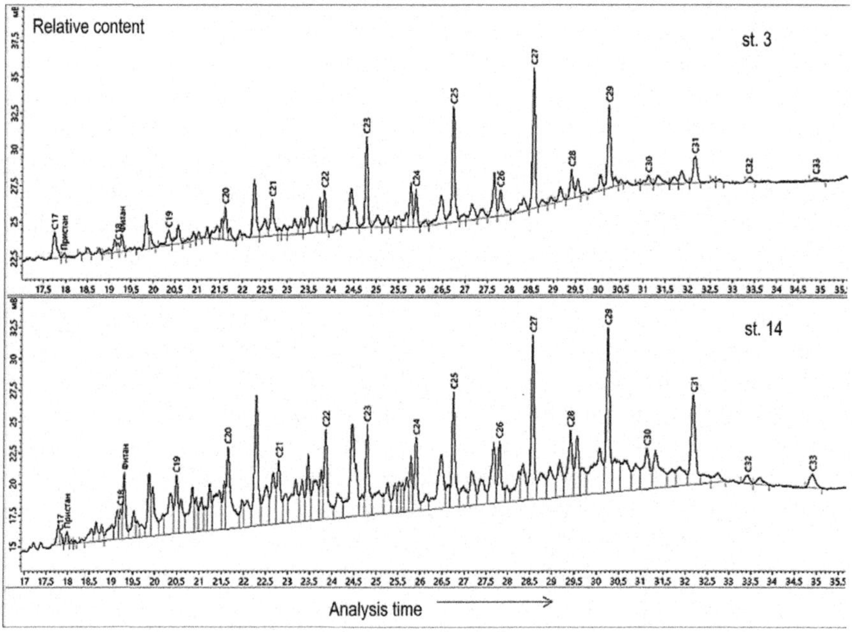

Fig. 3 Typical chromatograms of n-alkanes from sea bottoms sediments of the northeastern part of the Kara Sea, 2022

Table 2 Content of *n*-alkanes and diagnostic indices of bottom sediments of the northeastern part of the Kara Sea

Station number	D	C	LWH/HWH	Paq	TAR	ACL	TMD	Pr/Ph	$\Sigma C_{25-35}/\Sigma C_{15-21}$	CPI_1	CPI_2	C_{31}/C_{29}
1	16–33	21,24	0,45	0,625	1,73	27,45	1,36	1,15	1,55	1,23	2,01	0,31
2	17–33	28,09	0,26	0,46	4,61	27,99	2,12	0,08	2,82	0,9	1,61	0,58
3	17–33	11,45	0,23	0,625	4,28	27,31	2,01	0,23	3,23	2,05	3,59	0,42
4	17–33	43,23	0,18	0,5	11,32	27,83	2,82	0,12	4,32	1,17	3,22	0,6
5	16–33	13,69	0,39	0,45	8,85	27,87	2,72	0,6	1,85	0,4	2,12	0,46
6	17–32	12,6	0,16	0,62	4,29	27,12	2,42	0,14	4,72	3,86	2,81	0,47
7	17–32	18,4	0,24	0,66	7,69	27,09	1,78	0,09	3,01	1,49	2,17	0,36
8	17–33	21,33	0,25	0,53	6,66	27,6	2,05	0,04	3,09	1,5	1,91	0,52
9	17–33	19,12	0,32	0,54	4,69	28,15	2,04	0,48	2,4	0,9	1,89	0,64
10	17–33	26,71	0,063	0,37	8,4	28,25	4,38	0,07	12,6	7,04	1,84	0,56
11	17–33	27,56	0,18	0,51	13,81	27,61	3,09	0,25	4,16	0,85	2,43	0,48
12	17–33	14,38	0,38	0,47	4,55	27,94	1,87	0,09	1,94	1,07	3,43	0,55
13	17–33	21,38	0,48	0,52	2,13	27,77	1,28	0,05	1,51	1,69	2,25	0,6
14	17–33	46,18	0,25	0,4	7,33	28,3	2,66	0,15	3,11	0,8	2,11	0,7
15	17–33	22,66	0,25	0,43	4,8	27,88	2,66	0,22	3,26	0,97	1,99	0,46
16	17–33	30,66	0,17	0,38	10,7	28,3	3,29	0,04	4,6	1,41	2,99	0,58

Note D—range of identified *n*-alkanes; C—concentration of identified *n*-alkanes, mg/kg

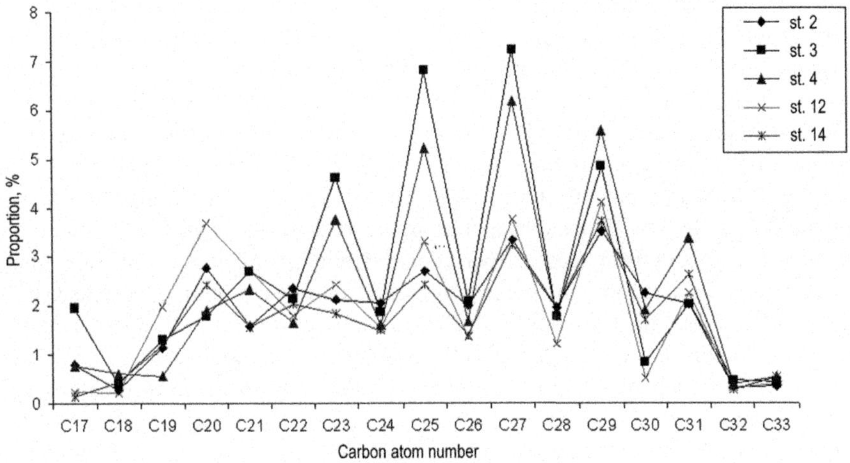

Fig. 4 Distribution of *n*-alkanes in sea bottom sediments at individual stations in the northeastern part of the Kara Sea

The ratio ($\Sigma C_{25-35}/\Sigma C_{15-21}$) above one indicates a strong influence of the terrigenous substance. In this work, the values of this marker fluctuated in the range of 1.55–12.65 (Table 2). At all stations, the index values significantly exceeded one with a maximum at station 10 ($\Sigma C_{25-35}/\Sigma C_{15-21} = 12.65$), which showed the dominance of allochthonous organic substances in the samples.

The ratio of isoprenoid alkanes (pristane to phytane), which marks the ratio of autochthonous (pristane, Pr) to petroleum (phytane, Ph) organic substances, ranged from 0.04 to 1.15 (Table 2). Judging by significant fluctuations in the ratio of the main isoprenoids, microbiological processes of organic matter transformation occur intensively in the upper layer of the marine sediments (Nemirovskaya et al. 2021b).

Another parameter that allows us to estimate the ratio of autochthonous and allochthonous *n*-alkanes is the TAR index. The values of this marker varied between 1.73 and 13.81 (Table 2), averaging 6.61. The value had obtained showed the predominance of allochthonous matter coming from land or brought with the flow of rivers, because such large rivers as the Ob and Yenisei have a direct impact on the water area (Nemirovskaya 2013).

When identifying hydrocarbons, such an important indicator as CPI is calculated. It indicates the ratio of the sum of *n*-alkanes with an odd number of carbon atoms to even ones (Nemirovskaya 2013; Valiev et al. 2020). The calculation of this index is based on the principle of predominance of odd, high-molecular weight *n*-alkanes, the source of which are plant objects. It is believed that with the predominance of petroleum hydrocarbons, the CPI value approaches one. Groups of *n*-alkanes with short chains are designated CPI_1, and those with long chains are designated CPI_2. The CPI_1 odd index calculated for lighter *n*-alkanes ranged in a wide range of 0.4–7.04 (Table 2), with an average value of 1.7. The value of the CPI_1 marker was low at stations 5, 11 and 14, at stations 2, 9, 15 the odd index approached 1, and at other stations the CPI_1 value was higher than 1. The high content of low molecular weight even *n*-alkanes in these areas of the water area may be the result of microbiological destruction of organic substances (Nemirovskaya 2013; Nishimura and Baker 1986).

The CPI_2 marker is more informative, since it is known that during the transformation of oil, light *n*-alkanes disappear first, and the content of more stable high-molecular compounds increases. Therefore, the ratio of *n*-alkanes in the high molecular weight region is more informative for determining oil pollution (Nemirovskaya 2013). CPI_2 values ranged from 1.61 to 3.59 (Table 2), averaging 2.39. At all stations, the values of this index exceeded one, therefore, it can be concluded that no oil pollution was detected in the upper layer of bottom sediments. So, in the northeastern part of the Kara Sea, the origin of hydrocarbons was biogenic. Biogenic sources of HC intake into bottom sediments are the processes of vital activity and metabolism of organisms (Nemirovskaya 2013).

Woody and grassy vegetation of the land has a great influence on the formation of the qualitative composition of hydrocarbons. The main peak associated with herbaceous vegetation falls on homologue C_{31}, and with woody vegetation—mainly on C_{29} (Huang et al. 2011). The ratio of these indicators (C_{29}/C_{31}) varied in the range of 0.31–0.7 (Table 2), averaging 0.52. Based on the data obtained, it can be concluded that woody vegetation prevailed in the bottom sediments as part of the allochthonous components.

Another index characterizing the dominance of a particular vegetation type is ACL (the average length of the carbon chain of *n*-alkanes). High values of the ACL marker indicate the predominance of herbaceous vegetation in the formation of HC, and low ACL values are characteristic of HC of woody plants (Huang et al. 2011). Low ACL values may also be an indicator of fresh oil pollution (Jeng 2006). This indicator ranged from 27.06 to 28.3 (Table 2), averaging 27.74, which reflected approximately the same contribution of woody and herbaceous plants to the formation of organic matter of sea bottom sediments of the water area. However, it should be borne in mind that this approach is applicable only to "fresh" organic matter (Nemirovskaya 2013).

The Paq index allows to find out which type of vegetation prevails in sea bottom sediments: terrigenous, submerged, floating or fresh macrophytes (Ficken 2000). This index ranged from 0.37 to 0.66 (Table 2), averaging 0.5. The calculated indicator indicates the presence of degraded vegetation in the sea bottom sediments. *N*-alkanes had formed as a result of the transformation of macrophytes were found at almost all stations.

Summarizing the above, it can be concluded that the values of the indices calculated for the sea bottom sediments of the northeastern part of the Kara Sea indicate the predominance of allochthonous hydrocarbons in the upper layer of sea bottom sediments of the water area. The presence of oil pollution was not detected at any of the investigated stations.

4 Conclusions

1. The total amount of HC in the sea bottom sediments of the northeastern part of the Kara Sea varied from 25.8 to 152.4 mg/kg, on average—68.21 $\pm$ 31.04 mg/kg. According to the classification (Uvarova 1989), sea bottom sediments were moderately polluted—37% and polluted—62%. According to the so-called «Dutch lists», the content of HC in sea bottom sediments exceed the permissible level (50 mg/kg) in 68% of samples.
2. The composition of the *n*-alkanes in sea bottom sediments was mainly dominated by high molecular weight homologues. The odd, most stable *n*-alkanes *n*-C_{25}–C_{29} predominated, which are markers of allochthonous organic matter.
3. The calculated diagnostic indices indicated the allochthonous origin of the organic matter in sea bottom sediments, the source of which could be a river runoff, which has a direct impact on the water area. Variability in the distribution of hydrocarbons in sea bottom sediments of the northeastern part of the Kara Sea was mainly due to natural processes.

Acknowledgements The sea cruise was carried out with the support of the Ministry of Science and Higher Education of the Russian Federation within the framework of: the plan-program of expeditionary research of Lomonosov Moscow State University on the topic "Features of Quaternary sedimentogenesis, relief formation and natural fluid discharge on the seabed in the northeastern part of the Kara Sea" and "Training through research on the Arctic shelf"; state assignment of the Moscow Institute of Physics and Technology (National Research University) "Additional support of the system in the field of marine sciences is the training of a young personnel reserve under the scientific and educational program "Floating University" on the basis of comprehensive studies of the seas of Russia and the World Ocean, the state assignment of the IO RAS "Ensuring scientific research, as well as experimental developments". The sample processing and chemical analysis was carried out within the framework of IBSS state research assignment "Study of biogeochemical patterns of radioecological and chemoecological processes in the ecosystems of water bodies of the Sea of Azov–Black Sea Basin in comparison with other areas of the World Ocean and individual aquatic ecosystems of their drainage basins to ensure sustainable development in the southern seas of Russia" (No. 1023032000047-8-1.6.19).

References

Aksenov PV, Ivanov VV (2018) "Atlantification" as a probable reason for the reduction of sea ice in the Nansen basin in the winter season. Probl Arctic Antarctic 64(1):42–54 (115)

Alekseeva TA, Serovetnikov SS, Frolov SV, Sokolov VT (2018) Ice conditions of swimming in the Arctic basin in the summer of 2018. Russian Arctic, No. 2, pp 31–40

Belyaev NA, Ponyaev MS (2022) Organic matter of the boundary zones of the Kara Sea. The ecosystem of the Kara Sea—new data from expeditionary research: conference proceedings. Moscow: APR, pp 227–231

Blumer M, Guillard RRL, Chase T (1971) Hydrocarbons of marine phytoplankton. Mar Biol 8(3):183–189. https://doi.org/10.1007/BF00355214

Duan Y (2000) Organic geochemistry of recent marine sediments from the Nansha Sea, China. Org Geochem 31(2–3):159–167. https://doi.org/10.1016/S0146-6380(99)00135-7

Drugov YS, Rodin AA (2020) Environmental analyses in oil and petroleum product spills: a practical guide. M.: Laboratory of Knowledge, 273 pp

Ficken KJ (2000) An *n*-alkane proxy for the sedimentary input of submerged floating freshwater aquatic macrophytes. Org Geochem 31:745–749

Flint MV, Poyarkov SG, Rimskii-Korsakov NA, Miroshnikov AY (2019) Ecosystems of the Siberian Arctic Seas 2018 (cruise 72 of the R/V Akademik Mstislav Keldysh). Oceanology 59(3):460–463

Flint MV, Poyarkov SG, Rimsky-Korsakov NA et al (2020) Ecosystems of the seas of the Siberian Arctic—2019: spring processes in the Kara Sea (76th voyage of the scientific research vessel Akademik Mstislav Keldysh). Oceanology 60(1):154–157

Flint MV, Poyarkov SG, Rimsky-Korsakov NA, Miroshnikov AY (2021) Ecosystems of the seas of the Siberian Arctic—2020: the Kara Sea (81st voyage of the scientific research vessel Akademik Mstislav Keldysh). Oceanology 61(2):330–333

Huang X (2011) Significance of long chain iso and anteiso monomethyl alkanes in the Lamiaceae (mint family). Org Geochem 42(2):156–165. https://doi.org/10.1016/j.orggeochem.2010.11.008

Jeng WL (2006) Higher plant *n*-alkane average chain length as an indicator of petrogenic hydrocarbons contamination in marine sediments. Mar Chem 102(3–4):242–251. https://doi.org/10.1016/j.marchem.2006.05.001

Kazennov AY, Nersesov BA, Rimsky-Korsakov NA (2018) Expeditionary studies of the ecology of the seas of the Russian Arctic M.: FSBI "Analytical Center" of the Ministry of Education and Science of Russia, p 307

Kontorovich AE (2015) Ways of developing oil and gas resources of the Russian Arctic sector. Bull Russian Acad Sci 85(5/6):420–430

Kontorovich VA, Kontorovich AE (2019) Geological structure and prospects of oil and gas potential of the Kara Sea shelf. Rep Acad Sci 489(3):272–276

Kotova VE (2020) Determination of petroleum products components in bottom sediments by high-performance liquid chromatography with spectrofluorimetric detection: abstract of the dissertation for the degree of Candidate of Chemical Sciences. Voronezh: Publishing House of the Federal State Budgetary Institution "Hydrochemical Institute", 21 pp

Kravchishina MD, Novigatsky AN, Savvichev AS, Pautova LA, Lisitsyn AP (2019) Investigation of sedimentation systems of the Barents Sea and the Norwegian-Greenland basin in the 68th voyage of the scientific research vessel Akademik Mstislav Keldysh. Oceanology 59(1):167–169

Kuznetsova IA, Dzyuban AN (2006) Bacteriobenthos communities under the influence of oil pollution of reservoirs. Hydrobiol J 42(5):88–92

L(u) X, Zhai S (2008) The distribution and environmental significance of n-alkanes in the Changjiang River estuary sediments. Acta Sci Circum 28(6):1221–1226

Mironov OA, Mironov OG (2020) Modern data on pollution of the coastal waters of the Azov-Black Sea region of Russia with petroleum hydrocarbons. South Russia Ecol Dev 15(3):77–85. https://doi.org/10.18470/1992-1098-2020-3-77-85

Nemirovskaya IA (2013) Oil in the ocean (pollution and natural flows). In: Nemirovskaya IA, Lisitsyn AP (eds) Scientific World, Moscow p 432. https://www.rfbr.ru/rffi/ru/books/o_1917147

Nemirovskaya IA (2015) Variability of hydrocarbon concentrations and composition in the frontal zones of the Kara Sea. Oceanology 55(4):552–562

Nemirovskaya IA (2021) Distribution and origin of hydrocarbons in the transarctic section through the seas of Siberia. Oceanology 61(2):209–219

Nemirovskaya IA, Rejepova ZYu, Lisitsyn AP (2019) Hydrocarbons of surface waters in the Transarctic section. Rep Acad Sci 483(3):345–349 (DAY). https://doi.org/10.31857/S0869-56524863345-34

Nemirovskaya IA, Khramtsova AV, Khalikov IS, Koltovskaya EV, Solomatina AS (2021a) Hydrocarbons in water and sediments of the Norwegian and Barents Seas. In: Proceedings of the Karelian Scientific Center of the Russian Academy of Sciences, no. 4, pp 94–107. https://doi.org/10.17076/lim1382

Nemirovskaya IA, Titova AM, Khramtsova AV (2021b) Hydrocarbons in the water column and bottom sediments of the Barents Sea. Barents Sea system.—M.: GEOS pp 531547. https://doi.org/10.29006/978-5-6045110-0-8/(35)

Nishimura M, Baker EW (1986) Possible origin of n-alkanes with a remarkable even-to-odd predominance in recent marine sediments. Geochimica Cosmochimica Acta 50(2):299–305. https://doi.org/10.1016/0016-7037(86)90178-X

Sanitary and biological studies of coastal waters of the southwestern Crimea at the beginning of the XXI century (2018) In: Mironova OG, Alimova SV (eds) Institute of Marine Biological Research named after A.O. Kovalevsky Ranu—Simferopol: IT "ARIAL", 276 pp

Silliman JE, Schelske CL (2003) Saturated hydrocarbons in the sediments of Lake Apopka, Florida. Org Geochem 34(2):253–260. https://doi.org/10.1016/S0146-6380(02)00169-9

Soloveva OV, Tikhonova EA, Gurov KI, Kotelyanets EA (2022) Hydrocarbons composition of sea bottom sediments (Balaklava Bay, Black Sea). Int J Environ Sci Technol. https://doi.org/10.1007/s13762-022-04167-y

Tikhonova EA (2021) Organic matter of bottom sediments of the Crimean and Caucasian coasts (Azov and Black Seas). Environ Safety Coastal Offsh Zones Sea 3:52–67. https://doi.org/10.22449/2413-5577-2021-3-52-67

Tikhonova EA et al (2020) Sanitary and biological characteristics of the coastal water area of the landscape reserve "Laspi" (Black Sea). In: Environmental safety of the coastal and shelf zone of the sea, vol 3. pp 95–106. https://doi.org/10.22449/2413-5577-2020-3-95-106

Tikhonova EA, Soloveva OV, Burdiyan NV (2021) Assessment of the pollution by organic substances of water and sea bottom sediments of the Kerch Strait and the adjacent Azov-Black

Sea water area. Progress in GeoMedia. Cham, Switzerland. Springer Nature Switzerland AG, vol 3, pp 285–293. https://doi.org/10.1007/978-3-030-69040-3_27

Uvarova VI (1989) The current state of the pollution level of waters and soils of the Ob-Irtysh basin. Collection of scientific papers of the State Research Institute, issue 305, pp 23–33

Valiev VS, Ivanov DV, Shagidullin RR (2020) Petroleum hydrocarbons in bottom sediments: composition, identification, mechanisms of transformation (review). Russian J Appl Ecol 1(21): 41–51

Zhang S et al (2014) An analysis of organic matter sources for surface sediments in the central South Yellow Sea, China: evidence based on macroelements and *n*-alkanes. Mar Pollut Bull 88(1–2):389–397. https://doi.org/10.1016/j.marpolbul.2014.07.064

Temperature Regime of the Snow Surface of Ski Slopes on the Example of the Ski Resort "Mountain Air" (Yuzhno-Sakhalinsk)

G. A. Rzhanitsyn and D. M. Frolov

Abstract The snow layer of a ski slope is a dynamic system that changes its structure depending on the temperature gradient. The spring period is characterized by large temperature changes, which affects the condition of the snow. Compacting snow on ski slopes is a complex task that requires significant funds. Recommended snow density parameters for safe use of pistes are 300–600 kg/m^3. The temperature of the snow is controlled to plan work on the slopes. Observations using loggers have shown that temperatures strongly depend on the depth of snow and underlying soils. Temperature gradients in snow horizons affect its thermal resistance. Simple mechanical compaction of snow is advisable in the evening, and soaking the slope with water in the morning. The use of water to harden snow must be regulated so as not to disturb the temperature field of the snow.

Keywords Snow sheet · Ski slope · Compaction · Flow · Density · Temperature regime · Temperature gradient · Thermometry

The snow column is a very dynamic system, which, depending on environmental conditions, changes its structure in all horizons, from the upper to the lower horizons. The temperature gradient has the greatest influence on the processes occurring in snow (Fierz et al. 2009; Heilig et al. 2008, 2009, Kotlyakov et al. 2017, Marshall et al. 2007, Rototaeva et al. 2019, Sokratov and Frolov 2023). In natural conditions, the spring period is most often characterised by large daily temperature differences in the atmospheric air. Accordingly, the snow layer is exposed to periodic cold waves at night and heat waves during the day. Monitoring of snow temperature regime is important for planning works on slopes.

Compaction of snow on the slopes of winter resorts for sporting events is a complex task that requires considerable resources from the organisers. The spring period is the most comfortable for the use of snow-covered slopes, but it is not easy to maintain the snow surface in the parameters permissible for use. The recommended snow

G. A. Rzhanitsyn · D. M. Frolov (✉)
Department of Geography, Lomonosov Moscow State University, Moscow, Russia
e-mail: denisfrolovm@mail.ru

T. Chaplina (ed.), *Processes in GeoMedia—Volume VIII*, Springer Geology,
https://doi.org/10.1007/978-981-97-6627-7_18

Fig. 1 Location of the slopes of the ski resort "Mountain Air"

parameters for safe use of snow slopes are mainly related to snow density. For event organisers and resort operators, the recommended values are 300–600 kg/m^3. Such values can be achieved by using snow compaction equipment and by adding water from snowmaking systems. Changing snow temperature in the daily regime can significantly change the result of the work carried out, both positively and negatively.

In order to control daily temperature dynamics in the upper horizons of the snow cover, field observations were carried out using loggers taking readings every minute. The work was carried out from 5 to 17 March 2024 on the slopes of the ski resort "Mountain Air" (Yuzhno-Sakhalinsk) (Fig. 1). Two measurement points were chosen at the summit of Mt Bolshevik (601 m a.s.l.). One on natural snow cover (loggers 2, 3) and the other on a compacted section of the ski track (loggers 4, 5, 6). Air temperature was recorded in parallel (logger 1).

By the beginning of the observation period, the thickness of natural snow cover was 53 cm and varied within 5–6 cm depending on precipitation, temperature regime and wind load. At Point 2, the thickness of compacted snow was 1.5 m (Fig. 2).

Loggers at point 1 were set at depths of 20 cm (no. 2) and 40 cm (no. 3).

Loggers at point 2 were set at depths of 25 cm (no. 4), 40 cm (no. 5) and 50 cm (no. 6).

The data of logger temperature values are presented in Fig. 3

The average air temperature during the observation period was – 7.54 °C. The maximum daily amplitude of air temperatures was 14.2 °C from 09 to 10 March

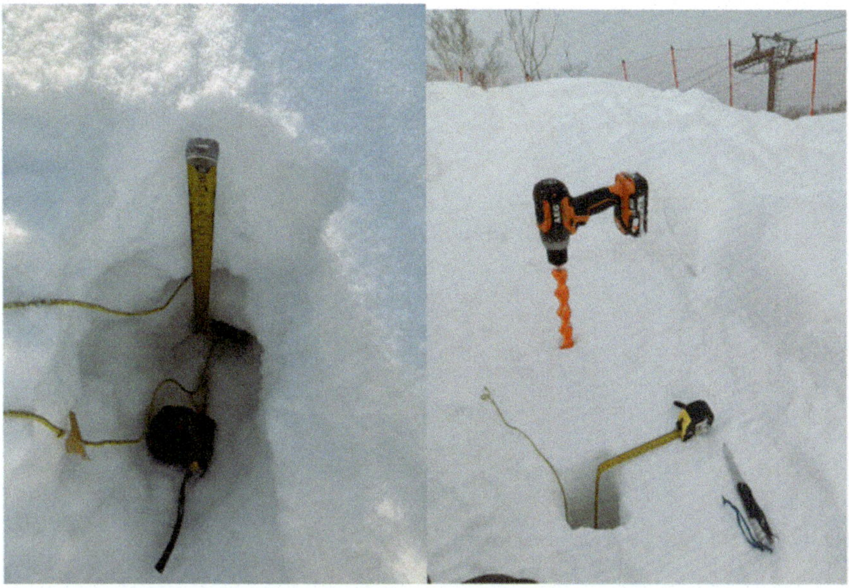

Fig. 2 Location of loggers in the snowpack

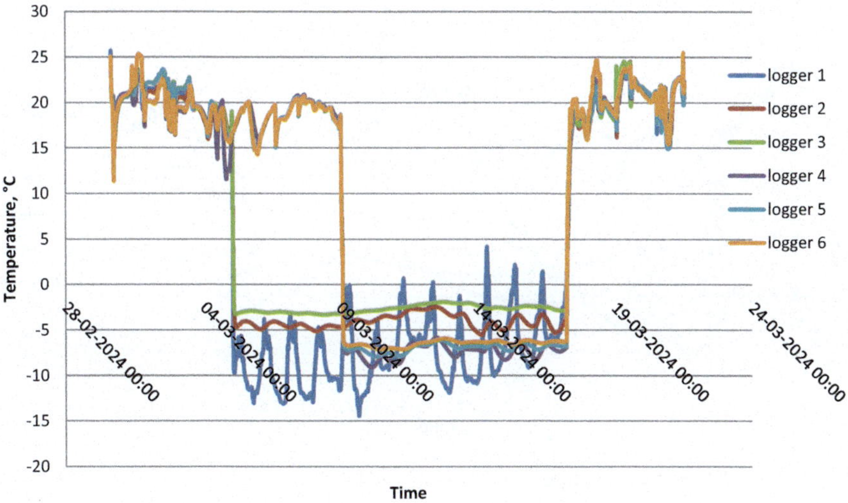

Fig. 3 Data of logger temperature values

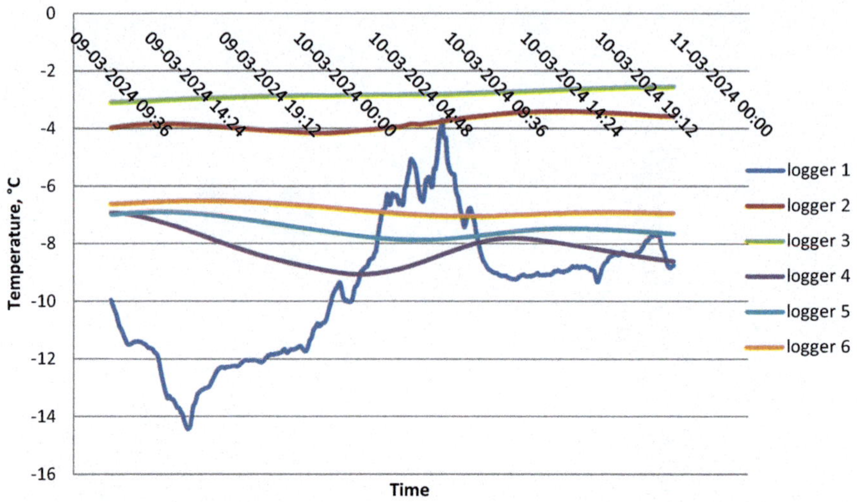

Fig. 4 Data of logger temperature values 9–10.03.2024

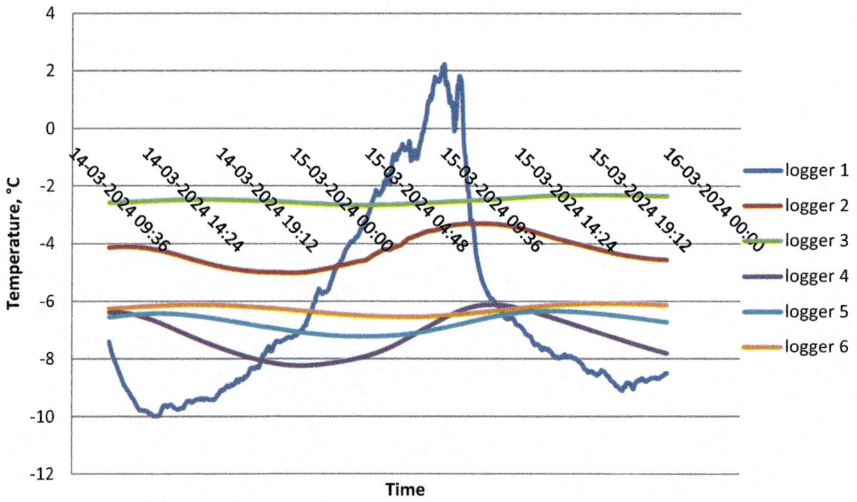

Fig. 5 Data of logger temperature values 14–15.03.2024

(Fig. 4). At stable transition of daytime temperatures through the plus mark, we also noted large amplitudes. Thus, 14.25 °C was recorded from 14 to 15 March (Fig. 5).

Daily air temperature fluctuations are known to penetrate into the snow thickness to a depth of 30–40 cm, while the temperature regime of deeper snow layers is determined mainly by long-period fluctuations of various natural factors, such as the movement of atmospheric fronts, insolation regime, cloudiness, etc.

The density of snow directly affects its thermal resistance. Snow density at site 1 was 275 kg/m^3, compacted snow at site 2 was characterised by values of 483 kg/m^3. Temperature gradients at horizons from 20 to 40 cm were 0.069 deg/cm and 0.028 deg/cm, respectively.

As the snow cover thickness increases, the temperature gradient decreases naturally, except for the very near-surface snow layers, whose temperature state is determined by daily fluctuations in air temperature and humidity, as well as by the radiation balance regime.

The gradient value of 0.2 deg/cm is considered to be a threshold value, above which mass transfer processes with loosening of horizons and formation of deep frost actively develop in the snow thickness. In our case, we neglect the consideration of these processes due to their insignificance in the stability of artificial snow cover.

The 20–40 cm deep snow horizon carries the main load, constituting the "Base"—a horizon that must have increased strength properties to resist mechanical loads. The layer above 0–20 cm is too variable in the daily regime, it is transformed daily both under the load of natural and anthropogenic factors (skis, transport mechanisms). The layer below > 40 cm is difficult to reach without specialised equipment.

The temperature regime of the Base Layer is of the greatest interest when performing methods of its artificial compaction.

It can be seen that temperatures strongly depend on the thickness of the snow column. The influence of the underlying soils is strong. The total cold reserve for the stages preceding the measurements forms a temperature field that varies only slightly in the diurnal regime.

Observing the dynamics of the cold wave of temperatures in the snow thickness, one can see that it follows the atmospheric air thermometer value with a delay of 10–12 h. The minimum values in the Base layer are observed around 8–12 a.m. of the next day.

The cold wave is more pronounced in the compacted stratum. The amplitude at a depth of 25 cm is about 2 °C (Fig. 6).

The warm wave carries the maximum daily temperatures to the Base layer with a delay of 3–6 h. The maximums are observed around 20–23 h.

Accordingly, simple mechanical compaction is advisable in the evening hours, while slope flooding using water is better to be done in the morning, before lunchtime. The onset of daytime plus temperatures shows us more active dynamics of horizon temperatures. At the same time, it is necessary to follow the time schedule of snow works more strictly.

In addition, the use of water for snow hardening should be regulated. Uncontrolled shedding, even with water temperatures of 2–4 °C, can completely destroy the snow temperature field. Recovery of the "Base" horizon due to short-term night minuses, especially in the spring period, is difficult.

The work was carried out in accordance with the state budgetary theme "Evolution of cryosphere under climate change and anthropogenic impact" (121051100164-0), "Danger and risk of natural processes and phenomena" (121051300175-4).

Fig. 6 Compaction of the snow bed by slope shedding

References

Fierz C, Armstrong RL, Durand Y, Etchevers P, Greene E, Mcclung DM, Nishimura K, Satyawali PK, Sokratov SA (2009) The international classification for seasonal snow on the ground. IACS Contribution No. 1 International Association of Cryospheric Sciences. 83 IHP-VII Technical Documents in Hydrology

Heilig A, Schneebeli M, Fellin W (2008) Cold Reg Sci Technol 51(2–3):178–190. https://doi.org/10.1016/j.coldregions.2007.06.003

Heilig A, Schneebeli M, Eisen O (2009) Cold Reg Sci Technol 59(2–3):152–162. https://doi.org/10.1016/j.coldregions.2009.07.008

Kotlyakov VM, Macheret YY, Sosnovsky AV, Glazovsky AF (2017) Ice Snow 57(1):45–56. https://doi.org/10.15356/2076-6734-2017-1-45-56

Marshall HP, Schneebeli M, Koh G (2007) Cold Reg Sci Technol 47(1–2):108–117. https://doi.org/10.1016/j.coldregions.2006.08.008

Rototaeva OV, Nosenko GA, Kerimov AM, Kutuzov SS, Lavrentiev II, Nikitin SA, Kerimov AA, Tarasova LN (2019) Ice Snow 59(1):5–22. https://doi.org/10.15356/2076-6734-2019-1-5-22

Sokratov SA, Frolov DM (2023) Geophysical research in the preparation of ski slopes for competitions: case study of the Garabashi glacier (Elbrus, Caucasus). In: XV international scientific conference “INTERAGROMASH 2022”, vol 575, pp 2686–2693

Experimental Study of the Rotation of Ice Markers on the Water Surface

T. O. Chaplina and V. P. Pakhnenko

Abstract Laboratory experiments confirmed the phenomenon of spontaneous rotation of melting ice on the water surface. It was found that the rotation speed increases with the height of the liquid. Using the technique of underpainting, the fact of vortex flow occurrence directly under the ice block in experiments on placing the ice block on the free surface of the resting liquid was established. Due to internal structural inhomogeneities of ice caused by various reasons, local areas called melting centres are formed on its surface, near which the melting process occurs more intensively than on average over the surface. It is shown that the presence of several melting centres can give the ice disc not only translational but also rotational motion. The results of the experiments will be important, first of all, for a better understanding of the formation of convective currents arising in the process of ice melting, and can be used to predict the thermal and dynamic regime of oceans, seas, and lakes.

Keywords Ice · Markers · Convection · Flows · Vortex · Experimental study

1 Introduction

The formation of flows and mixing processes in water bodies during different seasons is a crucial area of hydrophysical research. Recently, the study of ice melting dynamics has gained more attention due to the growth of oil production and transportation in high latitudes, such as the Arctic and Antarctic. A complete understanding of the formation of convective vortices caused by ice melt is still lacking, despite the large number of studies (Kouraev et al. 2016; Zyryanov et al. 2022; Heard 1977; Dorbolo et al. 2016a, b). Some hypotheses formulated in theoretical works have not yet been confirmed experimentally, due to the difficulty of describing them theoretically and carrying out the corresponding experiments.

T. O. Chaplina · V. P. Pakhnenko (✉)
Ishlinsky Institute for Problems in Mechanics of the Russian Academy of Sciences, Moscow, Russia
e-mail: pvpmos@gmail.com

T. Chaplina (ed.), *Processes in GeoMedia—Volume VIII*, Springer Geology,
https://doi.org/10.1007/978-981-97-6627-7_19

Fig. 1 Rotating ice disks on the surface of the water (https://www.newscientist.com/article/2191040-huge-spinning-ice-disc-in-river-provides-a-carousel-for-ducks/, https://www.cbc.ca/player/play/1684185667952)

An interesting phenomenon can be observed on frozen freshwater bodies—the rotation of ice blocks of various sizes, from small disks to huge panels that can reach hundreds of meters in diameter. Researchers report that the angular speeds of rotation of these ice formations are about several degrees per second (Chaplina and Glebova 2023). For many years, scientists have been paying special attention to this phenomenon, trying to understand its causes and mechanisms (Fig. 1).

The first documented observation of this phenomenon was made almost 130 years ago, in 1895, by a researcher named Bates. Since then, natural ice disks have been photographed on the planet several times in the twentieth century (Bates 1895). However, few researchers have detailed the conditions under which these disks were discovered, as well as their dimensions. Numerous studies present an analysis of the formation of convective currents that arise when ice melts. In one of the works carried out by S. N. Dikarev, laboratory experiments were carried out to study free small-scale convection under ice cover (Dikarev et al. 2004). The structure of this convection consists of isolated "salts" and jets that form stalactites and create a salt stratification of the water layer under the ice. Conditions for the formation of congelation and subaquatic ice have also been recorded. The results of laboratory experiments were applied to real observations of thermohaline processes.

The rotation mechanism of ice disks has not been thoroughly studied. Observations indicate that a distinct downward spiral flow forms below the ice-water interface when an ice block is placed on the water's surface. Subglacial hydrological measurements were conducted to determine the currents and thermal structure of water in the area of the rings. The results showed that in the geostrophic region of the lake under the ring, there is an anticyclonic eddy that delivers warmer deep water to the lower surface of the ice, leading to melting of the ice from below. However, the reason why the ice melts from below along the ring remains unknown.

Currently, there is no agreement on the mechanism behind the self-induced rotation of ice objects. This chapter presents the results of measuring the rotational speed of an ice disk on the surface of initially stationary fresh and salt water of varying depths.

2 Experimental Research Methodology

The study of flows resulting from the placement of an ice marker on the free surface of a resting fluid requires the consideration of a number of factors, such as the composition and mineralisation of ice and water, temperature, salinity, shape and size of the ice marker, roughness of its surface, available underwater and surface currents. By conducting the experiment under laboratory conditions, it is possible to exclude or minimise the influence of most of the weakly controlled factors and to study a simplified problem by reducing its thermodynamic part to the problem of free convection under a smooth interface. Experiments were carried out in containers of different shapes (cubic, cylindrical, triangular, square).

Ice pieces of different shapes and sizes were used as ice markers (Fig. 2).

Ice markers with frozen marker indicators were placed on the free surface of both fresh and salt water (35‰). All experiments were conducted at ambient water and air temperatures (20 °C), and the pattern of emerging movements was recorded using a video camera mounted on a tripod.

By placing an ice marker on the surface of both fresh and salty water, we can observe the rotation of the ice, the speed of which depends on both the depth of the water layer and the size of the marker. For example, in the experiments, the rotation of the ice block practically stopped at a depth of 2 cm. To visualise the flow in the liquid column beneath the block of ice placed on the surface of the water, a dye was added to the water during freezing—its release as the ice melts not only serves as an indicator of the flow (Fig. 3), but also reveals its structure—downward jets of cold meltwater under the action of gravity.

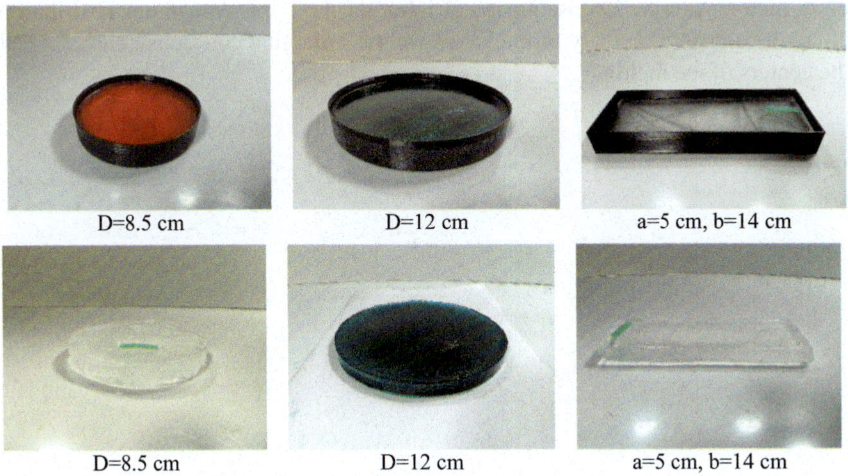

Fig. 2 Photographs of the ice markers used in the experiments

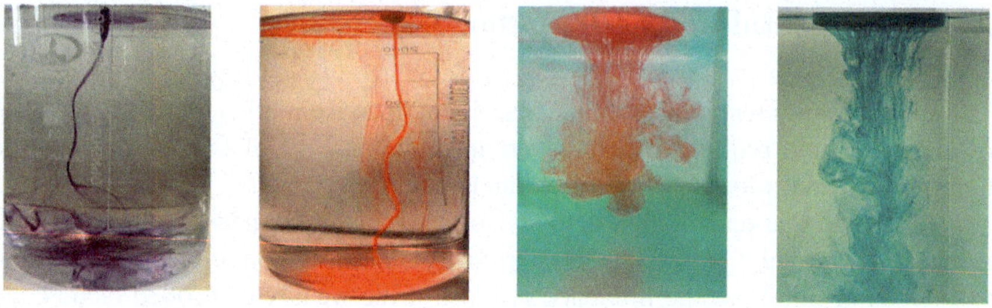

Fig. 3 Flow under the ice disc

3 Results and Discussion

Figures 4 and 5 present the results of experimental studies of the ice melting process and consequently the formation of vortex currents, when using cylindrical ice markers (Fig. 2) in a cylindrical container ($H_c = 50$ cm, $D_c = 30$ cm) Fig. 4 and cubic container ($A_c = 50$ cm) Fig. 5.

Thus, laboratory experiments confirmed the phenomenon of spontaneous rotation of melting ice on the water surface. It was found that the rotation speed increases with the height of the liquid. Using the technique of underpainting, the fact of vortex flow occurrence directly under the ice block in experiments on placing the ice block on the free surface of the resting liquid was established.

The rotation of an ice disc placed in resting water is caused by currents arising in water as a result of several physical effects. There are two mechanisms that can set the fluid in motion in the problem in the considered formulation: inhomogeneity of surface tension as a consequence of inhomogeneity of temperature distribution along the contact boundary of media, as well as emerging convective currents near the centers of ice melting.

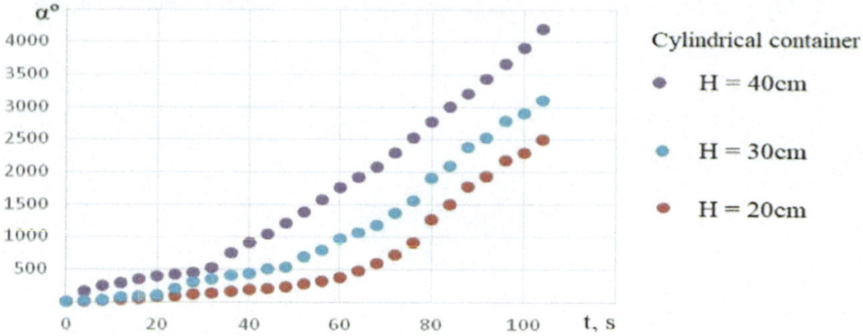

Fig. 4 Dependence of angular position of ice markers ($D = 8.5$ cm, $h = 1$ cm) on time at different depths in fresh water in a cylindrical container

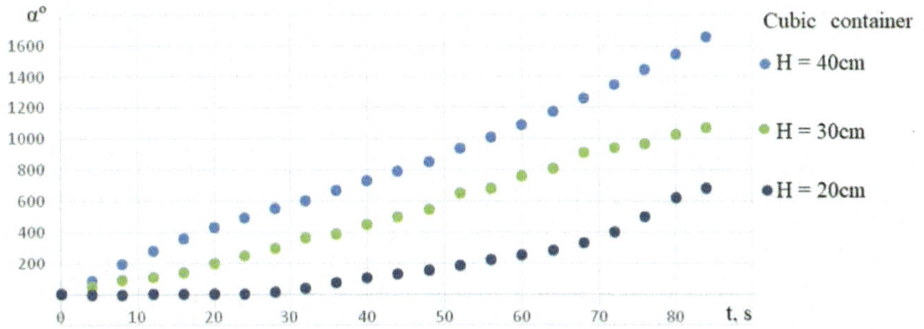

Fig. 5 Dependence of angular position of ice markers (D = 5.5 cm, h = 1 cm) on time at different depths in fresh water in a cubic container

$$z = h < 0 \; r = a \; b = c$$

Let's consider the ice marker to be perfectly smooth, the bottom of the marker is located at the level of $h = h_c(t) < 0$ (water level—$h = 0$), the location of the side wall of the ice marker will be determined by the diameter of the disc $D(t)$. We use a cylindrical coordinate system (a, φ, h), axis h coincides with the vertical axis passing through the centre of the disc (a—ice disc radius).

At initial moments of time, when the rotation of the disc around its own axis and its displacement along the water surface are just beginning to develop, it is acceptable to use approximate sticking conditions.

$$v|_{h=h_c(t)} = \dot{h}_c(t)e_h,\; v|_{a=A(t)} = \dot{A}(t)e_a,\; v = v_a e_a + v_\varphi e_\varphi + v_h e_h \tag{1}$$

where e_i, v_i—are unit orthants of the cylindrical coordinate system and the corresponding components of the velocity field, the point above the symbols denotes the time derivative.

Ice is solid and there is no motion of the medium in it, considering dynamic boundary conditions on the surface of viscous water, as well as approximate sticking conditions (1) on the bottom and on the side surface we can write down

$$\eta\frac{\partial v_r}{\partial h}\bigg|_{h=h_c} = -\frac{\partial\sigma}{\partial a},\; \eta\frac{\partial v_\varphi}{\partial h}\bigg|_{h=h_c} = -\frac{1}{a}\frac{\partial\sigma}{\partial\varphi},\; a \in [0, A(t)]$$
$$\eta\frac{\partial v_h}{\partial a}\bigg|_{a=A} = -\frac{\partial\sigma}{\partial h},\; \eta\frac{\partial v_\varphi}{\partial a}\bigg|_{a=A} = -\frac{1}{A}\frac{\partial\sigma}{\partial\varphi},\; h \in [h_c, h_m] \tag{2}$$

where z_m—level of the meniscus on the lateral surface of the ice, η—first dynamic viscosity of water, $\sigma = \sigma(T)$—surface tension coefficient at the "water–ice" boundary, depending on water temperature T.

If we consider that the total rotational momentum is determined only by the inhomogeneity of the surface tension coefficient distribution along the ice-water interface, we can write down

$$M = -e_h \left[\int_0^A \left[ae_a \times \frac{e_\varphi}{a} \right] I_\varphi ada + A \int_{h_b}^{h_m} \left[Ae_a \times \frac{e_\varphi}{A} \right] I_\varphi \, dh \right]$$

If the temperature distribution along the water–ice boundary is continuous, the surface tension distribution is also continuous.

$$I_\varphi = \int_0^{2\pi} \sigma'_\varphi d\varphi = 0, M = 0$$

Thus, the temperature dependence of the surface tension coefficient at the water–ice boundary can only lead to displacement (due to inhomogeneity of radial viscous stresses at the bottom), but not to rotation of the ice disc. That is, the rotation of the ice disc is caused by convection processes (Chaplina and Kistovich 2022).

Due to internal structural inhomogeneities of ice caused by various reasons, local areas called melting centres are formed on its surface, near which the melting process takes place more intensively than on the average of the surface. This fact is confirmed by experiments (Chaplina and Glebova 2023) investigating the melting of the lower surface of floating ice. As a result, the melting pattern that appears at the lower boundary of the disc after ice-water contact is a layer of meltwater and jets descending from it, generated by ice melting centres. These jets entrain the surrounding water in their motion, creating viscous stresses at this boundary that can, under certain conditions, set the ice disc in motion.

4 Conclusion

Thus, laboratory experiments confirmed the phenomenon of spontaneous rotation of melting ice on the water surface. It was found that the rotation speed increases with the height of the liquid. Using the technique of underpainting, the fact of vortex flow occurrence directly under the ice block in experiments on placing the ice block on the free surface of the resting liquid was established.

Due to internal structural inhomogeneities of ice caused by various reasons, local areas called melting centres are formed on its surface, near which the melting process occurs more intensively than on average over the surface.

It is shown that the presence of several melting centres can give the ice disc not only translational but also rotational motion. The direction of rotation depends on the sign of the total momentum, which is determined by the mutual location of the melting centres, and can take both positive and negative values.

The results of the experiments will be important, first of all, for a better understanding of the formation of convective currents arising in the process of ice melting, and can be used to predict the thermal and dynamic regime of oceans, seas, and lakes.

Acknowledgements This work was supported by the Russian Science Foundation, project No. 24-27-00029.

References

Bates JMA (1895) Revolving ice cake. Sci Am XXXVI(13)

Chaplina TO, Glebova OA (2023) Investigation of vortex flows formed when ice melting in multi-component liquids. In: Karev VI (ed) Physical and mathematical modeling of earth and environment processes—2022. PMMEEP 2022. Springer proceedings in earth and environmental sciences. Springer, Berlin, Cham. https://doi.org/10.1007/978-3-031-25962-3_62

Chaplina TO, Kistovich AV (2022) Self-induced rotation of an ice disk on the surface of water. Doklady Earth Sci 505(1):501–505

Dikarev SN, Poyarkov SG, Chuvilchikov SI (2004) Laboratory modeling of small-scale convection under growing ice cover in the winter arctic waters. Oceanology 44(1):70–79

Dorbolo S, Adami N, Dubois C et al (2016) Rotation of melting ice disks due to melt fluid flow. Phys Rev E 93:033112. https://doi.org/10.1103/PhysRevE.93.033112

Dorbolo S, Vandewalle N, Darbois Texier B (2016) Spontaneous rotation of an ice disk while melting on a solid plate. Phys Fluids 28(12):123601. https://doi.org/10.1063/1.4967399

Heard WB (1977) Steady-state convection with melting at a boundary. Phys Fluids 20(12):1993–1999

https://www.cbc.ca/player/play/1684185667952

https://www.newscientist.com/article/2191040-huge-spinning-ice-disc-in-river-provides-a-carousel-for-ducks/

Kouraev AV, Zakharova EA, Rémy F, Kostianoy AG, Shimaraev MN, Hall NM, Suknev AY (2016) Giant ice rings on Lakes Baikal and Hovsgol: inventory, associated water structure and potential formation mechanism. Limnol Oceanogr 61:1001–1014. https://doi.org/10.1002/lno.10268

Zyryanov VN, Kuraev AV, Kostyanoy AG (2022) Ice rings of Baikal: observations, hypotheses, theory. Water Resour 49(2):123–131. https://doi.org/10.31857/S0321059622020122

Changes in the Depth of Soil Thawing in the Russian Arctic in Recent Decades

D. M. Frolov, A. V. Koshurnikov, and V. E. Gagarin

Abstract The on-going climate warming has made 2023 the warmest year on record. According to the Japanese reanalysis JRA-552023, the air temperature anomaly in the Northern Hemisphere was + 0.72and + 1.09 °C in the Arctic compared to the same values in the baseline observation period 1991–2020. Previously, 2016 was considered the warmest year. This could not but bring changes in the values of ground thawing depth in the Russian Arctic in recent decades and in the overall permafrost thawing. The paper provides a quantitative assessment of these events.

Keywords Thawing depth · STS · Climate change

According to known data, about 65% of the territory of the Russian Federation and most of the Russian Arctic is occupied by permafrost. The ongoing climate change could not but bring changes in the values of ground thawing depths in the Russian Arctic in recent decades and in the general thawing of permafrost.

Variations in climatic conditions, snow cover thickness and associated ground freezing depth also lead to permafrost thawing and damage to structures and pipelines, and slope landslides. In particular, due to climate change over the last decade, at least in the Arctic, there has been an increase in ground temperature and an increase in seasonal thawing. The combined effect of changes in air temperature and snow cover thickness can result in a few degrees of additional heating (+ 2 + 3 °C). This is not critical for permanently frozen ground with temperatures of − 8–10 °C, but for permanently frozen ground with temperatures of − 3–5 °C and above, it can lead to thawing and an increase in the seasonally active layer.

According to temperature measurements in a borehole in the vicinity of Norilsk, the permafrost temperature at a depth of 6–18 m is − 2 °C.

This means that the permafrost temperature there is quite high and the permafrost there is quite vulnerable.

D. M. Frolov (✉) · A. V. Koshurnikov · V. E. Gagarin
Lomonosov Moscow State University, Moscow, Russia
e-mail: denisfrolovm@mail.ru

T. Chaplina (ed.), *Processes in GeoMedia—Volume VIII*, Springer Geology,
https://doi.org/10.1007/978-981-97-6627-7_20

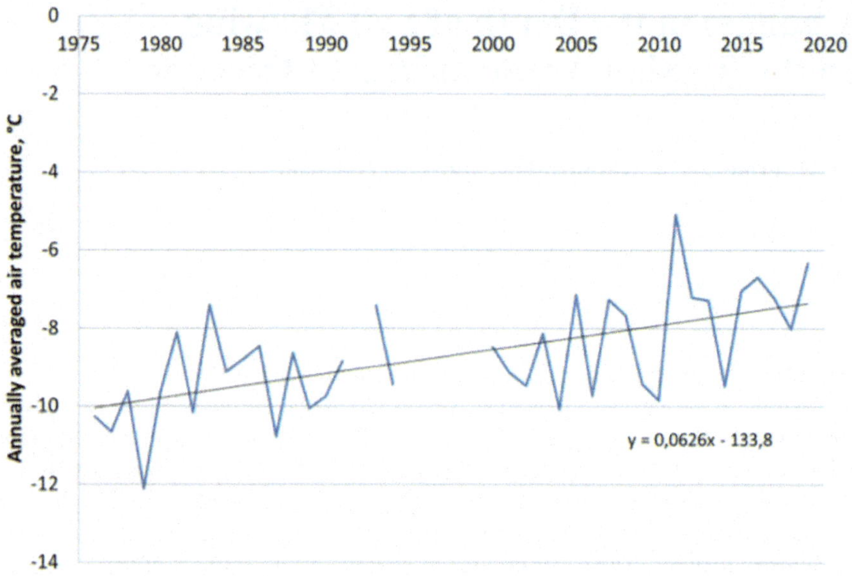

Fig. 1 The annually average air temperature of Norilsk for 1975–2020

The average annual air temperature according to the data of the Norilsk weather station (Meteorological and climatological website «Pogoda i klimat») for 1981–2020 was – 8.4 °C and had a positive trend. The average air temperature in the warm months (May–August) for 1931–2020 was 10.4 °C, and in the cold months (September-April) – 17 °C, respectively (Fig. 1).

For the study, the data obtained at the CALM monitoring sites and freely available on the website of this organization (https://www2.gwu.edu/~calm/data/north.htm) (https://www2.gwu.edu/~calm/data/north.htm) were taken. Also at the CALM site (Talnakh) (according to https://www2.gwu.edu/~calm/data/north.htm) there, the active layer for the time interval from 2005 to 2019 had an average value of 95 cm with an average trend of 1.3 cm increase per year, amounting to 90 cm at the beginning of the period and more than 100 cm at the end. Changes in thaw depth or seasonally thawed layer (STL) at the reference observation sites in Igarka and Talnakh (Norilsk region) are presented in Fig. 2.

The changes in thawing depth or seasonally thawed layer (STL) at the reference observation sites in Igarka and Talnakh, in particular, presented in Fig. 2 allow us to see that due to climate change the STL is increasing by about 1.3—3 cm per year, which amounted to about 20–30 cm over 20 years of observations. The general course of annual changes in ground temperature and thawing depth at observation sites in the Russian Arctic is presented in Fig. 3.

Thus, due to the 4–6 °C excess of air temperature in the winter-spring months (November–May) in Western and Central Siberia and the Russian Arctic compared to the long-term averages for 1981–2010, the amount of snow and the thickness of

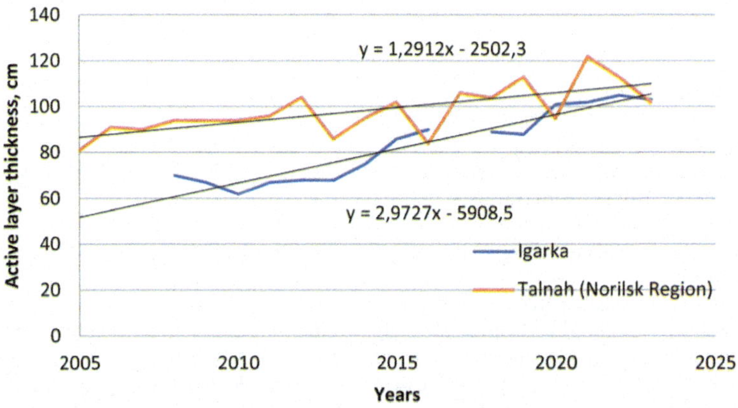

Fig. 2 Change of the thawing depth or seasonally thawed layer (STL) at reference observation sites in Igarka and Talnakh (Norilsk region)

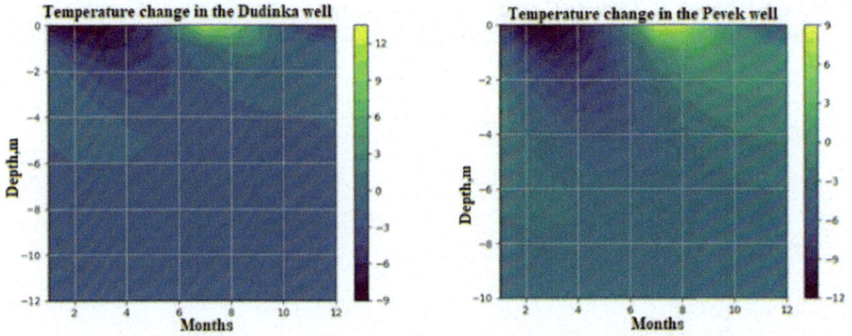

Fig. 3 Ground temperature changes at the observation site in Dudinka and Pevek

the snow cover have increased significantly over the last decade (Pulliainen et al. 2020). Also in Western and Central Siberia and the Russian Arctic, winter ground freezing becomes less intense and the thickness of the active layer increases (Frolov 2020, 2019; Kanevskiy et al. 2021; Jan and Painter 2020; Cao et al. 2020). This leads to a violation of the stability of foundations of buildings and structures and resulted, in particular, in damage to the structure of the oil tank in Norilsk on May 29, 2020 due to foundation instability and a catastrophic oil spill into the river. Moreover, due to the hot spring months, there is an increase in the number of wildfires.

The work was carried out in accordance with the state budgetary theme "Evolution of cryosphere under climate change and anthropogenic impact" (121051100164-0), "Danger and risk of natural processes and phenomena" (121051300175-4).

References

Cao B, Gruber S, Zheng D (2020) The ERA5-land soil temperature bias in permafrost regions. Cryosphere 14:2581–2595. https://doi.org/10.5194/tc-14-2581-202
Frolov D (2020) Peculiarities of weather and ground freezing conditions in Siberia and Russian Arctic in winter and spring period of 2019/2020. J Phys Conf Ser 1614:012078. https://doi.org/10.1088/1742-6596/1614/1/012078
Frolov DM (2020) Winter regime of temperature and snow accumulation as a factor of ground freezing depth variations E3S Web Conf 163(01005):1–5. https://doi.org/10.1051/e3sconf/202016301005
https://www2.gwu.edu/~calm/data/north.htm
Jan A, Painter SL (2020) Permafrost thermal conditions are sensitive to shifts in snow timing. Environ Res Lett 15:084026. https://doi.org/10.1088/1748-9326/ab8ec4
Kanevskiy M, Shur Y, Walker DA, Jorgenson T, Raynolds MK et al (2021) The shifting mosaic of ice-wedge degradation and stabilization in response to infrastructure and climate change, Prudhoe Bay Oilfield, Alaska, USA. Arctic Sci 8(2):498–530. https://doi.org/10.1139/as-2021-0024
Meteorological and climatological website «Pogoda i klimat». http://www.pogodaiklimat.ru/history/23078.htm
Pulliainen J, Luojus K, Derksen C et al (2020) Patterns and trends of northern hemisphere snow mass from 1980 to 2018. Nature 581:294–298. https://doi.org/10.1038/s41586-020-2258-0

Spatial and Temporal Interpolation of the Chlorophyll “a” Concentration Field in the Taganrog Bay Based on a Variational Assimilation Algorithm Using Remote Sensing Data

V. S. Kochergin and S. V. Kochergin

Abstract The aim of the work is to determine the concentration fields of chlorophyll “a” over a time interval of several days by integrating the transfer model, taking into account remote sensing data distributed over time and space. Satellite information is the most up-to-date, but it has gaps in its structure due to the scattering effect. Such omissions in the data are due to the presence, for example, of clouds or glare on the sea surface. Therefore, the construction of fields based on available information for the studied water area is an important and urgent task. In the numerical implementation of the transfer model and the variational method of assimilation of measurement data, the results of calculations based on the dynamic barotropic model of the Sea of Azov were used with wind exposure corresponding to the selected dates. In the iterative search for the extremum of the prediction quality functional, gradient methods are implemented, and the solution of the conjugate problem is used to construct the gradient of the functional in the parameter space. As a result of the calculations, the field of chlorophyll concentration “a” was obtained for the water area of the Taganrog Bay for the entire period of time of integration of the model consistent with the measurement data.

Keywords Chlorophyll concentration “a” · Variational assimilation algorithm · Conjugate problem · Taganrog Bay · Space–time interpolation

1 Introduction

The need to solve the environmental problems of the Azov-Black Sea basin requires the creation of systems that allow you to receive operational information about the ecological state of the reservoir. For such systems, first of all, numerical dynamic models (Marchuk and Sarkisyan 1988; Ivanov and Fomin 2008), models

V. S. Kochergin (✉) · S. V. Kochergin
Marine Hydrophysical Institute RAS, Kapitanskaya Street 2, 299011 Sevastopol, Russia
e-mail: vskocher@gmail.com

T. Chaplina (ed.), *Processes in GeoMedia—Volume VIII*, Springer Geology,
https://doi.org/10.1007/978-981-97-6627-7_21

of transportation of various suspended solids (Marchuk 1982), as well as procedures for assimilating reliable information about its concentration in the studied basin (Marchuk and Penenko 1979; Zalesny et al. 2016; Shutyaev et al. 2018; Kochergin et al. 2017; Kochergin and Kochergin 2017). In this paper, a variational assimilation algorithm is implemented to assimilate satellite information on the concentration of chlorophyll "a" in the upper layer of the Taganrog Bay. There are often omissions in such information, for example, due to clouds. The use of surface values of chlorophyll concentration in the transfer model makes it possible to obtain the missing information due to measurements for other time points and the model itself. The initial field of chlorophyll concentration is selected as the identifiable input parameter of the model. When implementing the procedure for searching for the extremum of the functional characterizing the deviations of model estimates from measurement data, the solution of the main, conjugate tasks and tasks in variations are used to construct the gradient of the functional and organize the iterative process. An extensive review of the methods of assimilation of measurement data is presented in Shutyaev (2019). The model in the algorithm used acts as a space–time interpolant (Penenko 1981) and the resulting solution to the problem in the time interval used is consistent not only with the model, but also with the measurement data.

2 Isotope Transfer-Diffusion Model

Let the model of the transfer-diffusion in σ—coordinates has the form:

$$\frac{\partial DC}{\partial t} + LC = 0, \tag{1}$$

with the conditions on the lateral boundaries:

$$\Gamma : \frac{\partial C}{\partial n} = 0, \tag{2}$$

boundary conditions on the surface and at the bottom:

$$\begin{aligned} \sigma = 0 : \frac{\partial C}{\partial \sigma} = 0, \\ \sigma = -1 : \frac{\partial C}{\partial \sigma} = 0, \end{aligned} \tag{3}$$

and initial data:

$$C(x, y, \sigma, 0) = C_0(x, y, \sigma). \tag{4}$$

In (1)–(4), the following designations are accepted:
$L = \frac{\partial DU}{\partial x} + \frac{\partial DV}{\partial y} + \frac{\partial W}{\partial \sigma} - \frac{\partial}{\partial x} A_H \frac{\partial D}{\partial x} - \frac{\partial}{\partial y} A_H \frac{\partial D}{\partial y} - \frac{\partial}{\partial \sigma} \frac{K}{D} \frac{\partial}{\partial \sigma}$,

$t \in [0, T]$—time. C—admixture concentration; D—dynamic depth; σ—dimensionless vertical coordinate that varies from -1 (on the bottom) to 0 (on the sea surface); x, y—admixture flow at the bottom; U, V, W—components of the speed field; A_H and K—coefficients of horizontal and vertical turbulent diffusion, respectively; n—normal to the side border; Γ—the border of the region M; $M_t = M \times [0, T]$.

3 Variational Assimilation Algorithm

To solve the problem of assimilation of measurement data it is necessary to find a minimum convex quadratic functional of forecast quality

$$I_0 = \frac{1}{2}(P(RC - C_{me}), (P(RC - C_{me})))_{M_t}. \tag{5}$$

where R is the operator for projecting to observation points, and P is the operator for filling in the field of forecast residuals with zeros in the absence of measurement data. Write (5) for linear relations (1)–(4) as:

$$I = I_0 + \left(\frac{\partial C}{\partial t} + LC, C^*\right)_{M_t} + \left(\frac{\partial C}{\partial n}, C^*\right)_{\Gamma_t} + \left(C - C_0, C^*\right)_M + \left(\frac{\partial C}{\partial \sigma}, C^*\right)_{\sigma_t^{-1}}. \tag{6}$$

We integrate the corresponding (6) expression for the variation of the functional in parts, taking into account the analog of the continuity equation and the boundary conditions. Chosen as the Lagrange multipliers for the solution of the following adjoint problem:

$$\frac{\partial DC^*}{\partial t} + L^* C^* = P(C_{me} - RC), \tag{7}$$

$$\Gamma : \frac{\partial C^*}{\partial t} = 0, \sigma = 0 : \frac{\partial C^*}{\partial \sigma} = 0, \sigma = -1 : \frac{\partial C^*}{\partial \sigma} = 0, \quad \mathrm{t = T} : C^* = 0, \tag{8}$$

where L^* is the formally adjoint operator to the operator L.

Then we have from the stationarity of the functional and the definition of its gradation

$$\nabla_{C^0} I = C(\sigma, x, y, 0).$$

The initial concentration distribution is searched iteratively

$$C_0^{n+1} = C_0^n + \tau \nabla_{C^0} I.$$

where τ is an iterative parameter, which is determined (Penenko 1981) taking into account the solution of the problem in variations, based on the minimum of the functional (5) by the following formula:

$$\tau = \frac{P(RC - C_{me}), (P(RC - C_{me}))}{(PR\delta C, PR\delta C)_{M_t}}$$

In (7) δC—the solution of the corresponding problem in variations.

4 Results of Numerical Experiments

Numerical experiments were carried out using a barotropic version of the model (Ivanov and Fomin 2008) for the water area of the Sea of Azov with a real bottom relief and wind action, which is characterized mainly by (http://dvs.net.ru/mp/data/201507vw.shtml) winds of the northern points with values of the order of 8 m/s. The calculated velocity fields and coefficients of the model were input information when integrating the passive impurity transfer-diffusion model for a period of 5 days. In the model, the values of steps in time sec and space km, km were set. Vertically, the model uses a calculated grid in coordinates with 15 horizons. In (Kochergin et al. 2017), the algorithm was tested on a test example and the results obtained were compared with satellite data.

Information on the concentration of chlorophyll-A was used as measurement data (http://dvs.net.ru/mp/data/201507vw.shtml) in the Taganrog Bay from September 7 to September 10, 2023. Often, certain areas may be partially, or completely covered by clouds. For example, 09.09.2023, almost the entire bay is covered with clouds. Using the concentration information for 09/10/2023 (Figs. 1 and 2), as well as other time points from the specified interval, we obtain fields for September 7 and subsequent time points, including for September 9 (Fig. 3).

Figure 1 shows the concentration of chlorophyll in the upper layer of the Taganrog Bay. The fields obtained as a result of the variational assimilation procedure are consistent with the measurement data and the model. The information used refers to the same points in time, so the daily course is not considered. Figure 2 characterizes the concentration of suspended matter in the upper layer of the sea on September 10. Based on such images, it is possible to judge the presence of clouds and maxima in the concentration field of suspended matter. Figures 1 and 2 show a correlation between data gaps off the northern coast of Taganrog and areas covered by clouds. Note that the absence of dense clouds does not guarantee a stable signal. The field shown in Fig. 3 is smoother than the measurement data due to the fact that it is obtained as a composite of solutions to conjugate problems with a certain viscosity. The field shown in Fig. 3 was found as a result of iterations while minimizing the functional (4) under the constraints imposed by the model itself.

In the first iterations, there is a fairly rapid drop in the normalized forecast quality functional. The rate of convergence is influenced by various factors—these are the

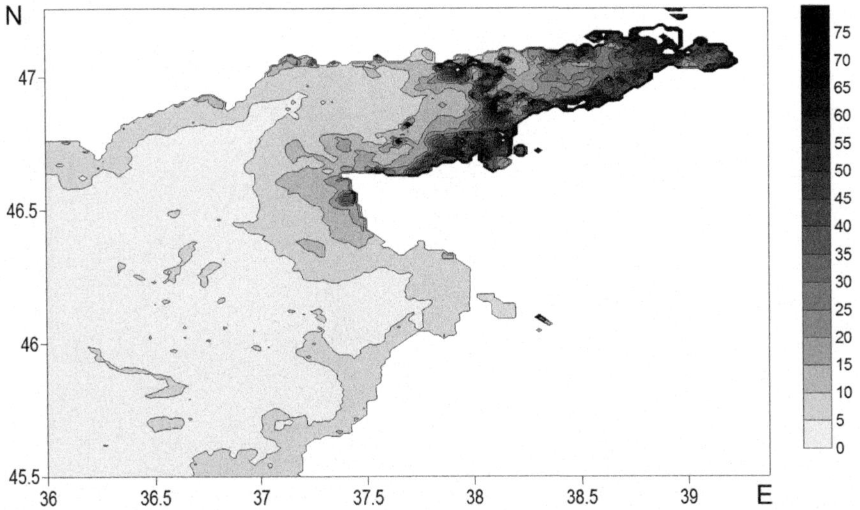

Fig. 1 The concentration of chlorophyll a on September 10, 2023 (mg/m^3)

Fig. 2 Concentration of suspended matter on September 10, 2023 (/worldview.earthdata.nasa.gov/?p=geographic&l=MODIS_)

coefficients of turbulent diffusion, and the time interval at which the problem is solved. Having an initial field consistent with the assimilated information and the model, it is possible to obtain fields for the entire time interval in which the problem is solved. Note that the approach used makes it possible to identify not only the initial fields, but also other parameters of the model, which can lead to an increase

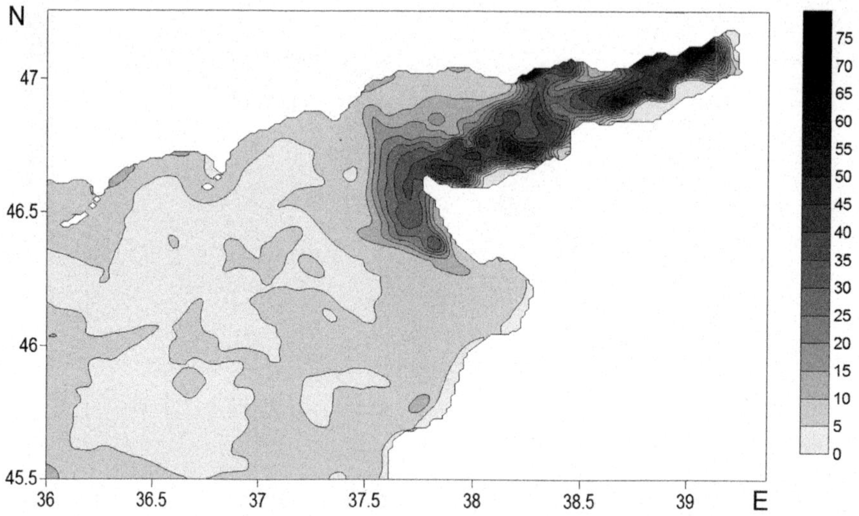

Fig. 3 The concentration of chlorophyll a on September 9, 2023 (mg/m^3)

in the adequacy of the model to processes occurring at sea. The conducted numerical experiments have shown the reliable operation of the variational algorithm for identifying the input parameters of numerical modeling. The results of the work can be used in solving environmental problems for the Azov-Black basin.

5 Conclusions

In the work, a variational assimilation of satellite information distributed over time and space was carried out, which made it possible to obtain a field of chlorophyll concentration "a" for the waters of the Taganrog Bay, taking into account incomplete coverage with observational data. The procedure can be used to determine the concentration fields of various suspended solids in the sea, taking into account the incomplete coverage of the studied water area with observational data for different time points.

Acknowledgements The work was carried out within the framework of the state task on the topic FNNN-2021-0005 "Complex interdisciplinary studies of oceanological processes that determine the functioning and evolution of ecosystems of the coastal zones of the Black and Azov Seas" (code "Coastal research").

References

Ivanov VA, Fomin VV (2008) Mathematical modeling of dynamic processes in the sea-land zone. In: ECOSI-hydrophysics, Sevastopol, 363 p

Kochergin VS, Kochergin SV (2017) Identification of the parameters of the instantaneous point pollution source in the Azov sea based on the adjoint method. Phys Oceanogr 1:62–67. https://doi.org/10.22449/1573-160X-2017-1-62-67

Kochergin VS, Kochergin SV, Stanichny VS (2017) Ispol'zovanie metoda sopryazhennykh uravnenii pri identifikatsii istochnikov zagryaznenii v Azovskom more (Identification of pollution sources in the sea of Azov using the adjoint equation method). Sovremennye Problemy Distantsionnogo Zondirovaniya Zemli iz Kosmosa 14(1):50–57

Marchuk GI (1982) Mathematical modeling in the problem of the environment. Nauka, Moscow

Marchuk GI, Sarkisyan AS (1988) Mathematical modeling of ocean circulation. Nauka, Moscow

Marchuk GI, Penenko VV (1979) Application of optimization methods to the problem of mathematical simulation of atmospheric processes and environment. In: Marchuk GI (ed) Modelling and optimization of complex systems. Springer, Berlin, pp 240–252. https://link.springer.com/chapter/10.1007%2FBFb0004167

Penenko VV (1981) Metody chislennogo modelirovaniya atmosfernyx pro-cessov. L.: Gidrometeoizdat, 350 p

Shutyaev VP (2019) Methods for observation data assimilation in problems of physics of atmosphere and ocean. Izv Atmos Oceanic Phys 55:17–31. https://doi.org/10.1134/S0001433819010080

Shutyaev VP, Le Dimet F-X, Parmuzin E (2018) Sensitivity analysis with respect to observations in variational data assimilation for parameter estimation. Nonlinear Process Geophys 25(2):429–439. https://doi.org/10.5194/npg-25-429-2018

Zalesny VB, Agoshkov VI, Shutyaev VP, Le Dimet F, Ivchenko VO (2016) Numerical modeling of ocean hydrodynamics with variational assimilation of observational data. Izv Atmos Ocean Phys 52(4):431–442

Calculation of Barotropic Currents in the Azov Sea with Different Wind Stress

V. S. Kochergin and S. V. Kochergin

Abstract The aim of the work is to calculate the circulation in the Sea Azov under intense wind stress of different directions, including cyclonic and anticyclonic nature. Numerical experiments were carried out using a three-dimensional barotropic model taking into account the real bottom relief, taking into account run-up phenomena, parametrization of variables in the space of turbulent diffusion coefficients, which, together with the flow fields, are input parameters for solving a wide class of problems of impurity transfer to the sea and solving problems related to the assimilation of measurement data, identification of pollution sources and their parameters.

Keywords Wind barotropic currents · The Sea of Azov · Surges · Cyclone · Anticyclone

1 Introduction

When solving a wide class of tasks related to the assessment of the ecological state of the Sea of Azov, it is necessary to analyze the possible consequences of the spread of various impurities. Such an analysis can be carried out on the basis of using mathematical modeling that takes into account the nature of their behavior in the aquatic environment, taking into account the ongoing dynamic processes. In recent years, interest in modeling dynamic processes in the Azov-Black Sea basin has grown significantly (Zalesny et al. 2013; Dianskij and Fomin 2017; Zalesny et al. 2012; Fomin et al. 2015a; Fomin and Diansky 2018) due to the need, in conditions of increasing anthropogenic pressure, to create reliable environmental monitoring systems. This work is devoted to modeling barotropic currents in the Sea of Azov, taking into account the real bottom relief and intense wind action. Numerical experiments were carried out using a three-dimensional hydrodynamic model (Fomin 2002; Ivanov and Fomin 2008) with constant density (without taking into account the thermohaline structure). The ideology of building the model code is laid down in the work

V. S. Kochergin (✉) · S. V. Kochergin
Marine Hydrophysical Institute RAS, Kapitanskaya Street 2, 299011 Sevastopol, Russia
e-mail: vskocher@gmail.com

T. Chaplina (ed.), *Processes in GeoMedia—Volume VIII*, Springer Geology,
https://doi.org/10.1007/978-981-97-6627-7_22

(Ivanov and Fomin 2008). A detailed description of the numerical implementation of the model used is presented in Fomin (2002), Ivanov and Fomin (2008). With its help, calculations were carried out to establish a model field of currents under the influence of intense constant wind of various directions. The obtained spatial distributions of velocity fields and turbulent diffusion coefficients were used as input information when integrating the passive impurity transfer model (Kochergin and Kochergin 2015).

2 Results of Numerical Experiments

Numerical experiments were carried out with the model (Fomin 2002; Ivanov and Fomin 2008) for the water area of the Sea of Azov. To implement the transfer model for this water area, calculations were carried out to establish model fields of currents with constant winds 10 m/s of various directions, as well as under the influence of a cyclone and an anticyclone with maximum velocities of the same order. An analytical representation of the wind effect is presented in Kochergin et al. (2022, 2023). On the southern border along the latitudinal section from cape Meganom to cape Utrish the condition of free passage for the meridional velocity of currents was used (Fomin et al. 2015a). The obtained velocity fields and turbulent diffusion coefficients were used in Kochergin and Kochergin (2015) as input parameters when integrating a passive impurity transfer model. At the same time, a step in time $\Delta t = 240$ s, a step in space $\Delta x = 0.78$ km, $\Delta y = 1.125$ km. Vertically, the model uses a calculated grid in coordinates with 15 evenly distributed horizons.

Figure 1 shows the velocity field on the sea surface under constant easterly 10 m/s wind action. The prevailing winds for the waters of the Sea of Azov are the winds of the eastern and north-eastern directions. The main features of the current fields in comparison with wind action are concentrated in the areas of influence of the spit and the Kerch Strait. The average depth of the sea is about 7–8 m, and the maximum does not exceed 13.5–14 m. Such values turn out to be significantly less than the depth of the Ekman layer, therefore (Kochergin 1978) the surface currents have the form shown in the figures. A significant influence on the structure of flows is provided by the accounting of run-up phenomena implemented in the dynamic model (Ivanov and Fomin 2008).

Figure 2 shows the level field calculated in the model for this wind effect. The greatest overburden phenomena are observed at the Arabatskaya Strelka, and the maximum values of overburden occur in the Taganrog Bay near the Don Delta.

Flooding or drainage of the delta occurs with more intense wind action, and at the same time, wind waves can make a significant contribution, which increases in the Taganrog Bay with sea level rise. This is due to a decrease in the effect of friction on the bottom with increasing depth (Fomin et al. 2015b). The elevation of the level is indicated by a dark color, and lighter tones characterize the driving phenomena. Zero is indicated by a white dashed line. Also, the wind of the north-easterly direction has

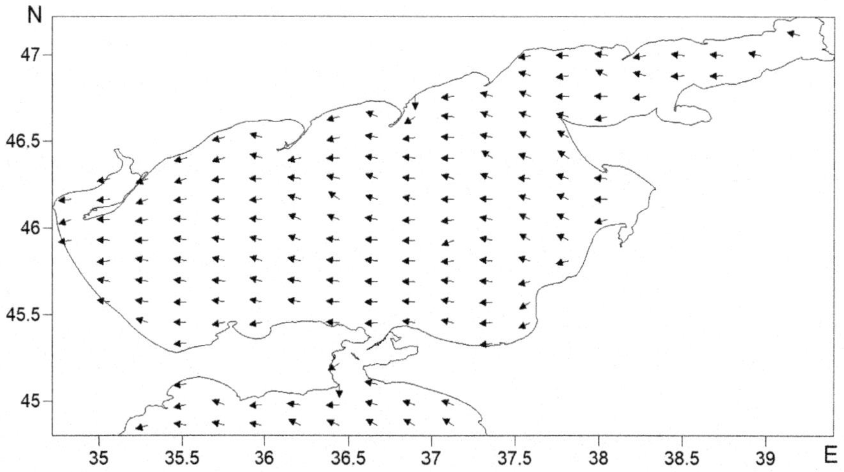

Fig. 1 Surface currents with an easterly wind

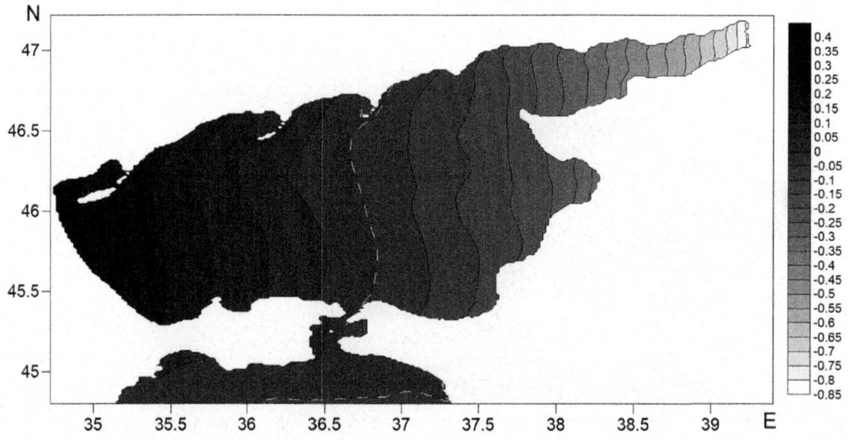

Fig. 2 The values of the model sea level (m) with an easterly wind

the greatest repeatability in the waters of the Sea of Azov. Figure 3 characterizes the surface circulation of waters under such a wind effect.

The main direction of movement of surface waters is also south-west. At the same time, the lowering of the level in the Taganrog Bay even exceeds the corresponding values obtained with the easterly wind effect (Fig. 4).

Accordingly, the south-westerly wind leads to more significant surges in the Taganrog Bay than with a westerly wind effect (Fig. 5).

Of particular interest is the circulation of waters and the behavior of the level surface during the passage of a cyclone (Figs. 6 and 7).

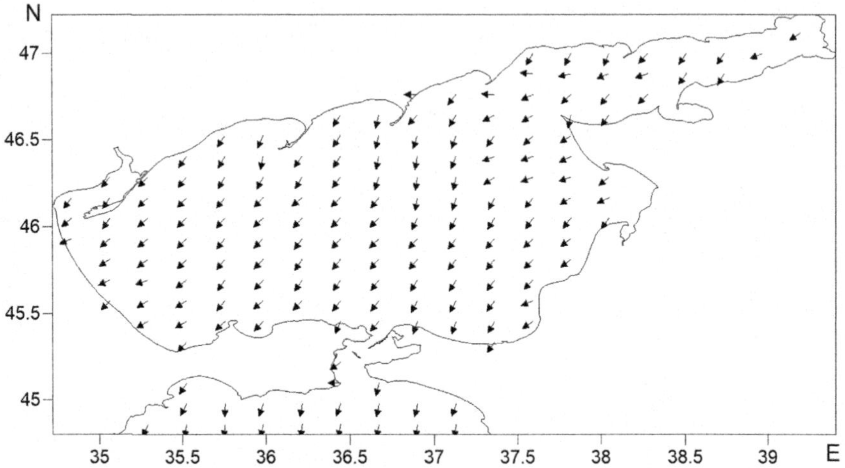

Fig. 3 Surface currents with a northeasterly wind

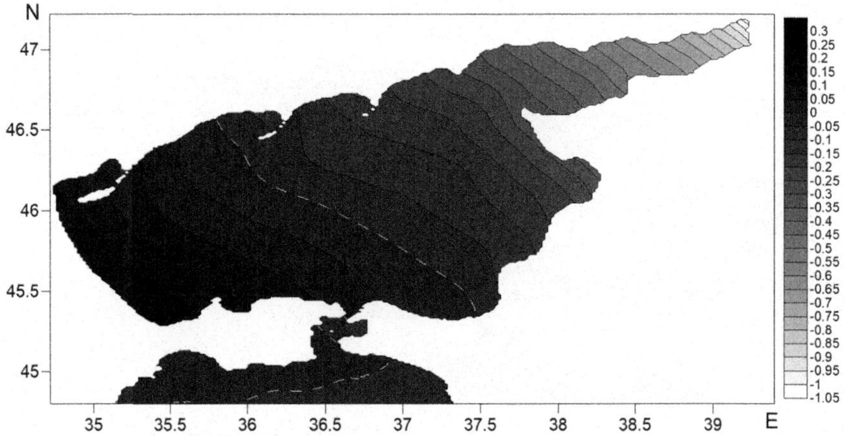

Fig. 4 The values of the model sea level (m) with a northeasterly wind

In addition, the level rises in the central part of the sea, and along the southern and northern shores there is a slight decrease (Figs. 8 and 9).

3 Conclusions

Thus, as a result of the calculations carried out, dynamic fields for the water area of the Sea of Azov were obtained, taking into account overburden phenomena that have a significant impact on the dynamics of waters in this basin. It is shown that under the

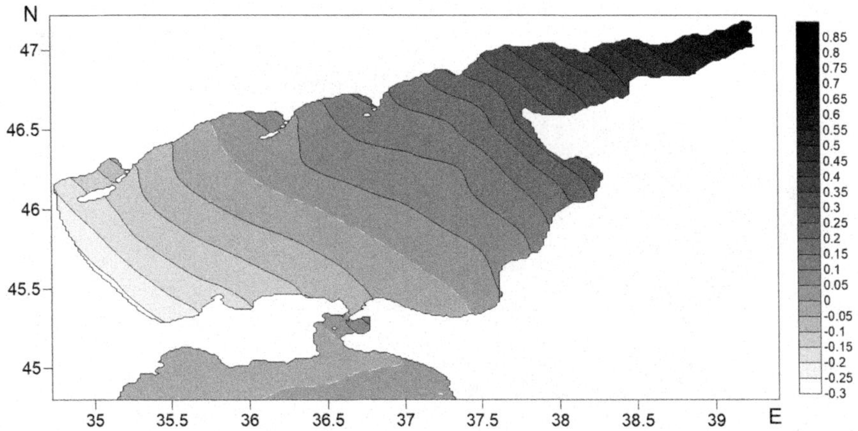

Fig. 5 The values of the model sea level (m) with a southwesterly wind

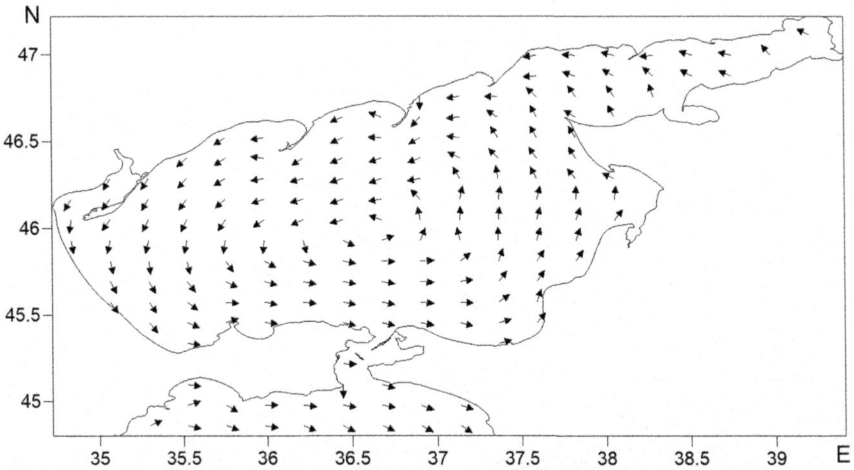

Fig. 6 Surface currents under cyclonic wind action

influence of the prevailing easterly and northeasterly winds in this area, a significant decrease in sea level occurs in the Taganrog Bay. The highest values are obtained with a north-easterly wind. A significant decrease in the level in the bay also occurs when a cyclone affects the formation of dynamic processes. On the contrary, with a cyclonic wind in the Taganrog Bay, the level rises. The results can be used to solve various environmental problems in the waters of the Sea of Azov.

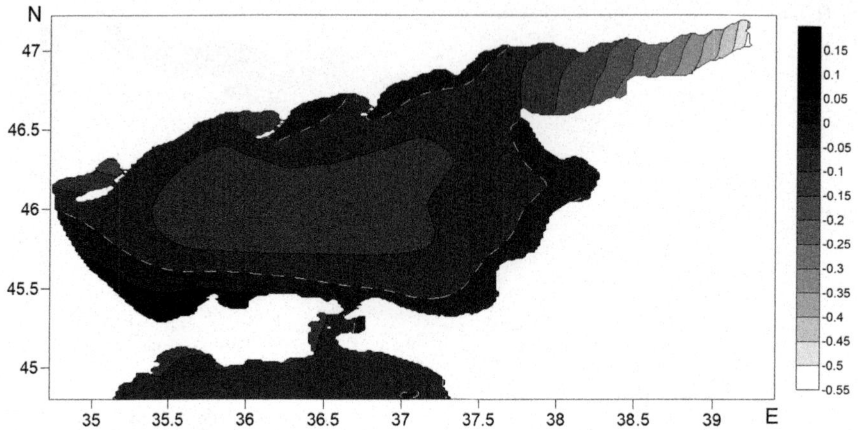

Fig. 7 The values of the model sea level (m) with cyclonic wind action

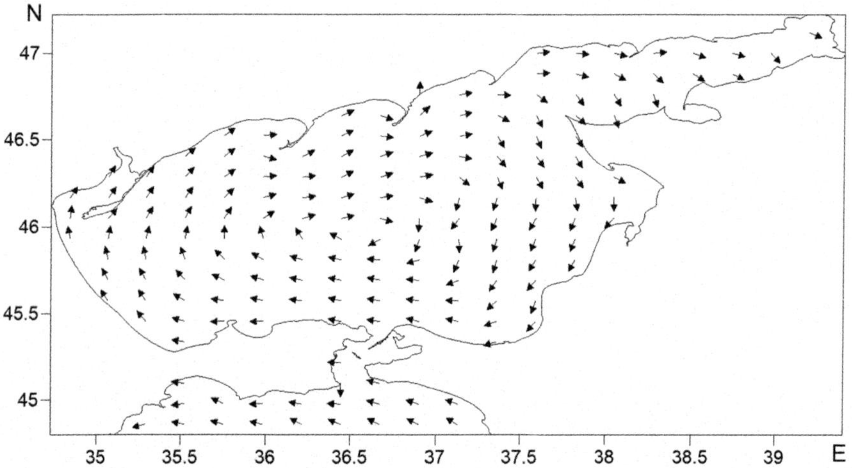

Fig. 8 Surface currents under anticyclonic wind action

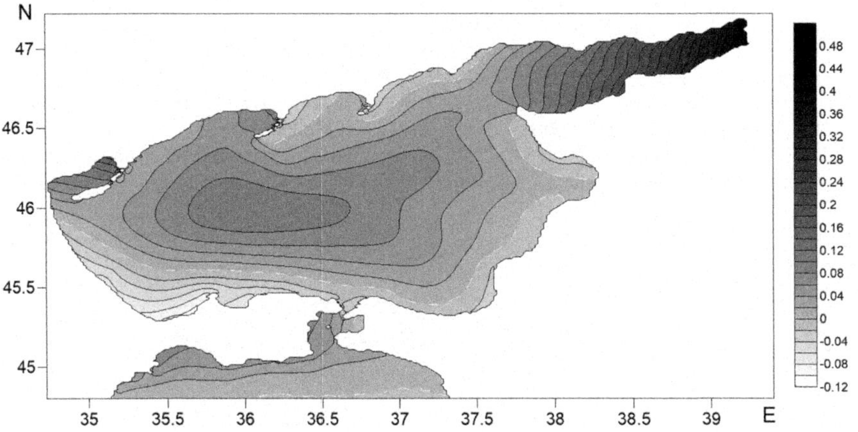

Fig. 9 The values of the model sea level (m) under anticyclonic wind action

Acknowledgements The work was carried out within the framework of the state task on the topic FNNN-2021-0005 "Complex interdisciplinary studies of oceanological processes that determine the functioning and evolution of ecosystems of the coastal zones of the Black and Azov Seas" (code "Coastal research").

References

Dianskij NA, Fomin VV (2017) Modelirovanie cirkulyacii Azovskogo morya i osobennosti vosproizvedeniya ekstremalnyx nagonov v Taganrogskom zalive. Trudy GOIN Tom 218:74–104

Fomin VV (2002) Chislennaya model cirkulyacii vod Azoskogo moray. Nauchnye Trudy UkrNIGMI. Vyp 249:S.246–255

Fomin VV, Diansky NA (2018) Simulation of extreme surges in the Taganrog bay with atmosphere and ocean circulation models. Russ Meteorol Hydrol [e-journal] 43(12):843–851. https://doi.org/10.3103/S1068373918120051

Fomin VV, Polozok AA, Fomina IN (2015a) Simulation of the Azov sea water circulation subject to the river discharge. Phys Oceanogr 1:15–26. https://doi.org/10.22449/1573-160X-2015-1-15-26

Fomin VV, Lazorenko DI, AlekseevD V, Polozok AA (2015b) Storm surge in the Taganrog bay and flooding of the Don delta. In: Environmental safety of the coastal and offshore zones of the sea. MHI, Sevastopol, issue 1, pp 74–82

Ivanov VA, Fomin VV (2008) Mathematical modeling of dynamic processes in the sea-land zone. In: ECOSI-hydrophysics, Sevastopol, 363 p

Kochergin VP (1978) Teoriya i metody rascheta okeanicheskix techenij. Nauka, Moskva, 127 p

Kochergin VS, Kochergin SV (2015) Identification of a pollution source power in the Kazantip bay applying the variation algorithm. Phys Oceanogr 2:69–76. https://doi.org/10.22449/1573-160X-2015-2-69-76

Kochergin VS, Kochergin SV, Sklyar SN (2002) Analytical solution of the equation for the stream function in the model of flows with variable wind action in space. Ecol Bull Res Cent Black Sea Econ Cooperation 19(1):16–24. https://doi.org/10.31429/vestnik-19-1-16-24

Kochergin VS, Kochergin SV, Sklyar SN (2023) Testing of numerical methods for solving the stream function problem in a model of stationary fluid motion. In: Processes in GeoMedia. Springer geology, vol VI, pp 513–521. https://doi.org/10.1007/978-3-031-16575-7_46
Zalesny VB, Diansky NA, Fomin VV, Demyshev SG (2012) Numerical model of the circulation of the Black Sea and Sea Azov. Russ J Numer Anal Math Model 27(1):95–112
Zalesny VB, Gusev AV, Moshonkin SN (2013) Numerical model of the Black Sea and Sea of Azov with variational initialization of temperature and salinity. Izv Atmos Ocean Phys 49(6):642–658

Study of the Slope Stability Using Analytical Method

N. S. Orlova and M. S. Bezuglov

Abstract It is known that rock collapse occurs due to loss of slope stability. The condition for loss of slope stability is expressed through the stability coefficient. The study of deterministic methods for calculating the stability coefficient and comparison of the results obtained using them (with a view to further selecting one of them for carrying out a probabilistic assessment of the rockfalls) is very important. We investigated the simplified analytical method for calculating the stability coefficient of slopes composed of coarse-grained, sandy or clayey soils with a homogeneous structure. The formula for calculating the slope stability coefficient is based on the method of circular cylindrical sliding surfaces. In addition to taking into account the own weight of the soil prism, it is possible to take into account the seismic impact during an earthquake. Using the analytical method, calculations of the slope stability coefficient were carried out without taking into account the influence of factors and taking into account the seismic impact. The calculation results obtained using the analytical method were compared with the data, obtained using various methods (including methods of dynamic analysis of accelerogram). The stability of the three geomechanical models with slope steepness of 20°, 30° and 40° were considered for a section of the northern slope of the Aibga ridge (Krasnaya Polyana, Sochi, Russia). In general, the calculation results using the analytical method and the calculated data, which were obtained using the various methods are in good agreement. The quantitative coincidence of the compared values (obtained using the analytical method and the methods of the pseudostatic analysis and the dynamic analysis) is observed (the relative difference in the values does not exceed 7%) at relatively low values of the adhesion (less than 15 kPa) and at the slope steepness greater than 20°.

Keywords Slope stability coefficient · Analytical method · Soils with a homogeneous structure · Seismic influence · Accelerogram data · Aibga Ridge · Slope steepness · Adhesion

---

N. S. Orlova (✉) · M. S. Bezuglov
Federal State Budgetary Educational University of Higher Education North Ossetian State University named after Kosta Levanovich Khetagurov, Vladikavkaz, Russia
e-mail: norlova.umi.vnc@gmail.com

T. Chaplina (ed.), *Processes in GeoMedia—Volume VIII*, Springer Geology,
https://doi.org/10.1007/978-981-97-6627-7_23

1 Introduction

It is known that rock collapse occurs due to loss of the slope stability. The condition for loss of the slope stability is expressed through the stability coefficient K_{st}. If the value of the stability coefficient is above 1, this means that the analyzed slope is stable. If the value of the stability coefficient is approximately equal to 1, this means that the analyzed slope is in a state of limiting equilibrium. If the value of the stability coefficient is less than 1, this means that the analyzed slope is not stable. The main causes of loss of the slope stability can be divided into natural and anthropogenic. The natural causes include changes in internal forces (an increase in the specific gravity of the soil with an increase in its humidity or, conversely, the influence of the suspended water pressure on the soil), the manifestation of hydrodynamic pressure and seismic forces. The anthropogenic causes include an increase in external load (construction of structures, storage of materials on the slope or near its edge), dynamic impacts of various kinds (for example, traffic, etc.).

To assess and predict the stability of the slopes, various methods are currently used (Tang et al. 2017; Fomenko et al. 2009; Morgenstern and Price 1965; Konyushkov 2020; Hoek and Brown 2019; Kang et al. 2018, 2021, 2021, 2022; Griffiths and Fenton 2004). These methods are based on deterministic and stochastic models. In deterministic models, the most widely used methods are the direct calculation of the slope stability coefficient. These methods are based on the conditions of limit equilibrium (Fomenko et al. 2009; Morgenstern and Price 1965; Konyushkov 2020), and calculations are carried out using formulas and graphical techniques of classical soil mechanics. The success of using these methods largely depends on the calculation schemes that are most successfully selected and correspond to the main features of the slope structure, as well as reflecting weakened zones. Most methods are implemented in commercial packages. It should be noted that numerical modeling of the slope stability is often carried out using the finite element method (Hoek and Brown 2019; Kang et al. 2020, 2021, 2022; Griffiths and Fenton 2004).

For a full analysis of the slope stability, it is recommended to use various calculation methods and then compare their results (Fomenko et al. 2009; Konyushkov 2020; Kang et al. 2018; Kang and Zerkal 2016). After choosing a deterministic method for calculating the stability coefficient, they proceed to a probabilistic analysis of the slope stability assessment (Hoek and Brown 2019; Kang et al. 2020, 2021, 2022; Griffiths and Fenton 2004; Kang and Zerkal 2016; Miao et al. 2017; Zerkal and Fomenko 2017). In this regard, the study of deterministic methods for calculating the stability coefficient and comparison of the results obtained using them (with a view to further selecting one of them for carrying out a probabilistic assessment of the collapse) is very important.

In this work, we investigated the simplified analytical method for calculating the stability coefficient of the slopes composed of coarse-grained, sandy or clayey soils with a homogeneous structure (Konyushkov 2020). In addition to taking into account the own weight of the soil prism, it is possible to take into account natural (such as the influence of the hydrodynamic effect of water and the seismic impact during an earthquake) and anthropogenic factors (such as the additional external load on the edge of the slope). Using the analytical method, calculations of the slope stability coefficient were carried out without taking into account the influence of factors and taking into account the seismic impact. We used input data (physical and mechanical properties of the soil, geometry of model slopes, etc.) from the article (Kang et al. 2018), in which the stability of the slopes under seismic influence was assessed using various methods (including methods of dynamic analysis of accelerogram data (Liu et al. 2003)). The calculation results obtained using the simplified analytical method were compared with data from the article (Kang et al. 2018).

2 Analytical Method for Slope Stability Calculation

We carried out the study of the simplified analytical method for calculating the stability coefficient of the slopes composed of coarse-grained, sandy or clayey soils with a homogeneous structure. This method was proposed by Konyushkov (2020). The formula for calculating the slope stability coefficient is based on the method of circular cylindrical sliding surfaces, which has become widespread for solving practical problems. The well-known technique of Maslov N. N. was used to determine the center of rotation of the sliding prism (Konyushkov 2020). Figure 1 shows this technique. Initially, we find point 1, which is located vertically at a depth of $2H$ (H is the height of the slope) from the edge of the slope and horizontally at a distance of $4.5H$ from the bottom of the slope. Then from point 1 we draw an inclined line passing through the edge of the slope until it intersects with a vertical line, located at a distance l/H (l is the length of the horizontal projection of the slope line) from the bottom of the slope. Point 2 is the center of rotation, relative to which a sliding arc of length L is constructed. In the case of heterogeneous water-saturated soils that make up the body of the slope, and weak soils at its base, the slip prism can be determined by other alternative methods.

Based on the calculation scheme (Fig. 1), a simplified analytical method for calculating the slope stability coefficient is recommended:

$$K_{st} = \frac{\left(\mathrm{tg}\varphi + \left(\frac{L \cdot c}{G \cdot \cos\theta}\right)\right)}{\mathrm{tg}\theta}, \quad (1)$$

$\mathrm{tg}\varphi$ is the tangent of the angle of internal friction of the slope soil (degrees); L is the arc length of the sliding prism (m); c is the slope soil adhesion (kPa); G ($G = A\gamma$) is the own weight of the sliding prism per linear meter in the transverse direction (kN/m); A is the sliding prism area (m^2); γ is the specific gravity of the sliding prism soil (kN/m^3); θ is the angle, obtained by constructing a straight line from the center (half) of the sliding arc to the intersection with the center of rotation of the prism (degrees). If sandy soils have specific adhesion $c = 0$, $\mathrm{tg}\theta$ is taken equal to $\mathrm{tg}\alpha$ (α is the slope angle).

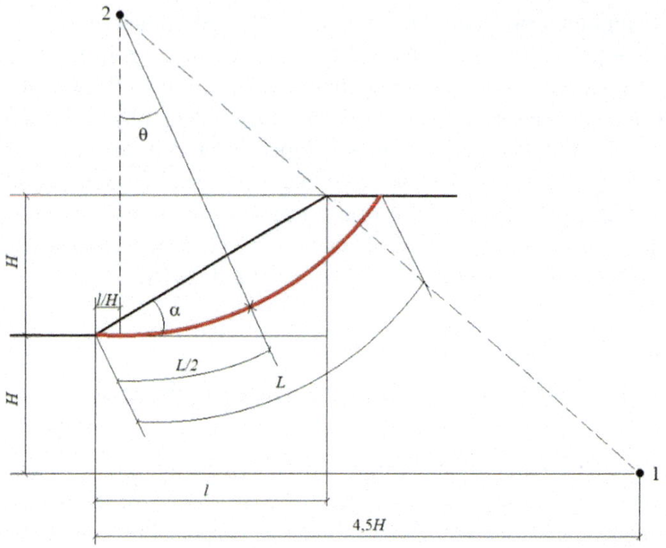

Fig. 1 The calculation scheme of the slope with the sequence of determining the sliding prism (the case of a homogeneous dry slope on a solid foundation without external loads and influences)

Formula (1) can be used to calculate the stability of slopes composed of coarse, sandy or clayey soils with a homogeneous structure. In the case of heterogeneous bedding of soils and significant differences in their properties, it is necessary to reduce the strength characteristics of soils to average values or determine the slope stability coefficient using more complex methods.

To take into account the seismic impact, formula (2) is used:

$$K_{st} = \frac{\left(\mathrm{tg}\varphi + \left(\frac{L \cdot c}{G \cdot \cos\theta}\right)\right)}{\mathrm{tg}\theta + \mathrm{tg}\delta}, \tag{2}$$

$$\delta = \mathrm{arctg}\left(\frac{a}{g}\right), \tag{3}$$

g is the acceleration of gravity ($g = 9.8$ m/s^2). a is the seismic acceleration, which takes the following values depending on the intensity of the earthquake on the MSK-64 scale:

$a = 0.1\ldots0.25$ m/s^2 with 7 points;
$a = 0.25\ldots0.5$ m/s^2 with 8 points;
$a = 0.5\ldots1.0$ m/s^2 with 9 points;
$a = 1.0\ldots2.5$ m/s^2 with 10 points.

The values of seismic angles δ, as a rule, take the following values depending on the intensity of the earthquake on the MSK-64 scale:

$\delta \approx 2°$ at 7 points;
$\delta \approx 3°$ at 8 points;
$\delta \approx 6°$ at 9 points;
$\delta \approx 14°$ at 10 points.

It is possible to take into account the influence of the hydrodynamic effect of water, as well as additional external load in formulas (1) and (2). Since these factors were not taken into account in further calculations, formulas for calculating the stability coefficient taking into account the influence of the hydrodynamic influence of water and additional external load are not given. A more detailed description of these formulas is presented in the paper (Konyushkov 2020).

Using formulas (1) and (2), calculations of stability coefficients were carried out at different adhesion values for the slopes of different steepness (at different slope angles). It was assumed that the soil has a homogeneous structure. The adhesion characterizes the strength of the soil. Three geomechanical models with slope steepness of 20°, 30° and 40° were considered. The height of all model slopes was assumed to be 25 m. The following soil properties were accepted in the models: soil density 2200 kg/m^3; internal friction angle 28°; adhesion 10, 15, 20, 25 kPa. We compared the calculation results obtained using the analytical method with the data from the article (Kang et al. 2018). For this we constructed the graphs of the dependence of the stability coefficient on the steepness of the slope at different values of adhesion.

3 Analysis of the Calculation Results

The stability of the slopes composed of mudstones was analyzed under possible seismic impacts for a section of the northern slope of the Aibga ridge (Krasnaya Polyana, Sochi, Russia) in the papers (Fomenko et al. 2009; Kang et al. 2018; Kang and Zerkal 2016; Zerkal and Fomenko 2017). The northern slope of the Aibga ridge in its middle and lower part is composed of Lower Jurassic mudstones, represented by both dense semi-rock varieties and zones with a high degree of weathering, which has a significant impact on the modern stability of the slopes on this territory. The region is characterized by high seismicity, which also affects the current stability of the slopes. The maximum magnitudes of possible seismic events in the area of the Aibga ridge can be up to 6.0–7.5 (Kang et al. 2018).

The seismic impacts in the slope stability quantification were assessed using pseudostatic and dynamic analyses. The Newmark method was used for the dynamic analysis (Kang et al. 2018). This method makes it possible to assess the dynamic stability of the slope, quantified as the critical threshold acceleration of the base (massif) necessary to overcome the static friction force and the beginning of movement. According to Newmark's approach, it is assumed that the development of displacements begins when the acceleration, caused by the seismic effect, exceeds the value of the critical acceleration. A more detailed description is presented in Kang et al. (2018).

The calculations of the slope stability coefficient were carried out using various methods in the paper (Kang et al. 2018). To calculate the stability coefficient without the influence of external factors (the seismic impact), the Morgenstern-Price method was used (close analogues are the Spencer method, the generalized limit equilibrium method), based on the equilibrium of forces and moments. To take into account the seismic impact, pseudostatic analysis and dynamic analysis methods were used. In the pseudostatic analysis, the effect of an earthquake was modeled by introducing an additional force determined by the ratio of the seismic acceleration to gravitational acceleration (more details in Kang et al. (2018)). The slope stability calculation was carried out taking into account the expected 8-point seismic impact.

In the dynamic analysis, the calculation of the slope stability is based on the use of accelerogram data, which allows the actual duration and intensity of the earthquake to be taken into account. The dynamic analysis makes it possible to determine the change in the stress–strain state of the massif during the passage of the seismic waves. The dynamic analysis is based on a finite element solution of the equation of motion (more details in Kang et al. (2018)). When carrying out the dynamic analysis, modified data (accelerograms) were used for the Rachin earthquake (magnitude 6.8), which occurred on April 29, 1991 (Kang et al. 2018).

In this section we compare the results of calculations, obtained using the analytical method, with those obtained from the pseudostatic analysis and the dynamic analysis (Kang et al. 2018). When carrying out the dynamic analysis, the method of the minimum value of the stability coefficient $K_{st}^{\min}$ (in which the minimum value is used as the resulting value of the stability coefficient (Kang et al. 2018; Su et al. 2003)) was used. The 0.65-method was also used (Kang et al. 2018; Liu et al. 2003). This method converts the minimum values $K_{st}^{\min}$, obtained when performing dynamic analysis.

The value of the stability coefficient without taking into account the seismic impact K_{st} and the minimum value of the stability coefficient $K_{st}^{\min}$, obtained from the dynamic analysis of accelerogram data, are used in the calculations by the 0.65-method. It is assumed that the slope instability does not necessarily occur instantaneously when the seismic loading is applied. The following formula is used:

$$K_{st}^{dyn} = K_{st} - 0.65 \times \left(K_{st} - K_{st}^{\min}\right), \quad (4)$$

A more detailed description of the methods is presented in the paper (Kang et al. 2018).

Figure 2 shows the curves of the dependence of the stability coefficient (without taking into account the seismic impact) on the slope angle for adhesion of 10 kPa (a) and 15 kPa (b). Curve 1 corresponds to the calculation results that were obtained using the analytical method, curve 2—to the calculated data by the Morgenstern-Price method from the article (Kang et al. 2018).

It can be seen (from Fig. 2) that the values of the stability coefficient, calculated using the analytical method, better coincides with the data by the Morgenstern-Price method when the value of the slope angle is increasing.

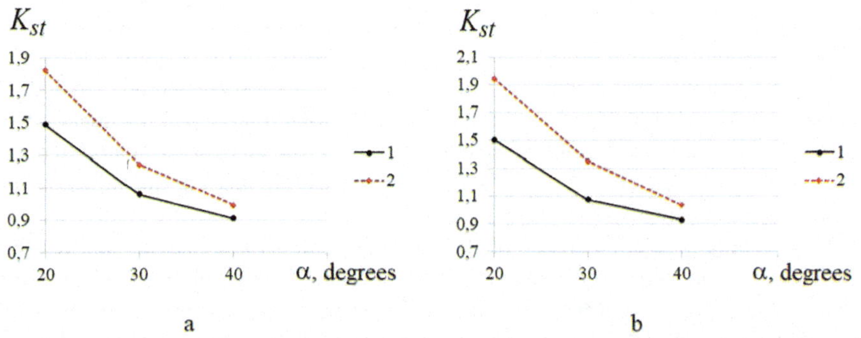

Fig. 2 The dependence of the slope stability coefficient (without taking into account the seismic impact) on the slope angle for adhesion of 10 kPa (a) and 15 kPa (b). Curve 1—the calculation results, obtained using the analytical method; curve 2—the calculated data by the Morgenstern-Price method

Figure 3 shows the curves of the dependence of the stability coefficient (taking into account the seismic impact) on the slope angle for adhesion of 10 kPa (a) and 15 kPa (b). Curve 1 corresponds to the calculation results that were obtained using the analytical method (taking into account the seismic impact); curve 2—to the calculated data, which were obtained using the pseudostatic analysis in Kang et al. (2018); curve 3—to the calculated data, which were obtained using the dynamic analysis of accelerograms (0.65-method) in Kang et al. (2018); curve 4—to the calculated data of the dynamic analysis (the minimum value of the stability coefficient $K_{st}^{\min}$) in Kang et al. (2018), in which the minimum value $K_{st}^{\min}$ is used as the resulting value of the stability coefficient.

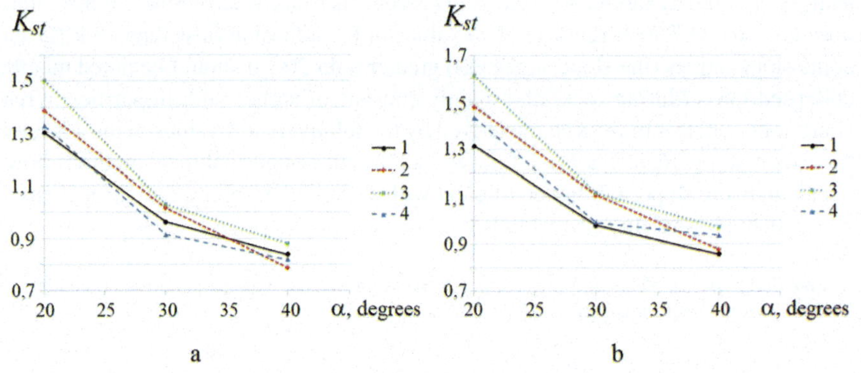

Fig. 3 The dependence of the slope stability coefficient (taking into account the seismic impact) on the slope angle for adhesion of 10 kPa (a) and 15 kPa (b). Curve 1—the calculation results, obtained using the analytical method; curve 2—the calculated data of the pseudostatic analysis; curve 3—the calculated data of the dynamic analysis (0.65 method); curve 4—the calculated data of the dynamic analysis (the minimum value of the stability coefficient $K_{st}^{\min}$)

It can be seen (from Fig. 3) that the results of calculations, obtained using the analytical method, qualitatively describe the dependence of the stability coefficient on adhesion and slope steepness, obtained based on various methods (including based on accelerogram processing data). It is important to note that the quantitative difference between the results of calculations using the analytical method and the data from the article (Kang et al. 2018) increases with increasing the values of the adhesion (at 20 and 25 kPa).

The quantitative coincidence of the compared values is observed (the relative difference in the values does not exceed 7%) at relatively low values of the adhesion (less than 15 kPa) and at the slope angles (the slope steepness) greater than 20°. It should be noted that the collapsed mass (during rockfalls) mainly consists of stones with impurities in real conditions, and this mass is characterized by the relatively low values of the adhesion (usually no more than 10 kPa). It follows that the analytical method allows to calculate the slope stability coefficient with sufficient accuracy, and it can be used to assess the probability of rockfalls (collapses).

4 Conclusions

In general, the calculation results using the simplified analytical method and the calculated data, which were obtained using the various methods (including the Morgenstern-Price method, the method of the pseudostatic analysis and the methods based on the dynamic analysis of accelerograms) are in good agreement. When we consider the case of the dependence of the stability coefficient on the slope angle (taking into account the seismic impact) the quantitative coincidence of the compared values (obtained using the analytical method and the methods of the pseudostatic analysis and the dynamic analysis) is observed (the relative difference in the values does not exceed 7%) at relatively low values of the adhesion (less than 15 kPa) and at the slope angles (the slope steepness) greater than 20°. It should be noted that the collapsed mass (during rockfalls) mainly consists of stones with impurities in real conditions, and this mass is characterized by the relatively low values of the adhesion (usually no more than 10 kPa). It follows that the simplified analytical method allows to calculate the slope stability coefficient with sufficient accuracy, and it can be used to assess the probability of rockfalls (collapses).

Acknowledgements This study was funded by Russian Science Foundation, project number 23-27-00264, https://rscf.ru/en/project/23-27-00264/.

References

Fomenko IK, Zakharov RG, Samarkin-Dzharsky KG, Sirotkina ON (2009) Taking into account seismic impact when calculating slope stability (using the example of the Krasnaya Polyana geodynamic site). Georisk 4:50–55 (in Russian)

Griffiths DV, Fenton GA (2004) Probabilistic slope stability analysis by finite elements. J Geotech Geoenviron Eng 130(5):507–518. https://doi.org/10.1061/(ASCE)1090-0241(2004)130:5(507)

Hoek E, Brown ET (2019) The Hoek-Brown failure criterion and GSI-2018 edition. J Rock Mech Geotech Eng 11(3):445–463. https://doi.org/10.1016/j.jrmge.2018.08.001

Kang K, Zerkal OV (2016) Application of probabilistic analysis in quantitative assessment of slope stability. Eng Geol 1:16–21

Kang K, Zerkal OV, Fomenko IK, Ponomarev AA (2018) Modern approaches to quantitative assessment of slope stability under seismic impact. Eng Geol 13(1–2):72–85 (in Russian). https://doi.org/10.25296/1993-5056-2018-13-1-2-72-85

Kang K, Fomenko I, Wang J, Nikolskaya OV (2020) Probabilistic assessment of rock slope stability in open pit mine Chaarat using the generalized Hoek–Brown criterion. J Min Sci 56(5):732–740

Kang K, Zerkal OV, Ponomarev AA, Fomenko IK (2021) Probabilistic slope stability assessment under seismic conditions based on the generalized Hoek-Brown criterion. Soil Mech Found Eng 58:223–229. https://doi.org/10.1007/s11204-021-09732-0

Kang K, Huang S, Liu W, Cheng H, Fomenko I, Zhou Y (2022) Sandstone slope stability analysis under wetting-drying cycles based on generalized Hoek-Brown failure criterion. Front Earth Sci 10:838862. https://doi.org/10.3389/feart.2022.838862

Konyushkov VV (2020) Comparison of the results of numerical modeling of slope stability in the Plaxis program with analytical calculations using a simplified method. Bull Civ Eng 3(80):108–115 (in Russian). https://doi.org/10.23968/1999-5571-2020-17-3-108-115

Liu H, Fei K, Gao Y (2003) Time history analysis method of slope seismic stability. Rock Soil Mech 24(4):553–556 (in Chinese)

Miao F, Wu Y, Xie Y, Yu F, Peng L (2017) Research on progressive failure process of Baishuihe landslide based on Monte Carlo model. Stoch Environ Res Risk Assess 31:1683–1696. https://doi.org/10.1007/s00477-016-1224-8

Morgenstern NR, Price VE (1965) The analysis of the stability of general slip surfaces. Geotechnique 15(1):79–93

Su C, Li J, Ren Q (2003) FEM-based dynamic stability analysis of abutment for high arch dams. J Hehai Univ (Nat Sci) 31(2):144–147 (in Chinese)

Tang H, Yong R, Ez Eldin MAM (2017) Stability analysis of stratified rock slopes with spatially variable strength parameters: the case of Qianjiangping landslide. Bull Eng Geol Environ 76:839–853. https://doi.org/10.1007/s10064-016-0876-4

Zerkal OV, Fomenko IK (2017) Influence of various factors on the results of probabilistic analysis of activation of landslide processes. Eng Geol (4):18–26 (in Russian). https://doi.org/10.25296/1993-5056-2016-1-16-21

Modelling of Deformation Processes in Structures of Sports Constructions

German Alekseev

Abstract At scientific and technical support of construction, it is necessary to model deformation processes both at the stage of construction and in the process of operation of the performed in order to develop a reconstruction project, long-term improvement of strength and deformation characteristics and to ensure regulatory requirements for limiting absolute and relative deformations of foundations. The results of modeling of deformation development in time of a unique sports facility located in the zone of increased geodynamic activity are presented.

Keywords Sports infrastructure structures · Loadings · Seismic loads · Deformations · Movements · Settlements · Operational loads

Ensuring the specified quality parameters of unique sports infrastructure facilities under construction, operation and reconstruction to ensure the safety of their operation during their entire life cycle is a very urgent task (Neguritsa et al. 2022; Oreste 2005; Oznamets and Tsvetkov 2018; Alekseev et al. 2023).

It is urgent to ensure safe operation of unique structural structures in conditions of increased geodynamic activity. For this purpose, it is necessary to conduct regular engineering and geological surveys, monitor the condition of structures, as well as take measures to strengthen and protect them from the negative impact of natural factors.

The complexity of the implementation of construction projects of unique sports infrastructure facilities lies in the very short timeframe for the entire complex of works, including survey, design, construction and operation of these facilities, in the complexity of construction conditions with a unique layout and the need to use unique architectural and engineering solutions (Kulikova 2018; Neguritsa 2018; Neguritsa et al. 2023). The shift in the timing of work on individual stages of work production, in the end, inevitably affects the quality of construction (Neguritsa et al. 2021; Zaitsev et al. 2015; Tereshin et al. 2016; Gribkova et al. 2016).

G. Alekseev (✉)
National Research Moscow State University of Civil Engineering, Moscow, Russia
e-mail: 1104780@gmail.com

T. Chaplina (ed.), *Processes in GeoMedia—Volume VIII*, Springer Geology,
https://doi.org/10.1007/978-981-97-6627-7_24

The chapter considers modeling of deformation of structures of a separate section of a sports grandstand of a unique structure located in a zone of increased geodynamic activity. The structural scheme of the section is a monolithic reinforced concrete grandstand frame with metal cover structures resting on it. The foundation of the building is a monolithic reinforced concrete slab and partially crossed belts 800 mm thick, made of concrete of class B25 with reinforcement of class A500 and A240. Monolithic walls 250 and 350 mm thick made of B25 class concrete; columns with dimensions 800 × 800 mm, 650 × 650 mm, 500 × 500 mm. Monolithic reinforced concrete beams 200 mm thick. Beams with dimensions 600 × 900 mm, 500 × 900 mm, 400 × 900 mm, 600 × 600 mm. Monolithic reinforced concrete stairs 200 mm thick.

The steel pavement structures are L-shaped cantilever frames with a 19.8 m outreach, which are supported on reinforced concrete structures along the B and D axes and on the foundation slab. Nodes of support on the foundation slab are rigid, on axis B and D—hinged. The frames are made of welded I-beams of variable cross-section. The purlins and connections are made of rolled sections. All structures are made of steel. The overall stability of the structures is ensured by the joint operation of the frame structures and the system of links with reinforced concrete walls, columns and floor discs.

Loads in the calculation of operational and seismic loads

Load 1: Weight of the MC;
Load 2: Weight of reinforced concrete;
Load 3: Floors;
Load 4: Roofing;
Load 5: Suspended chute and equipment;
Load 6: Partitions;
Load 7: Facade;
Load 8: Usable on slabs;
Load 9: Useful load on bleachers;
Load 10: Seismic of design earthquakes of the DE (X);
Load 11: Seismicity of design earthquakes of the DE (Y);

1 Building Calculation

Calculation of the main load-bearing elements of monolithic reinforced concrete structures and steel structures of the above-ground part of the building of the sports infrastructure facility, for the following types of impacts: operational loads; seismic impacts of the earthquake level with the calculated seismic impact used to check the seismic resistance of all structures located on this construction site—8 points, 3 category of soil on seismic properties. The calculation was carried out using the LIRA-SADP software package. The calculation is based on the finite element method in displacements.

The following node displacements are taken as the main unknowns:

- X linear along the X axis;
- Y linear along the Y axis;
- Z linear along the Z axis;
- UX angular around the X axis;
- UY angular around the Y axis;
- UZ angular around the Z axis.

The following element types are included in the design scheme:

- Type 10. Universal spatial rod-rod FE for modelling columns;
- Type 41. Universal rectangular shell FE for modelling floors and walls;
- Type 42. Universal triangular shell KE for modelling floors and walls;
- Type 44. Universal quadrangular KE shells for modelling floors and walls.

2 Interpretation of Calculation Results

Linear movements are considered positive if they are directed along the coordinate axes. Positive angular movements correspond to counterclockwise rotation when viewed from the end of the corresponding axis.

The movements are indexed as follows:

- X linear on the X axis;
- Y linear in the Y axis;
- Z linear in the Z axis;
- UX angular around the X axis;
- UY angular around the Y axis;
- UZ angular around the Z axis.

3 Indexing and Rules for Force Signs in Finite Elements

Type 10. Universal spatial rod-rod KE.

The end element absorbs the following types of forces:

N axial force; the positive sign corresponds to stretching.

M_k torque about the X_1 axis; the positive sign corresponds to the counterclockwise action of the torque, when viewed from the end of the X_I axis, on the section belonging to the end of the rod.

M_Y bending moment with respect to the Y_1 axis; the positive sign corresponds to the counterclockwise action of the moment, when viewed from the end of the Y_1 axis, on the section belonging to the end of the rod.

M_z bending moment about the Z_1 axis; the positive sign corresponds to the counterclockwise action of the moment, when viewed from the end of the Z_1 axis, on the section belonging to the end of the rod.

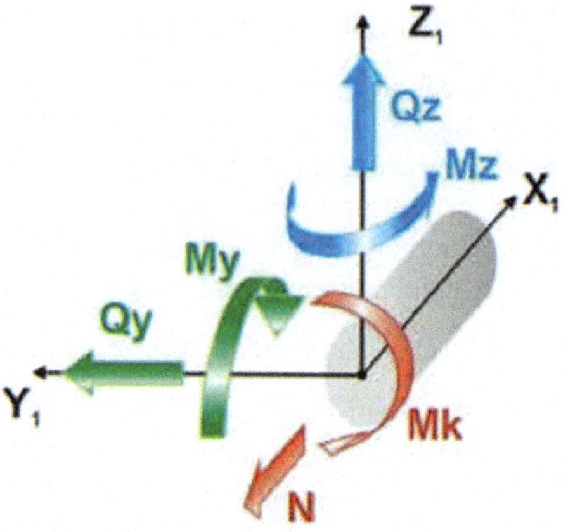

Fig. 1 Forces in rod cross-section Type 41, 42 and 44. Universal triangular and quadrilateral finite element shells

Q_y shear force along the Y_1 axis; a positive sign corresponds to the coincidence of the direction of the force with the Y_1 axis for the section belonging to the end of the rod.

Q_z shear force along the Z_1 axis; a positive sign corresponds to the coincidence of the force direction with the Z_1 axis for the section belonging to the end of the rod (Fig. 1).

The end element absorbs the following types of forces, stresses and reactions: N_x normal stress along axis X_I; positive sign corresponds to tension.

N_y normal stress along the Y_1 axis; positive sign corresponds to tension.

N_z normal stress along the Z_1 axis (in the case of plane deformation); the positive sign corresponds to tension.

TX_y shear stress parallel to the X_I axis and lying in a plane parallel to $X_{10}Z_1$; the direction coinciding with the direction of the X_I axis is taken as positive if NY coincides in direction with the Y_1 axis.

M_x the moment acting on the section orthogonal to the X_I axis; a positive sign corresponds to the stretching of the bottom fibre (with respect to the Z_1 axis).

M_Y moment acting on the section orthogonal to the Y_1 axis; positive sign corresponds to the stretching of the bottom fibre (with respect to the Z1 axis).

MX_y torque; positive sign corresponds to the curvature of the diagonal 1–4, directed convexly downwards (with respect to the Z_1 axis).

Q_x shear force in the section orthogonal to the X_I axis; the positive sign corresponds to the coincidence of the force direction with the Z_1 axis direction in the part of the element where node 1 is missing.

Q_y shear force in the section orthogonal to the Y_1 axis; a positive sign corresponds to the coincidence of the force direction with the Z_1 axis direction on the part of the element where node 1 is missing (Fig. 2).

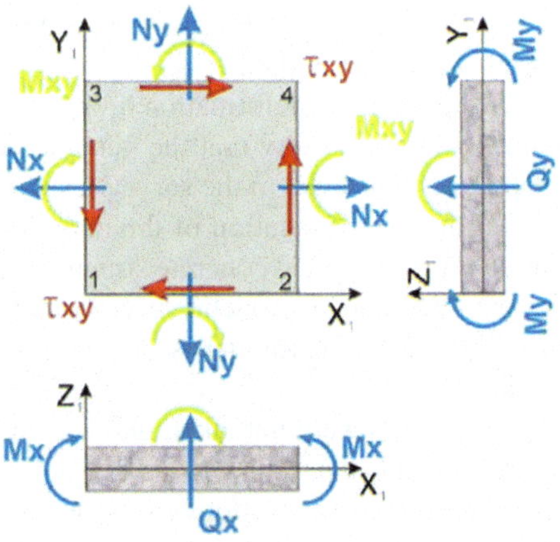

Fig. 2 Forces in the shell

4 Calculation Assumptions

Calculation for operational loads was carried out according to I and II group of limit states. Calculation for special loads was carried out according to Group I of limit states. Load reliability coefficients and combination coefficients are taken in accordance with the normative requirements (Fig. 3).

The reliability coefficient of responsibility is $\gamma_n = 1$.

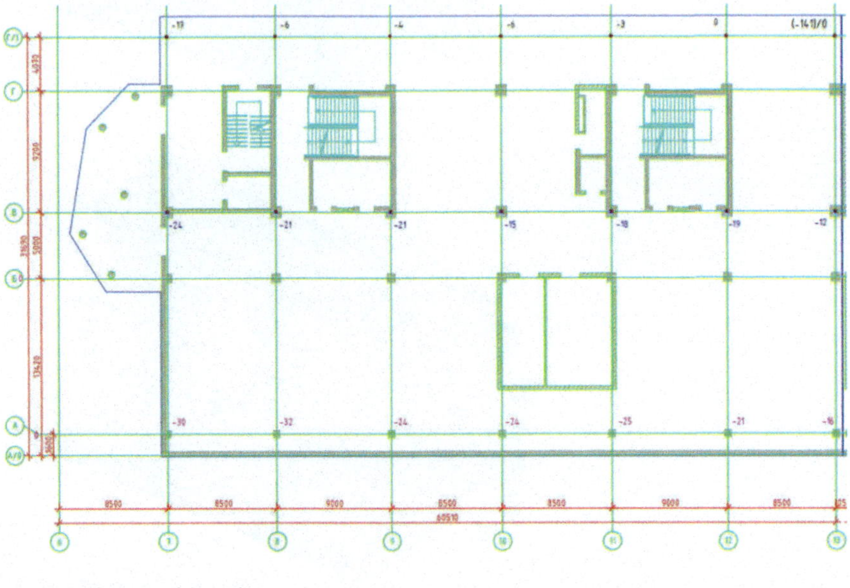

Fig. 3 Settlement of 1 section

The interaction of the structures with the foundation soil is set in accordance with the geodetic survey data.

The modulus of deformation of the foundation soils in the calculation model was selected in such a way that the general character of the foundation slab settlement isopoles corresponds to the survey results (Figs. 4, 5, 6, 7, 8, 9, 10, 11 and 12).

Verification calculation of the main load-bearing elements of the above-ground monolithic reinforced concrete structures of the sections was carried out, taking into account the data of geodetic survey for operational loads and seismic effects of the level of artificial earthquakes of 8 points, 3 category of soil on seismic properties (Table 1).

Snow loads were not taken into account in the calculations. The values of short-term uniformly distributed loads were taken with a reduction factor of 0.5. Horizontal displacements of the section cover structures do not exceed the limit normative values:

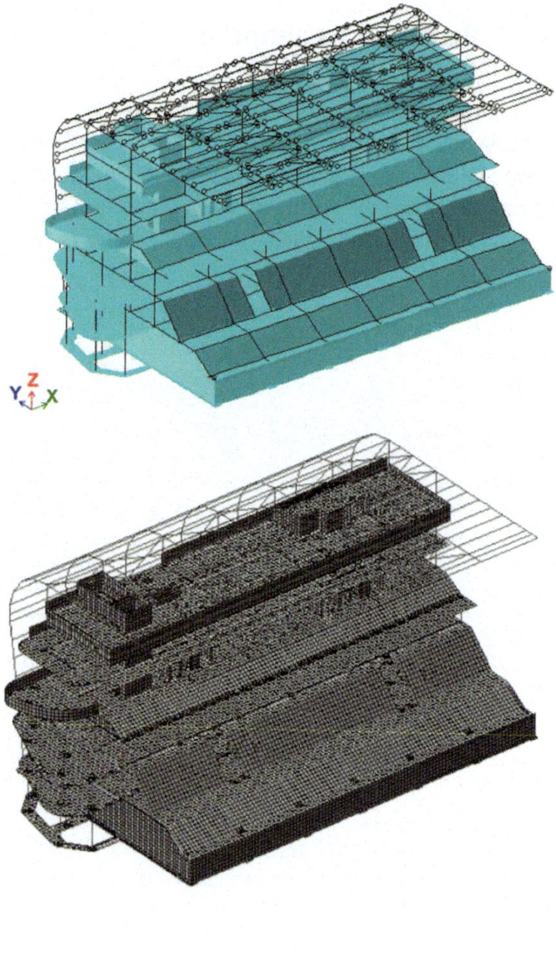

Fig. 4 Results of the calculation of the sports building for operational loads. Calculation scheme of the building

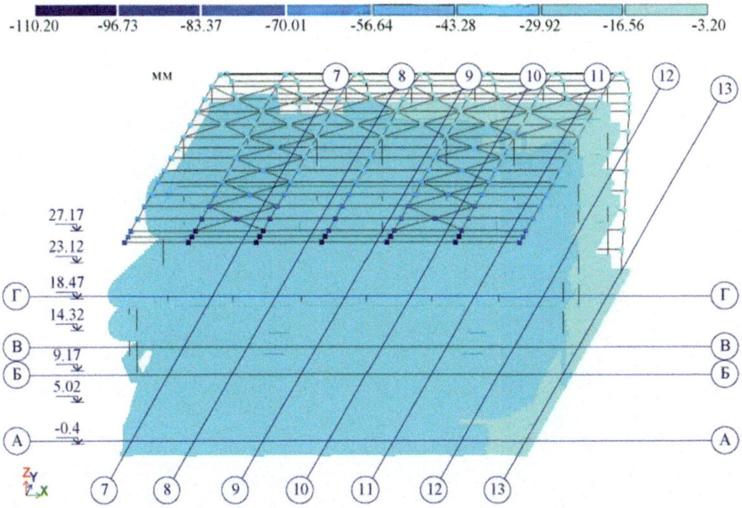

Fig. 5 Building displacements Z displacements, mm

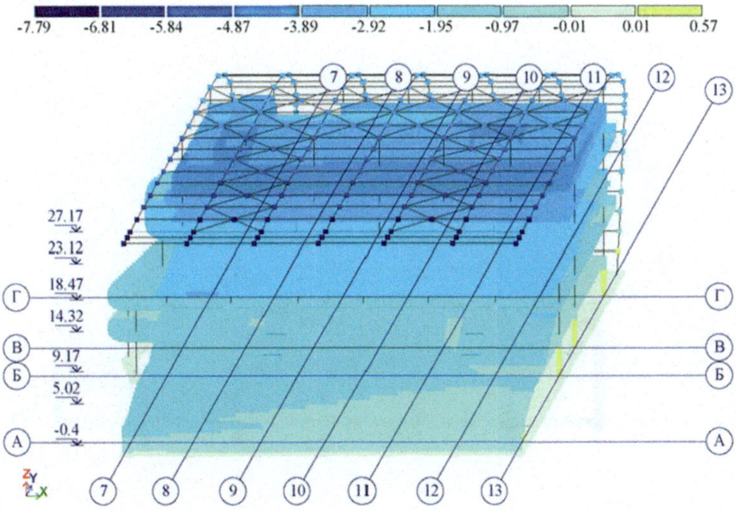

Fig. 6 Horizontal movements in the X axis

$$\Delta \max = 2.9 \text{ cm} < \Delta \lim = \frac{1}{500} 2800 = 5.6 \text{ cm}.$$

The maximum structural settlement of the foundations of Section No. 1 does not exceed the limit normative values:

$$Z\max = 29.6 \lim < Znp = 150 \text{ mm}.$$

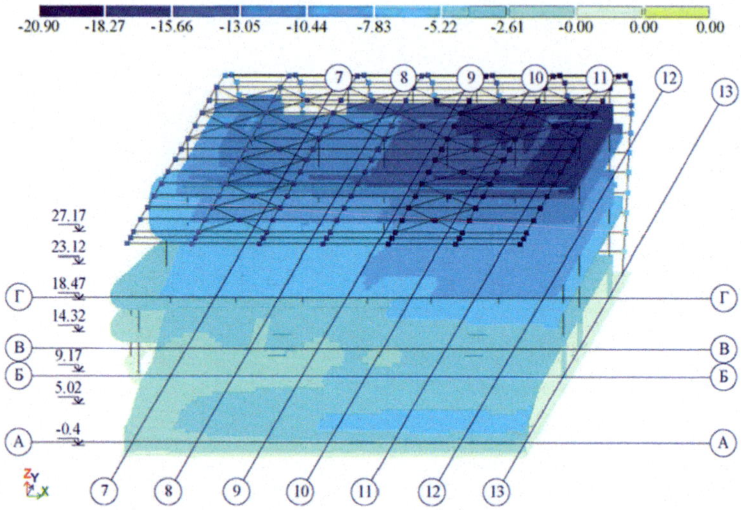

Fig. 7 Horizontal movements in the Y axis

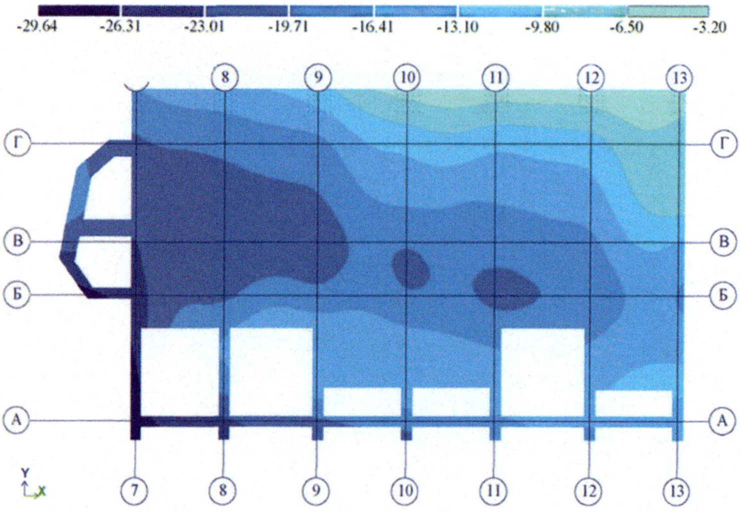

Fig. 8 Sedimentation

Relative difference in structural settlement of the foundations of the section:

$$(As/L) = (2.96 - 0.32)/6195 = 0.00043 < (\Delta s/L)_u = 0.003.$$

The deformability conditions are fulfilled.

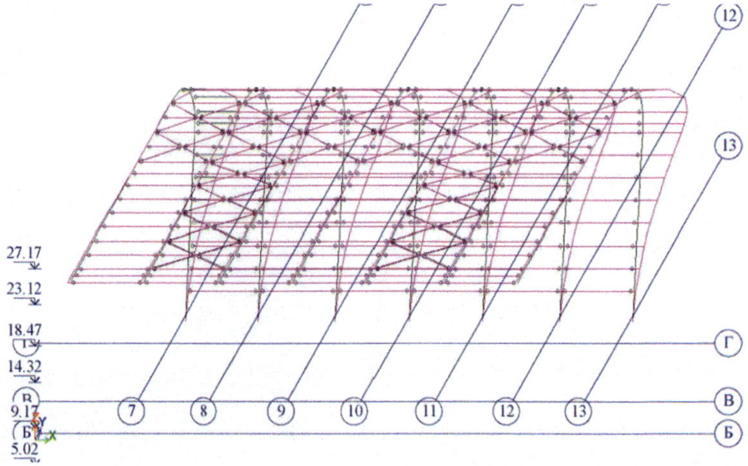

Fig. 9 Determination of the overall stability margin 1 form stability margin factor 4.76

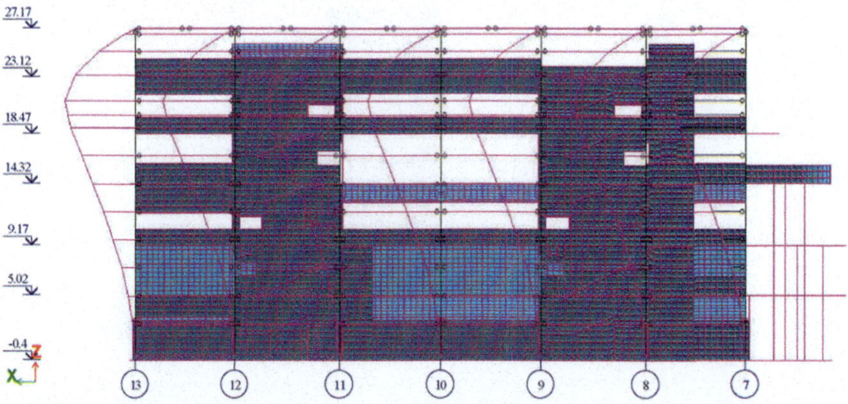

Fig. 10 Shapes and frequencies of building oscillation 1 shape

The modeling showed that the development of deformation processes in this part of the sports facility is within the established safe limits. The obtained results will be used in the organization of surveying and geodesic monitoring during the reconstruction of the facility.

Fig. 11 Shapes and frequencies of building oscillation 2 shape

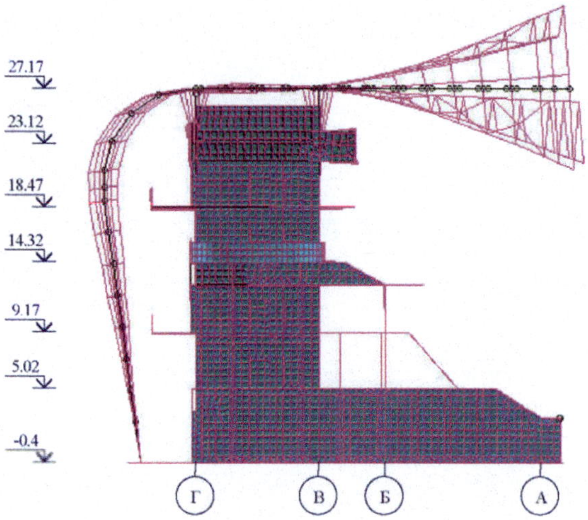

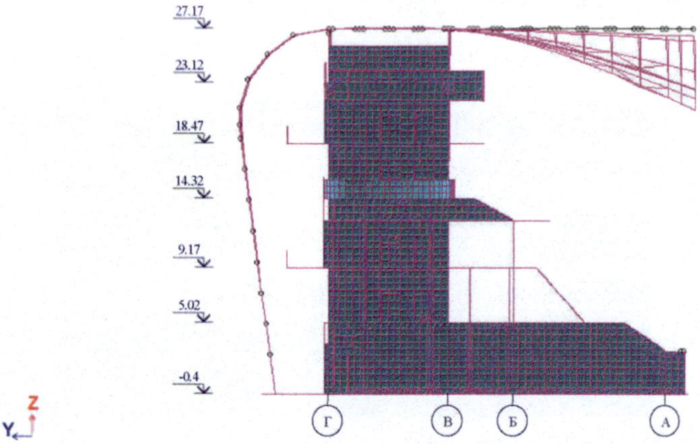

Fig. 12 Shapes and frequencies of building vibration 9 shape

Table 1 Frequencies of natural vibrations

№ forms	Own values	Frequencies		Period (c)
		Circle, frequency (rad/s)	Frequency (Hz)	
1	0.174394	5.73413	0.912615	1.095752
2	0.104621	9.558337	1.521257	0.657351
9	0.085252	11.729972	1.866883	0.535652

References

Alekseev G, Medvedev E, Neguritsa D (2023) Monitoring of deformation processes during scientific and technical support of construction. E3S Web Conf 410:02045. https://doi.org/10.1051/e3sconf/202341002045

Gribkova LA, Skripkina IA, Shabanova IG, Lychagin DV, Gorbenko VV, Salnikov YaO (2016) Geodetic deformation monitoring of the technical condition of unique buildings and structures. Branch Sci Appl Res Earth Sci 2:104–108

Kulikova EY (2018) Defects of urban underground structure and their prediction. IOP Conf Ser Mater Sci Eng 012108. https://doi.org/10.1088/1757-899X/451/1/012108

Neguritsa D (2018) The problems of monitoring the deformation processes in the integrated development of the underground space of metropolitan cities. E3S Web Conf 56:02027. https://doi.org/10.1051/e3sconf/20185602027

Neguritsa DL, Alekseev GV, Medvedev EA, Tereshin AA (2021) Geomechanical support of the multifunctional centre construction in megapolis conditions. Izv Tula State Univ Earth Sci 3:286–296 (in Russian)

Neguritsa DL, Alekseev GV, Tereshin AA, Medvedev EA, Slobodin KM (2022) Deformation processes modelling throughout underground construction within megapolis limits. In: Processes in GeoMedia. Springer geology, vol IV. https://doi.org/10.1007/978-3-030-76328-2_23

Neguritsa DL, Alekseev GV, Medvedev EA, Tereshin AA (2023) Assessment of the impact of the construction of underground structures with the use of special methods on underground pedestrian passengers. In: Springer proceedings in earth and environmental sciences, pp 641–646. https://doi.org/10.1007/978-3-031-25962-3_64

Oreste P (2005) Back-analysis techniques for the improvement of the understanding of rock in underground constructions. Tunn Undergr Space Technol 20(1):7–21. https://doi.org/10.1016/j.tust.2004.04.002

Oznamets VV, Tsvetkov VY (2018) Geomonitoring: state and development. MAKS Press LLC, Moscow, 112 p

Tereshin AA, Neguritsa DL, Alekseev GV (2016) Monitoring of the building deformations by geodetic methods during the reconstruction of its foundation. Bull Peoples' Friendship Univ Russ Ser Eng Res 4:84–90

Zaitsev AK, Simonyan VV, Shmelin NA (2015) Geodesic monitoring of buildings and structures. NIU MSCU, Moscow, 144 p

Charts Resultant Interpretation of Geochemical Data

G. A. Ponomareva

Abstract A new method of processing a multidimensional analytical material based on the use of charts resultant is offered. This tool opens up wide opportunities for rapid interpretation of hidden processes that determine the nature of the investigated geochemical data.

Keywords Resultant chart points · Diagnostic measuring grid · Multidimensional data

1 Introduction

One of the main tasks of analytical results processing is a reasonable grouping of observations according to the composition of analyzed components. It is the basis of numerous classifications of numerical material necessary for its taxonomic recognition.

The recognition procedure itself is usually reduced to the projection of the studied samples from the feature space into other integrated coordinate systems of truncated dimensionality. In geological subject areas they can be described in terms of mineral and petrochemical associations, formational, geographical or other affiliation. An indirect attribute of classification success is the identification of explicable from the researcher's point of view groupings of figurative points, which gives serious grounds for various predictive ideas and hypotheses.

An important point of the described interpretation on scheme is the need for a meaningful interpretation of each of the formal classifications of analytical material. As a rule, it is based on the speculative study of some set of point diagrams illustrating the distribution of the studied samples in certain slices of the attribute space. However, its multiaxiality seriously complicates this procedure, since for the completeness of the experiment it should be performed at least for all orthogonal perspectives. In

G. A. Ponomareva (✉)
Orenburg State University, 13 Pobeda Avenue, Orenburg 460018, Russia
e-mail: galy.ponomareva@mail.ru

T. Chaplina (ed.), *Processes in GeoMedia—Volume VIII*, Springer Geology,
https://doi.org/10.1007/978-981-97-6627-7_25

complex cases, the number of chart planes to be studied can quickly exceed the limit of sense perception (Nikiforov 2014; Nikiforov et al. 2014).

This paper describes a new approach to the meaningful classification of geochemical samples, which has proven itself positively when processing the results of sampling a wide range of geological formations and analytical data. It is based on the generation of an chart resultant, which is a two-dimensional map of figurative points, the position of each of which corresponds to the complex composition of the sample.

Thus, if the chart is constructed correctly, geochemical samples of close composition must necessarily neighbor each other forming accumulations and clusters. A possible exception are cases when objectively different between themselves content of analyzed components are represented by the same numbers, for example, due to the peculiarities of the analytical method.

However, this problem equally complicates any way of classifying numerical material.

2 Methods and Results

The algorithm for constructing the chart resultant is described in detail in Nikiforov (2014), Nikiforov et al. (2014). It is explained in Fig. 1, which shows the scheme of positioning the figurative point of the two-dimensional chart resultant for a six-dimensional attribute space.

Here, for clarity, the radial vectors are signed by the content of attributes in the analyzed sample, and the broken line shows the imaginary trajectory or track of the figurative point movement from zero to the final position ($X = -6.35$; $Y = 1.64$).

Each track segment is labeled with the name of the vector to be summarized, and its ending indicates the intermediate resultant of all vectors preceding it. The described actions are very simple to implement in any algorithmic language.

The application of the resultant chart is especially productive at the initial stage of interpretation of multidimensional geologic data. Since this (in essence—graphical technique) does not provide complex statistical calculations, it allows simultaneous processing of analytical data of different scale and degree of accuracy. To illustrate the above, let us consider an example of studies of the metal contentы of oil from a number of fields in the Orenburg region, in order to clarify their nature and possible sources of metals.

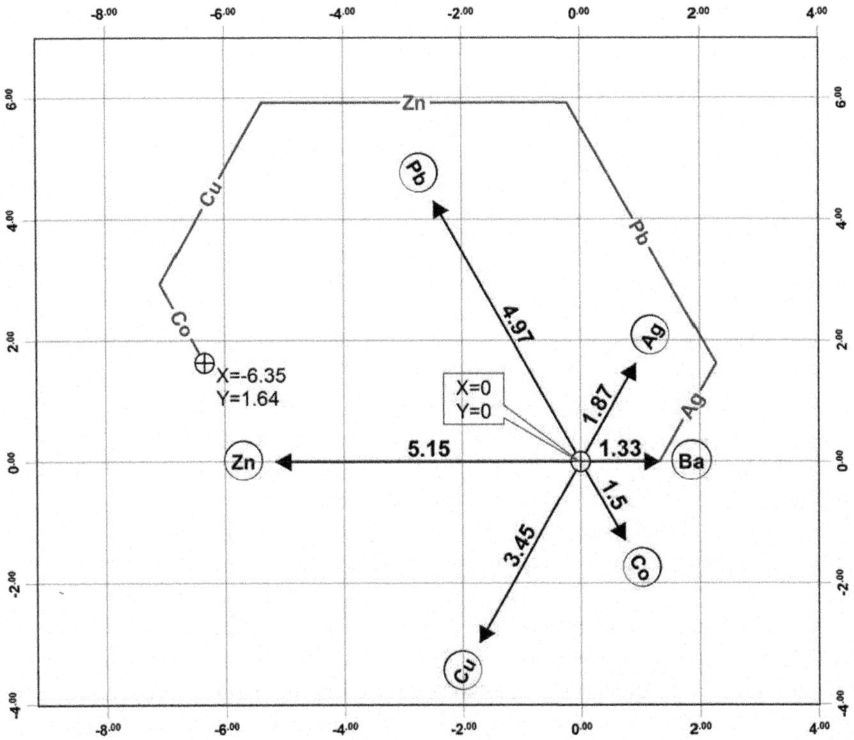

Fig. 1 Example of calculation resultant for six attributes (metal concentrations)

3 Geochemical Interpretation of Data

The laboratory of physical research methods of the Orenburg State University has been conducting systematic studies of oil metallogeny from fields in the western part of the Orenburg region for a number of years. Oil samples were taken by reservoir tester in terrigenous and carbonate reservoirs of different but carefully controlled stratigraphic and structural-tectonic affiliation. Analyses of geochemical samples for noble metals were performed by atomic absorption spectrometry with electrothermal atomizer (ETA AAS) on the spectrometer MGA-915 using the patented method (Russian Federation Patent No. 2409810) for determination of platinoids in carbonaceous media (Ponomareva 2013). For diagnostics of other elements semi-quantitative spectral analysis was used.

The accumulated analytical material (more than 500 elemental determinations) was investigated to reveal hidden regularities that shed light on the nature of metals in hydrocarbons.

To select informative features from the initial set (Au, Ag, Pd, Pt, Co, Ni, Cu, Cu, Pb, Cr, V, Ti, Mo, Zr), the most available methods of nonparametric statistics were used. The attributes were ranked, and as a non-strict criterion of their informativeness, a simple number of ranks was used, especially high for noble metals. Randomization of ranks by random permutations (Gaek 1971) was used to exclude repetitions arising due to quantization of spectral analysis results. Elements with a rank range below the average were excluded from the informative set.

Additional steps of geochemical material preparation were:

- recalculation of the initial results of the analysis of informative attributes into the scale of standard deviations to eliminate the effect of variability;
- harmonization of the obtained results to positive values (Nikiforov et al. 2014; Ponomareva 2013).

The main parameters of the statistical distribution of metals of the final geochemical sample are presented in Table 1.

The obtained data allowed us to construct a six-axis chart resultant in the space of attributes: Au, Ag, Pd, Pt, Co, Ni. In addition, the so-called diagnostic palette was created on its basis, which is a mosaic of Thiessen polygons merged by the most important attributive characteristics into similar loci.

Figure 2 shows the chart resultant combined with the diagnostic palette, the loci of which reflect the lithologic belonging of the sampled reservoirs (Nikiforov et al. 2014).

At first glance, the analysis of the figure confirms the genetic connection of oil metalliferousness with the type of reservoirs. Indeed, the diagnostic palette is divided into two dominant loci of terrigenous and carbonate composition, which may indicate hydrocarbon contamination of metals from the host rocks. However, the test of the hypothesis about the difference in statistical distributions of the studied components in the rock material of terrigenous and carbonate reservoirs was not confirmed.

The author suggests that the solution to this issue is related to the influence of an additional factor that can simultaneously affect the composition of hydrocarbons and lithology of different-aged reservoirs.

Table 1 Statistical characteristics of metals content in oil (mg/t)

Parameters of statistical distribution	Au	Ag	Pd	Pt	Co	Ni
Average	12.49	15.71	7.51	17.66	9.67	8987.15
Standard error	4.00	3.27	1.42	3.67	2.22	2212.32
Median	6.00	11.00	6.00	10.00	5.00	5080.00
Mode	1.00	0.10	6.00	18.00	3.00	180.00
Standard deviation	24.95	20.40	8.85	22.89	13.87	13,815.92
Excess	27.82	4.70	15.90	5.23	7.37	18.43
Asymmetry	4.96	2.10	3.61	2.29	2.77	3.84
Minimum	0.30	0.10	0.10	0.40	1.00	100.00
Maximum	153.00	90.00	51.00	102.00	63.00	79,870.00

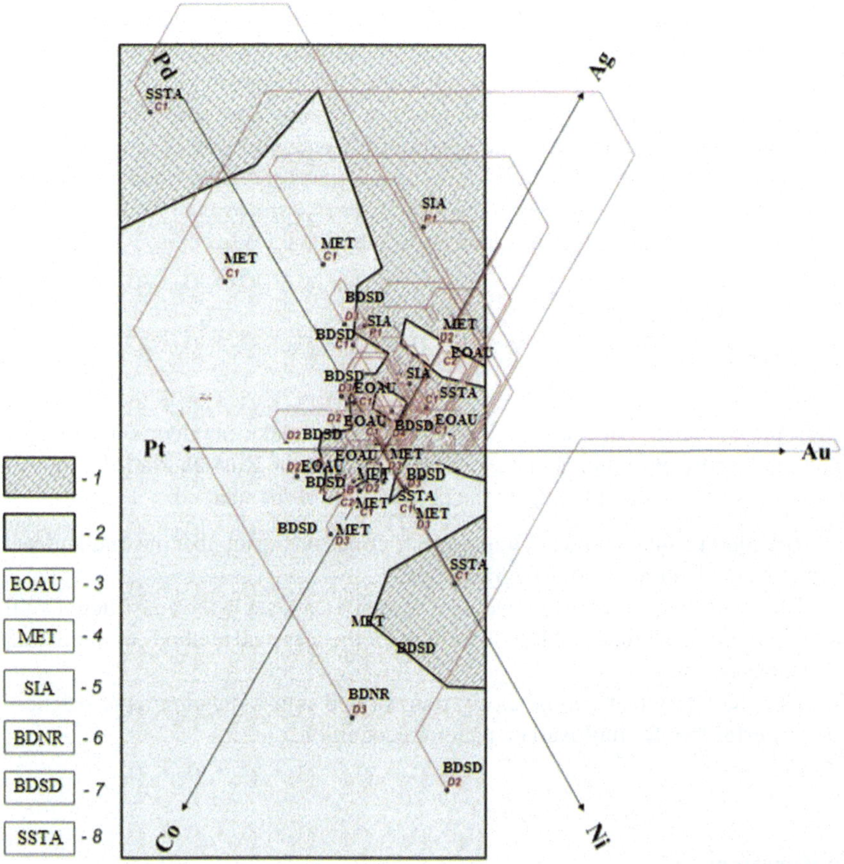

Fig. 2 Chart resultant metal contents of oils. 1—carbonate reservoir; 2—terrigenous reservoir; 3 EOAU—East Orenburg arch uplift; 4 MET—Mukhano-Erokhovsky trough; 5 SIA—Sol-Iletsky arch; 6 BDNR—Buzuluk depression: northern region; 7 BDSD—Buzuluk depression: southern dive; 8 SSTA—southern slope of the Tatar arch

Each figurative point in Fig. 2 is signed with the abbreviation of the structural-formational zone in which the tested reservoir is located.

It can be seen that figurative points falling in the terrigenous field are, with few exceptions, associated with negative tectonic elements, while carbonate reservoirs, on the contrary, are associated with positive structures.

It is not excluded that tectonic processes controlling the structural, structural-facial and lithological diversity of the geological section in some cases were accompanied by deep hydrocarbon emanations of mantle nature. Such circumstances could occur during the inversion of phases of geotectonic cycles, confidently fixed by the accumulation of clastic material in depressions.

In this case, judging from Fig. 2, the main contribution to the metal-bearing component of oil fluids is made by Pd, Pt and Co, which noticeably shift the "terrigenous" figurative points to the left half of the diagnostic palette.

The areas of the chart with predominance of carbonate reservoirs are associated with relatively stable tectonic conditions. Concentrations of analyzed metals within the corresponding structural-formational zones are determined by dispersion processes and are close to the average value ones (Nikiforov et al. 2014).

However, confirmation of this idea needs additional research.

4 Conclusion

Thus, the application of chart resultant allowed us to make an important assumption about the nature of oil fluids in the eastern margin of the Russian Platform. In many respects it became possible due to the peculiarities of the method:

1. The chart resultant does not impose restrictions on the number of simultaneously projected features on the diagram plane 2.
2. The requirements for the preparation of analytical data in the construction of the diagram of equalities are less strict than for the classical methods of multivariate statistics.
3. The chart resultant can be easily transformed into a diagrammatic palette—a powerful tool for multivariate pattern recognition.

References

Gaek Ya (1971) Shidak 3. Theory of rank criteria. Transl from English. M.: Nauka

Nikiforov IA (2014) Geochemical classifications using the method of mapping resultants. Geochemistry (4):361–369

Nikiforov IA, Ponomareva GA, Nikiforov IA (2014) Interpretation of geochemical data using resultant diagrams. International Independent Institute of Mathematics and Systems "MiS". Materials of the VIII international scientific and production complex "Topical issues of development of innovative activity in the new millennium", Novosibirsk, no 8, pp 104–108

Ponomareva GA (2013) Regional regularities of platinoids distribution in the Orenburg part of the Southern Urals. Autoref Dis Kand Geol-Min Nauki: 25.00.11. Ekaterinburg, 23 c

Critical Analysis of Remote Sensing Data of Chlorophyll-a in the Surface Waters of Lake Onego

V. N. Baklagin and I. S. Novikova

Abstract The work is devoted to a comparative analysis of processed satellite data of chlorophyll-a concentration for the period 1998–2022 and field data collected during expeditions by the Laboratory of Hydrobiology of the Northern Water Problems Institute of the Karelian Research Centre of the Russian Academy of Sciences and published in scientific periodicals. GlobColour datasets provided by the Copernicus Marine Environmental Monitoring Service and datasets of the Ocean Color Climate Change Initiative project of the European Space Agency were used. The comparison results showed that GlobColour data are more consistent with field data of the Lake Onego than OC-CCI data: average absolute error values for the period 1998–2020 are 2.97 μg/l (n = 509) and 5.24 μg/l (n = 33) respectively.

Keywords Lake Onego · Chlorophyll-a · Satellite data · Field data

1 Introduction

A large freshwater body on the European continent, Lake Onego is one of the most significant landscape elements in the North-West Russia in many aspects. Water exchange intensity is 16 year^{-1}. More than 150 rivers flow into Lake Onego and its drainage is the Svir River, which flows into Lake Ladoga (Filatov et al. 2015; Kalinkina et al. 2017; Filatov 2010). The northern basin of the lake is characterized by a rugged shoreline with numerous bays and cliffs, while the southern shoreline is flat and weakly indented. Traditionally, the main limnic areas of Lake Onego are distinguished (Fig. 1): Poveneckij bay, Zaonezhskij bay, Kondopozhskaya bay, Bol'shoe Onego, Maloe Onego, Petrozavodskaya bay, Central'noe Onego, Yuzhnoe Onego (Filatov et al. 2015).

V. N. Baklagin (✉) · I. S. Novikova
Northern Water Problems Institute Karelian Research Centre Russian Academy of Sciences, Petrozavodsk, Republic Karelia, Russia
e-mail: slava.baklagin@mail.ru

T. Chaplina (ed.), *Processes in GeoMedia—Volume VIII*, Springer Geology,
https://doi.org/10.1007/978-981-97-6627-7_26

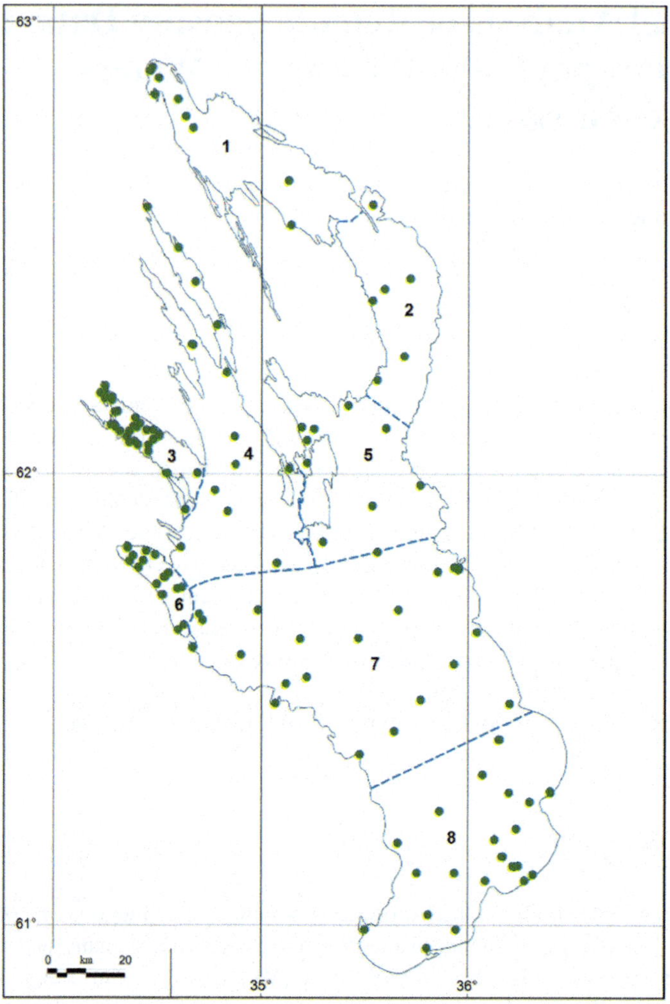

Fig. 1 Location of sampling stations on Lake Onego for 1960–2022: 1—Poveneckij bay, 2—Zaonezhskij bay, 3—Kondopozhskaya bay, 4—Bol'shoe Onego, 5—Maloe Onego, 6—Petrozavodskaya bay, 7—Central'noe Onego, 8—Yuzhnoe Onego

The complex structure of the lake basin, its morphometric composition, uneven distribution of inflows and anthropogenic pressure in total determine the spatiotemporal originality of hydrological, hydrochemical and hydrobiological processes in the water area of Lake Onego.

Lake Onego, being an important water body for economic multi-purpose activities in the North-West of the country, requires regular and in-depth monitoring of the ecological state and its dynamics. The main factors that can lead to changes in the state of the Lake Onego water are: anthropogenic impacts, biological pollution and climate change (Filatov et al. 2015).

Along with such parameters as the concentration of suspended particles and colored dissolved organic matter, the concentration of chlorophyll-a is one of the most important indicators of the trophic status and water quality of a reservoir (Pozdnyakov and Filatov 2021; Dekker et al. 1995; Bridget et al. 2021). Take for example, there are a lot of US federal and regional water quality monitoring programs (Nutrient Criteria Technical Guidance Manual: Lakes and Reservoirs. EPA 822-B-00-001, National Lakes Assessment: A Collaborative Survey of the Nation's Lakes. EPA 841-R-09-001), which are more focused on obtaining regular chlorophyll-a concentration data as a priority information.

The most common approach to get chlorophyll-a concentration data is in situ water sampling using watercraft. This method provides high quality obtained data, but it has significant drawbacks arising due to irregular spatiotemporal data. The problem is especially acute for large water bodies, such as Lake Onego, that makes regular, statistically reliable and required time resolution water sampling of the chlorophyll-a concentration (as well as other water quality parameters) expensive and logistically difficult, and in many cases even impossible in the water area of the lake (Kondratyev and Filatov 1999; Papenfus et al. 2020). Although observations of the parameters characterizing the ecological state of Lake Onego have been regularly carried out for several decades by the Northern Institute of Water Problems of the Federal Research Center "Karelian Scientific Center of the Russian Academy of Sciences" during expeditions—sampling from research vessels (Filatov et al. 2015). Individual or limited in number and time sampling from reservoir does not allow for a complete assessment of the reservoir entire condition (Lesht et al. 2018). Although time constraints can be partially overcome by continuously operating probes, the spatial distribution of data remains undetailed (Suarez et al. 2019).

Remote sensing methods using for estimating chlorophyll-a concentration in lakes significantly complement field data, since this approach solves the problems of spatiotemporal limitations of the sampled data. In addition, remote sensing data are effectively used to assess long-term system changes over wide spatial and temporal scales, as well as to observe rapidly evolving (short-term) events, such as episodic algal blooms (Pozdnyakov and Filatov 2021; Bridget et al. 2021; Pozdnyakov et al. 2013).

The study of the dynamics of water quality parameters of Lake Onego (including chlorophyll-a concentration) was also carried out using satellite data, in particular, the visible range (Filatov et al. 2015). The study used data obtained in the optical range by the MERIS spectrometer (English: The Medium Resolution Imaging Spectrometer) for 2011. Based on these data, high-resolution spatial distributions of chlorophyll-a concentration (300 m) were obtained for the first time, allowing to estimate the seasonal dynamics of chlorophyll-a concentration and in the central part of Lake Onego, as well as in large bays. In addition, there was developed a more detailed classification of trophic level for individual parts of this water area as a result of studies assessing the trophic level of Lake Onego conducted in 2021 (Korosov et al. 2021). The classification is based on such indicators as the amount of saprophytic bacterioplankton, biomass of crustacean zooplankton, chlorophyll-a concentration and water transparency.

However, despite the advantages of using remote sensing data for water bodies, achieved due to temporal and spatial regularity, they require statistically confirmed conclusions about comparability with field observation data due to errors that arise during detection, especially for such reservoirs as Lake Onego, which has the rugged shoreline and complex morphometric structure.

In this regard, the purpose of this study is comparative analysis chlorophyll-a concentration from remote sensing of Lake Onego surface water for the period 1998–2022 and field observation data obtained during expeditions.

2 Materials and Methods

In this work we used daily chlorophyll-a concentration data at a 4 × 4 km spatial resolution, available for the period from September 1997 to December 2022 from the GlobColour project of the CMEMS Marine Environment Monitoring Service (Copernicus Marine Environment Monitoring Service) (https://data.marine.copernicus.eu/products). The GlobColour dataset was created by combining data from different satellite sensors: the Sea-viewing Wide Field-of-view Sensor (SeaWiFS), the Moderate Resolution Imaging Spectroradiometer (MODIS) on board the satellites Aqua and Terra, the Medium Resolution Imaging Spectrometer (MERIS), a set of visible and IR resolution sensors VIIRS-SNPP (Visible Infrared Imaging Radiometer Suite—Suomi National Polar-orbiting Partnership) and JPSS1 (Joint Polar Satellite System), the sensor for recording ocean and land optical characteristics—Ocean and Land Color Instrument (OLCI) of satellites Sentinel-3A and Sentinel-3B. One of the characteristics of this set is the presentation of multi-year time series data, that allows to be used with greater accuracy for scientific purposes, because it is applied a more advanced primary processing algorithm to them (Beckers and Rixen 2003; Volpe et al. 2007, 2012, 2019) than for real-time data. These data have a fourth level of processing (L4), characterized by the combination of several detection algorithms for chlorophyll-a concentration: CI for oligotrophic waters, the general approach OCx (Ocean Color) (OC3, OC4 or OC4Me depending on sensor) for mesotrophic waters and the algorithm OC5 for coastal waters (Garnesson et al. 2019), as well as spatiotemporal interpolation of combined arrays of chlorophyll-a concentrations for each sensor (Saulquin et al. 2019), that minimized missing data located outside the sensor swaths, as well as missing data associated with the presence of clouds.

To compare the results of different satellite data, the OC-CCI (Ocean Color Climate Change Initiative) set of the European Satellite Agency (ESA) version 6.0 was also used in this work. This dataset is represented of monthly chlorophyll-a concentration data at a 4 × 4 km spatial resolution for the period from September 1997 to December 2022 (https://climate.esa.int/en/projects/ocean-colour/data/). The data is based on data from MERIS, SeaWiFS, MODIS, VIIRS, Sentinel 3A and 3B-OLCI. The OC-CCI represents multi-year time series dataset, that are necessary for assessing climate trends. Chlorophyll-a concentrations were calculated using the OCI, OCI2, OC2, OC3, OCx (OC3/OC4 ratio) and OC5 algorithms, which take into account different types of natural water (Jackson et al. 2017; Moore et al. 2009).

The GlobColour and OC-CCI datasets are generated differently (Garnesson et al. 2019). Thus, when forming the GlobColour dataset, the chlorophyll-a concentration value is calculated for each sensor taking into account its characteristics (spectral range and resolution), and then obtained values are combined. In addition, in the OC-CCI dataset, obtaining chlorophyll-a concentration value is preceded by combining reflectance values from different sensors.

Calculations and additional data processing were performed using software developed in the Python programming language within the framework of the Russian Science Foundation project No. 22-17-00193. This software is integrated with the information and analytical system "Lake Onego and its catchment" for calibration and verification of a three-dimensional model of the Lake Onego ecosystem, as well as for visualization and presentation of calculation results and satellite observation data, including the concentration of chlorophyll-a.

To compare the results of processing satellite data of chlorophyll-a concentrations and field data, we used data collected from research vessels of Lake Onego water as a result of expeditionary research by the Institute of Water Problems of the North for the period 1998–2020 (Filatov et al. 2015; Kalinkina et al. 2017, 2021; Tekanova et al. 2018). This data was structured and collected in the database "chlorophyll-a" of Lake Onego water (Sabylina et al. 2018). Over the period 1998–2020, 2174 measurements of surface chlorophyll-a concentrations were collected from May to October at 133 sampling stations (Fig. 1).

Comparison of series of individual values of chlorophyll-a concentrations obtained during expeditions and from satellite observations was carried out ensuring spatiotemporal synchronicity of the compared values. There was created a program in the Matlab and Python programming languages for this purpose, that carries out the procedure for selecting GlobalColour and OC-CCI satellite datasets most closely correspond to sampling stations finding values by sampling date, and also latitude and longitude indices for lake surface waters only (at the depth of less than 1 m).

The average value of absolute errors was determined by the formula:

$$\overline{\delta} = \frac{\sum_{i=1}^{n} |chl_{insitu\,i} - chl_{sat\,i}|}{n},$$

where $chl_{insitu\,i}$—is the i-th value of chlorophyll-a concentration obtained as a result of direct measurements, $chl_{sat\,i}$—is the i-th value of chlorophyll-a concentration reconstructed from satellite data, n—is the sample volume.

3 Results

Analysis of the satellite data of chlorophyll-a concentration showed that the GlobColour dataset provides significantly greater spatiotemporal coverage than OC-CCI dataset. GlobColour data provided 85% coverage of the average chlorophyll-a concentration for the period 1998–2022, while in the case of OC-CCI—only for the 9-year period 2002–2011.

It can be noted that there is a moderate correlation between the time series of GlobColour and OC-CCI data (the pair correlation coefficient of average seasonal values is 0.37), however, the values of chlorophyll-a concentration in the surface waters of Lake Onego according to OC-CCI data are higher values (on average 3 times) than those obtained from GlobColour data.

The results of comparison of satellite data and available published field data showed that GlobColour data are more consistent with Lake Onego field data (Filatov et al. 2015; Kalinkina et al. 2017, 2021; Tekanova et al. 2018; Sabylina et al. 2018) than the OC-CCI data (Table 1)—the average absolute error values for the period 1998–2020 are 2.97 μg/l ($n = 509$) and 5.24 μg/l ($n = 33$) respectively.

In this regard, we believe that the assessment of interannual variability of chlorophyll-a concentration in the surface waters of Lake Onego should be based primarily on GlobColour data.

The choice of using data is also supported by the fact that the obtained values of chlorophyll-a concentration during expeditions in August 2017 are in the range of 2.0–3.2 μg/l for central and southern areas of Lake Onego water (Tekanova et al. 2018) and in close agree with the estimated value of this parameter (2.47 μg/l) for the same year and month of GlobColour data in August 2017. Averaged chlorophyll-a concentration in June based on collected field data of the central area of Lake Onego waters for the period 2001–2019 also have values similar to those obtained from GlobColour data and are 0.3–0.9 μg/l and 0.89 μg/l, respectively (Kalinkina et al. 2021).

The maximum (more than 3 μg/l) values of the seasonal concentration of chlorophyll-a were observed in 2002, 2003, 2016 and 2018 in the surface waters of Lake Onego according to GlobColour data. At the same time, the maximum monthly

Table 1 Average values of absolute errors ($\overline{\delta}$, μg/l) of satellite data for retrieving chlorophyll-a concentrations for the periods 1998–2022 and 2002–2011, respectively, according to GlobColour and OC-CCI data, based on comparison with the results of expeditions for the period 1998–2022

Regions	GlobalColour	OC-CCI
Yuzhnoe Onego	2.11	4.36
Central'noe Onego	2.09	4.71
Bol'shoe Onego	3.23	5.76
Maloe Onego	3.72	6.31
Petrozavodskaya bay	3.40	7.34
Kondopozhskaya bay	3.56	6.16
Zaonezhskij bay	2.69	–
Poveneckij bay	2.75	9.60

Table 2 Comparison of MERIS/ENVISAT and GlobColour satellite data on average monthly values of chlorophyll-a concentration (μg/l) for August 2011 in some areas of Lake Onego

Regions	MERIS/ENVISAT	GlobColour
Unitskaya bay	4.0–5.0	3.0–6.0
Poveneckij bay	4.0–5.0	3.0–4.0
South part	3.0–3.5	2.5–3.0

average value of chlorophyll-a concentration was noted in September 2003 (more than 6 μg/l); the minimum values of chlorophyll-a concentration were noted in 2012, 2013 (less than 0.6 μg/l). The maximum average chlorophyll-a concentration was noted in the central area of Lake Onego in June 2020 (3.22 μg/l). It fits well with the chlorophyll-a concentration value (2.48 μg/l) for this period, obtained through field studies in frame of the research Lake Onego ecosystem in the spring–summer period to abnormally high air temperatures in the winter of 2019–2020.

It should be noted that the concentration of chlorophyll-a is distributed evenly over the lake area, however, in June, July, August, higher values can be observed in Unitskaya Bay (3.0–6.0 μg/l), Poveneckij Bay (3.0–4.0 μg/l), as well as in the southern area of Lake Onego (2.5–3.0 μg/l). In September and October, high chlorophyll-a concentrations are observed in the Zaonezhskij Bay (4.0–4.5 μg/l). These results are consistent with the spatial distributions of chlorophyll-a concentrations obtained from MERIS/ENVISAT (Environmental Satellite) satellite data (Filatov et al. 2015): according to both data sources, high chlorophyll-a concentrations were recorded in some bays of Lake Onego (Table 2).

4 Discussion of Results and Conclusions

Despite the preference of using GlobColour satellite data of chlorophyll-a concentration in comparison with similar OC-CCI data, the results obtained for this parameter in the surface waters of Lake Onego are not free from inaccuracies, especially in the case of small bays where waters are characterized by high levels of turbidity and coloration. In particular, satellite data for such areas as Lake Onego water do not image the high values of chlorophyll-a concentration that were revealed during onboard studies.

However, in the case of large bays, such as Petrozavodskaya and Kondopozhskaya bays, satellite data of chlorophyll-a concentrations in surface waters are quite adequate.

In general, we can approve that today satellite observations of the Lake Onego trophic status provide unique opportunities today to identify the most characteristics of the distribution and spatiotemporal dynamics of chlorophyll-a concentration today, being important indicator of the trophic state of a large and multi-purpose economic activities reservoir, that is statistically confirmed by the results of expeditional data collected during years.

Acknowledgements The work was supported by the RSF project No. 22-17-00193. The authors thank Dr., Prof., D. V. Pozdnyakov and Corresponding member of RAS, Dr., Prof., N. N. Filatov for useful comments during the preparation of the manuscript.

References

Beckers JM, Rixen M (2003) EOF calculations and data filling from incomplete oceanographic datasets. J Atmos Oceanic Tech 20:1839–1856. https://doi.org/10.1175/1520-0426(2003)020%3c1839:ECADFF%3e2.0.CO;2

Bridget N, Seegers P, Werdell J, Vandermeulen RA, Salls W, Stumpf RP, Schaeffer BA, Owens TJ, Bailey SW, Scott JP, Loftin KA (2021) Satellites for long-term monitoring of inland U.S. lakes: the MERIS time series and application for chlorophyll-a. Remote Sens Environ 266:112685. https://doi.org/10.1016/j.rse.2021.112685

Dekker AG, Malthus TJ, Hoogenboom HJ (1995) The remote sensing of inland water quality. In: Advances in environmental remote sensing. Wiley, Chichester, pp 123–142

Filatov NN (ed) (2010) Lake Onego. Atlas. Karelian Scientific Center of the Russian Academy of Sciences, Petrozavodsk, p 151 (in Russian)

Filatov NN, Kalinkina NM, Kulikova TP, Litvinenko AV, Lozovik PA (eds) (2015) The largest lakes and reservoirs in the North-West of European Russia: current state and changes in ecosystems under climatic and anthropogenic influences. Karelian Scientific Center of the Russian Academy of Sciences, Petrozavodsk, p 375 (in Russian)

Garnesson P, Mangin A, Fanton DO, Demaria J, Bretagnon M (2019) The CMEMS GlobColour chlorophyll a product based on satellite observation: multi-sensor merging and flagging strategies. Ocean Sci 15(3):819–830. https://doi.org/10.5194/os-15-819-2019

Jackson T, Sathyendranath S, Mélin F (2017) An improved optical classification scheme for the ocean colour essential climate variable and its applications. Remote Sens Environ 203:152–161. https://doi.org/10.1016/j.rse.2017.03.036

Kalinkina NM, Tekanova EV, Syarki MT (2017) Ecosystem of Lake Onego: unusual aquatic communities on anthropogenic factors and climate change. Water Manag Russ Probl Technol Manag 1:4–18 (in Russian). https://doi.org/10.35567/1999-4508-2017-1-1

Kalinkina NM, Tekanova EV, Efremova TV, Palshin NI, Nazarova LE, Baklagin VN, Zdorovennov RE, Smirnova VS (2021) Response of Lake Onego ecosystem in the spring–summer period to anomaly high air temperature in winter 2019/2020. Izv Ross Akad Nauk Ser Geogr 85(6):888–899 (in Russian). https://doi.org/10.31857/S2587556621060078

Kondratyev KY, Filatov NN (1999) Limnology and remote sensing: a contemporary approach. Praxis Publishing Ltd., London, p 406

Korosov AV, Kalinkina NM, Tekanova EV, Syarki MT, Isakova KV (2021) Development of trophic indices for Lake Onego using of the ecological information system. In: Proceedings of the international conference "InterCarto. InterGIS. GI support of sustainable development of territories, vol 27, MSU, Faculty of Geography (Part 3), Moscow, pp 256–273 (in Russian). https://doi.org/10.35595/2414-9179-2021-3-27-256-273.

Lesht BM, Barbiero RP, Warren GJ (2018) Using satellite observations to assess the spatial representativeness of the GLNPO water quality monitoring program. J Great Lakes Res 44(4):547–562. https://doi.org/10.1016/j.jglr.2018.05.001

Moore TS, Campbell JW, Dowell MD (2009) A class-based approach to characterizing and mapping the uncertainty of the MODIS ocean chlorophyll product. Remote Sens Environ 113(11):2424–2430. https://doi.org/10.1016/j.rse.2009.07.016

Papenfus M, Schaeffer BA, Pollard AI, Loftin KA (2020) Exploring the potential value of satellite remote sensing to monitor chlorophyll-a for US lakes and reservoirs. Environ Monit Assess 192(12):1–22. https://doi.org/10.1007/s10661-020-08631-5

Pozdnyakov DV, Filatov NN (2021) Interannual water quality variations in Lake Ladoga in spring during 2016 and 2017: satellite observations. Fundam Appl Hydrophys 14(1):79–85 (in Russian). https://doi.org/10.7868/S2073667321010081

Pozdnyakov DV, Korosov AA, Petrova NA, Grassl H (2013) Multi-year satellite observations of Lake Ladoga's biogeochemical dynamics in relation to the lake's trophic status. J Great Lakes Res 39(1):34–45. https://doi.org/10.1016/j.jglr.2013.05.002

Sabylina AV, Tekanova EV, Kalinkina NM (2018) Chlorophyll-a in the water of Lake Onego. Certificate of state registration of the database No. 2018621068 dated 06/13/2018 (in Russian)

Saulquin B, Gohin F, Fanton d'Andon O (2019) Interpolated fields of satellite-derived multi-algorithm chlorophyll-a estimates at global and European scales in the frame of the European Copernicus-Marine Environment Monitoring Service. J Oper Oceanogr 12(1):47–57. https://doi.org/10.1080/1755876X.2018.1552358

Suarez EL, Tiffay M-C, Kalinkina N, Tchekryzheva T, Sharov A, Tekanova E, Syarki M, Zdorovennov RE, Makarova E, Mantzouki E, Venail P, Ibelings BW (2019) Diurnal variation in the convection-driven vertical distribution of phytoplankton under ice and after ice-off in large Lake Onego (Russia). Inland Waters 9(2):193–204. https://doi.org/10.1080/20442041.2018.1559582

Tekanova EV, Kalinkina NM, Zdorovennov RE, Makarova EM (2018) Results of the study of Lake Onego ecosystem in the period of summer stratification based on measured data from a 2017 expedition. Trans Karelian Res Center Russ Acad Sci Limnol Oceanol 9:44–53. https://doi.org/10.17076/lim753

Volpe G, Santoleri R, Vellucci V, Ribera DM, Marullo S, D'Ortenzio F (2007) The colour of the Mediterranean Sea: Global versus regional bio-optical algorithms evaluation and implication for satellite chlorophyll estimates. Remote Sens Environ 107(4):625–638. https://doi.org/10.1016/j.rse.2006.10.017

Volpe G, Colella S, Forneris V, Tronconi C, Santoleri R (2012) The Mediterranean ocean colour observing system—system development and product validation. Ocean Sci 8(5):869–883. https://doi.org/10.5194/os-8-869-2012

Volpe G, Colella S, Brando V, Forneris V, La Padula F, Di Cicco A, Sammartino M, Bracaglia M, Artuso F, Santoleri R (2019) The Mediterranean ocean colour level 3 operational multi-sensor processing. Ocean Sci 15(1):127–146. https://doi.org/10.5194/os-15-127-2019

Numerical Modeling of Long-Wave Oscillations in Two Coupled Bays

Yu. V. Manilyuk, S. V. Dovgaya, V. V. Fomin, and D. I. Lazorenko

Abstract Within the framework of the theory of long waves (based on nonlinear Boussinesq equations without dispersion), using the numerical hydrodynamic model *FUNWAVE* (Fully Nonlinear WAVE model), the dynamics of water in a system of two coupled model rectangular bays of constant depth was reproduced. The sequence of incident monochromatic waves, having periods within the range corresponding to the lowest modes of the basin's natural oscillations, was specified as a disturbance. Numerical experiments were performed for the basin with parameters corresponding to the laboratory experimental apparatus (Nakano and Fujimoto 1987). The modes of forced and free oscillations have been studied. It was possible qualitatively numerically repeat the free level fluctuations in tops of the bays that took place during laboratory experiments in (Nakano and Fujimoto 1987): with and without beats. But in all our numerical experiments, the oscillations in coupled bays were co-phasic. Even in the case of beats, free oscillations did not turn into contra-phasic over time. Only a slight phase shift was observed, which disappeared over time and the oscillations became co-phasic.

Keywords Boussinesq equations · Numerical model FUNWAVE · Coupled bays · Seiches

Consideration of oscillations in systems of coupled bays is of interest, when studying the dynamics of waters near sea coasts. There are a fairly large number of works on this topic. In Liu et al. (2003) you can find a description of the phenomenon and an extensive bibliography. In Manilyuk et al. (2019), the generation of seiches by long-wave disturbances in coupled model bays is studied, in Rabinovich (2009) the interaction of the bays of Platja Gran and Ciutadella, located on the island of Menorca, is investigated, and in Manilyuk et al. (2020)—the Sevastopol and Krantinnaya bays,

Yu. V. Manilyuk (✉) · S. V. Dovgaya · V. V. Fomin · D. I. Lazorenko
Marine Hydrophysical Institute of RAS, Sevastopol, Russian Federation
e-mail: uvmsev@yandex.ru

V. V. Fomin
e-mail: v.fomin@mhi-ras.ru

T. Chaplina (ed.), *Processes in GeoMedia—Volume VIII*, Springer Geology,
https://doi.org/10.1007/978-981-97-6627-7_27

in Kovalev et al. (2022) seiche oscillations in Terpeniya Bay are considered, in Manilyuk et al. (2022) the resonant response of the Sevastopol Bay to long-wave disturbances coming from the open sea is studied. The work Nakano (1932) indicates that in the two neighboring bays of Kozioro and Moroiso, located on the Miura Peninsula at the mouth of Tokyo Bay, sea level fluctuations in the form of beats are observed, resulting from the superposition of several seiche modes. In Nakano (1932), based on the analytical solution of the Lagrange equation, a theoretical explanation of this phenomenon is given. In this case, two adjacent equal rectangular bays with equal depths are considered. Coupling between them is carried out through their inlets. In Nakano and Fujimoto (1987), to verify the theoretical results obtained in Nakano (1932), experiments were carried out on a laboratory installation, which showed satisfactory agreement with Nakano (1932). At the same time, it has been experimentally established that in a system of coupled bays two modes of oscillations are possible: co-phasic and contra-phasic, which occur at periods different from the own period of one bay. It is of practical interest to perform a numerical simulation of the excitation of waves in adjacent bays, the process of free oscillations, and compare them with the results of laboratory experiments conducted in Nakano and Fujimoto (1987). This is the main goal of this work.

1 Materials and Method of Research

We used the numerical hydrodynamic model *Fully Nonlinear WAVE model* (*FUNWAVE*) (https://fengyanshi.github.io/build/html/basics.html; Kennedy et al. 2000; Shi et al. 2012), which is based on solving the nonlinear Boussinesq equations (Chen 2006). Let us briefly consider the main equations of the model (Shi et al. 2012). The depth-integrated volume conservation equation is given by formula

$$\eta_t + \nabla \cdot \mathbf{M} = 0,$$

where

$$\mathbf{M} = H\{\mathbf{u}_\alpha + \overline{\mathbf{u}}_2\} \tag{1}$$

is a full stream. Here $H = h + \eta$—total depth at a point in the water area (at the same time h—local depth at rest, and η—local displacement of the free surface), $\mathbf{u}_\alpha$—velocity at a certain reference level $z = z_\alpha$, $\overline{\mathbf{u}}_2$—depth-average contribution of order $O(\mu^2)$ in the field of the horizontal velocity component $\mathbf{u}_2(z)$ (μ—a parameter characterizing the ratio of depth to wavelength and is assumed to be small). In this case, the total horizontal speed looks like:

$$\mathbf{u} = \mathbf{u}_\alpha + \mathbf{u}_2(z),$$

where

$$\mathbf{u}_2(z) = (z_\alpha - z)\nabla \cdot A + \frac{1}{2}\left(z_\alpha^2 - z^2\right)\nabla \cdot B \tag{2}$$

and $\mathbf{u}_2(z)$ is a depth-dependent correction at $O(\mu^2)$. In (2) A, B and z_α are given as:

$$A = \nabla \cdot (h\mathbf{u}_\alpha),$$
$$B = \nabla \cdot \mathbf{u}_\alpha,$$
$$z_\alpha = \zeta h + \beta\eta,$$

where ζ and β are constants (Kennedy et al. 2001).

In (1) the term $\overline{\mathbf{u}}_2$ is determined by the expression:

$$\overline{\mathbf{u}}_2 = \frac{1}{H}\int_{-h}^{\eta} \mathbf{u}_2(z)dz = \left(\frac{z_\alpha^2}{2} - \frac{1}{6}\left(h^2 - h\eta + \eta^2\right)\right)\nabla B + \left(z_\alpha + \frac{1}{2}(h-\eta)\right)\nabla A.$$

The equation for horizontal momentum averaged over depth is written as

$$\mathbf{u}_{\alpha,\mathrm{t}} + (\mathbf{u}_\alpha \cdot \nabla)\mathbf{u}_\alpha + g\nabla\eta + \mathbf{V}_1 + \mathbf{V}_2 + \mathbf{V}_3 + \mathbf{R} = \mathbf{0},$$

where g is the acceleration of free fall; $\mathbf{R}$—diffusion and dissipative components, including bottom friction and horizontal turbulent mixing; $\mathbf{V}_1$, $\mathbf{V}_2$ and $\mathbf{V}_3$ are dispersion terms given by the expressions:

$$\mathbf{V}_1 = \left\{\frac{Z_\alpha^2}{2}\nabla B + z_\alpha \nabla A\right\}_{\mathrm{t}} - \nabla\left[\frac{\eta^2}{2}B_{\mathrm{t}} + \eta A_{\mathrm{t}}\right],$$
$$\mathbf{V}_2 = \nabla\left\{(z_\alpha - \eta)(\mathbf{u}_\alpha \cdot \nabla)A + \frac{1}{2}\left(z_\alpha^2 - \eta^2\right)(\mathbf{u}_\alpha \cdot \nabla)B + \frac{1}{2}[A + \eta B]^2\right\}.$$
$$\mathbf{V}_3 = \omega_0 \mathbf{i}^{\mathbf{z}} \times \overline{\mathbf{u}}_2 + \omega_2 \mathbf{i}^{\mathbf{z}} \times \mathbf{u}_\alpha,$$

where

$$\omega_0 = (\nabla \times \mathbf{u}_\alpha) \cdot \mathbf{i}^{\mathbf{z}} = v_{\alpha,x} - u_{\alpha,y}, \tag{3}$$

$$\omega_2 = (\nabla \times \overline{\mathbf{u}}_2) \cdot \mathbf{i}^{\mathbf{z}} = z_{\alpha,x}\left(A_y + z_\alpha B_y\right) - z_{\alpha,y}(A_x + z_\alpha B_x). \tag{4}$$

In (3) and (4) $v_{\alpha,x}$, $z_{\alpha,x}$ are the derivatives of v_α and z_α with respect to x, $u_{\alpha,y}$, $z_{\alpha,y}$ are the derivatives of u_α, z_α with respect to y.

A detailed description of the model is given in Toro (2009), Tonelli and Petti (2009), Gotlieb et al. (2001).

2 Scheme of Numerical Experiments

Numerical experiments were carried out for the basin with parameters corresponding to the laboratory experimental apparatus (Nakano 1932). Figure 1 shows a schematic representation of the model domain, reproducing the laboratory apparatus from Nakano (1932), where d is the distance between the bays, l_1 and l_2—the lengths of bays A and B, respectively, b_1 and b_2—the widths of these bays. The following parameter values were set: $l_1 = l_2 = 51.2$ cm, $b_1 = b_2 = 15$ cm, $d = 5$ cm, when carrying out the calculations. The depth of the basin is constant $h = 8.7$ cm.

The model integration step in time is $\Delta t = 0.02$ s, in space—$\Delta x = \Delta y = 1$ cm, the computational domain consists of 726×312 cells.

The numerical experiments carried out consisted of two stages. At the first stage, forced oscillations were modeled, excited by a disturbance representing a sequence of incident sinusoidal waves with an amplitude of 0.87 cm, perpendicular to the entrances to the bays. At the second stage, free oscillations occurring after the disturbance ceased were modeled. The duration of the disturbance is 280 s, the duration of the interval of free oscillations is 280 s. The no-flow condition was used at all solid basin boundaries (Fig. 1). Absorbing layers on the lateral boundaries were not specified in the modification of the model used in the calculations.

Two series of numerical experiments were carried out. The first series is devoted to the study of oscillations excited by disturbances with resonant periods specified in Nakano and Fujimoto (1987): 2.57 s—the period of co-phasic oscillations, 3.08 s—contrai-phasic oscillations, 2.67 s—the natural period of one bay, and 2.22 s—theoretical value of the Helmholtz mode period of bay A/B, calculated using the Merian formula: $T = 4l_1(gh)^{-1/2}$. Second series—calculations of oscillations in bays caused by disturbances with periods corresponding to the most energy-bearing periods, established on the basis of the first series of numerical experiments, were carried out.

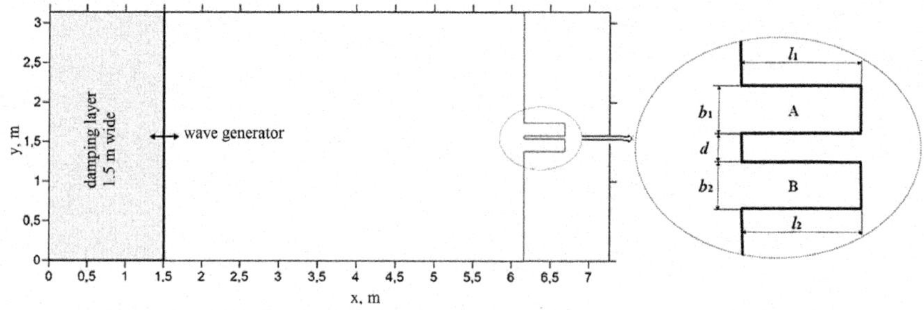

Fig. 1 Scheme of the calculation area (A and B—bays under consideration)

3 Observational Data from Nakano and Fujimoto (1987)

Figure 9 in Nakano and Fujimoto (1987, p. 129) shows sample records of level oscillations obtained on a laboratory installation, when the disturbance period was equal to 3.08 s (the period of co-phasic oscillations) and for a period of 2.57 s (the period contra-phasic oscillations). In Nakano and Fujimoto (1987) an analysis of the oscillations presented in this figure is given. It is shown that when waves are generated during the period of co-phasic oscillations (3.08 s, Fig. 9a (Nakano and Fujimoto 1987)), damped co-phasic oscillations without beats arise in both bays. When waves are excited by a disturbance having a period of contra-phasic oscillations (2.57 s, Fig. 9b (Nakano and Fujimoto 1987)), co-phasic damped oscillations appear in the bays, accompanied by beats. Level oscillations gradually become contra-phasic, the phase of which then does not change over time. The work Nakano and Fujimoto (1987, p. 130) indicates that the values of the natural periods of each of the bays are equal to 2.67 s.

4 Results and Discussion

Two series of numerical experiments were carried out to simulate forced oscillations excited by a sinusoidal disturbance, turning into free oscillations after the termination of the disturbance.

4.1 The First Series of Numerical Experiments

The calculation of waves in the model domain (Fig. 1) was carried out at the following disturbance periods: 2.57; 3.08; 2.67; 2.22 s. Depending on the period of disturbances, there are different types of oscillations reaching a stationary regime (Fig. 2).

As can be seen from Fig. 2, it takes a long time for the oscillations to reach a stationary mode (compared to the period of the bays' own oscillations)—about 1.5 min. In the process of establishing oscillations, their amplitude can monotonically increase (perturbation period 2.57 s, Fig. 2b). Or, increasing, reaches a certain level and begins to decrease (Fig. 2a, c). Moreover, after the end of the transition process the amplitude of level oscillations decreases by more than 2 times, when the disturbance has a period of 3.08 s.

Figure 3 shows the energy spectra of level oscillations. As the analysis of Fig. 3 showed, the oscillation periods are 2.57; 3.08; 2.67 s are not resonant for the model region. The highest oscillations energy occurs at periods of 2.81; 3.16 and 3.62 s. There is also a period of 24.86 s (apparently corresponding to the beats), which is present in all spectrograms, except for the case when the period of the disturbance is 3.08 s (Fig. 3b).

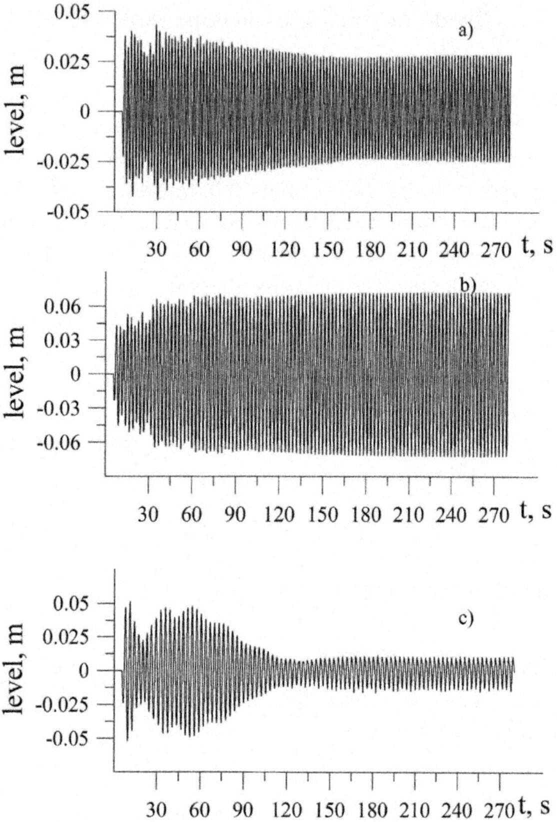

Fig. 2 Forced waves in Bay A, point 1, caused by disturbances with periods: **a** 2.22 s; **b** 2.57 s; **c** 3.08 s

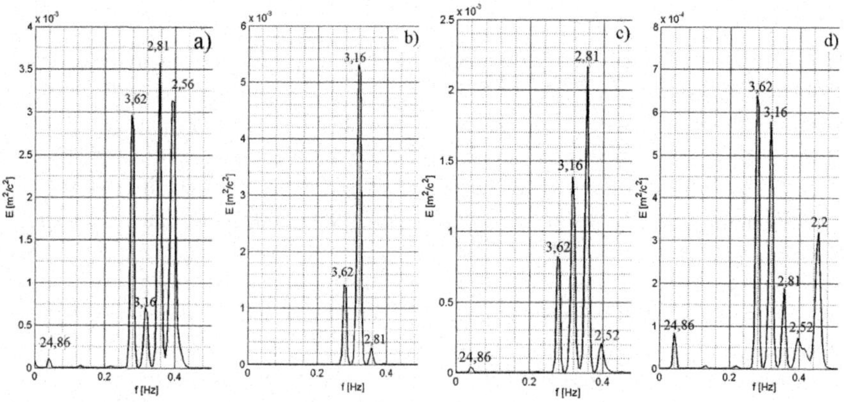

Fig. 3 Energy spectra of free level oscillations at point 1, occurring after the cessation of disturbances with periods equal to: **a** 2.57 s; **b** 3.08 s; **c** 0.67 s; **d** 2.22 s

Note that for point 2, located in Bay B (Fig. 1), the results of calculating level fluctuations are identical to the results for point 1. Therefore, it is not considered in this section.

From the theory of oscillations (Rabinovich and Trubetskov 1984) it is known that partial periods (the case when there is no connection between the resonators, the entrance to one bay is closed) lie between normal periods (occurring when there is a connection between the resonators, one normal period corresponds to contra-phasic oscillations, the other is co-phasic, the period of contra-phasic oscillations is less than that of co-phasic ones). Thus, we can assume that in the oscillatory system considered here there are two normal periods: 2.81 s corresponds to contra-phasic oscillations, 3.62 s to co-phasic oscillations, and a period equal to 3.16 s is partial.

4.2 The Second Series of Numerical Experiments

Waves were excited in the computational domain by disturbances with periods of 2.81; 3.62 and 3.16 s. The processes of forced and free oscillations in bays A, B (Fig. 1) were studied.

Analysis of the results of calculations of forced oscillations showed the following. If the disturbance periods T are equal to 2.81 and 3.62 s, then the oscillation charts are similar: first there is a short transient process, lasting for about 30 s, then the oscillations go into a stationary mode. If $T = 2.81$ s, then the waves are symmetrical, the heights of the crests and soles are about 8 cm, if $T = 3.62$, then the waves are asymmetrical, the height of the crest is 11 cm, the soles are 9 cm. In the case when $T = 3.16$ s there is a longer transient process consisting of two phases. In the first phase, lasting about 15 s, the amplitude of oscillations quickly increases to 6 cm, then over the next 15 s it decreases to 1 cm and the second phase begins when the amplitude of oscillations monotonically increases to 11 cm over 60 s, and then it practically does not change. The waves have an asymmetrical structure: the maximum heights of the wave crests are 11 cm, and the maximum heights of the wave bottoms are 9 cm.

Let us consider in more detail the consideration of free oscillations that arise after the stopping of the action of disturbances. Figure 4 shows the energy spectra of free oscillations that occur after the stopping of disturbances with periods equal to 2.81 s (a), 3.16 s (b) and 3.62 s (c).

As can be seen, in all the cases considered, there is a clearly defined peak at the resonant period at which the waves were generated. For the case when the perturbation period is equal to 2.81 s, two more well-defined peaks can be traced at periods of 3.16 and 3.62 s, which, as shown above, are resonant for the oscillatory system under consideration.

Figure 5 shows the model level progress in the free oscillation mode. As can be seen from this figure, level fluctuations caused by a disturbance with a period of 2.81 s, corresponding to the period of contra-phase oscillations, are accompanied by beats. A similar picture was recorded in laboratory experiments (Fig. 2b (Nakano and Fujimoto 1987)). When the period of disturbance corresponds to the period of

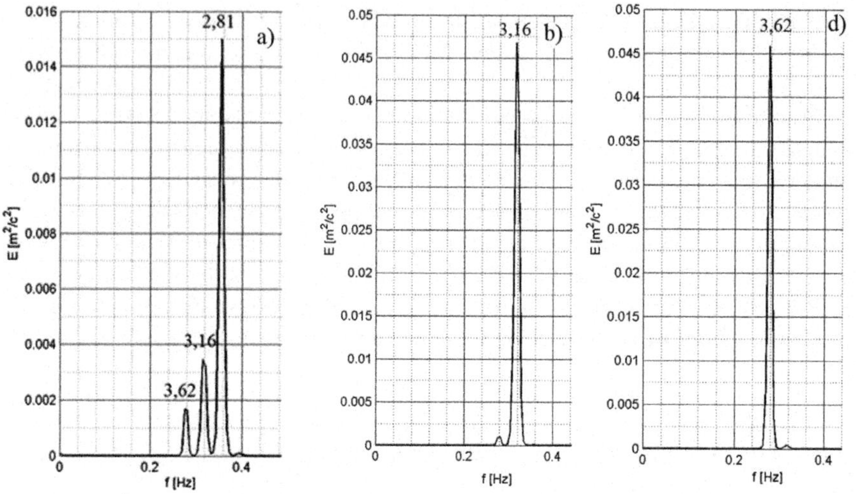

Fig. 4 Energy spectra of free level oscillations at point 1, occurring after the cessation of disturbances with periods equal to: **a** 2.81 s; **b** 3.16 s; **c** 3.62 s

co-phasic oscillations, no beat is observed, both in laboratory (Fig. 2a (Nakano and Fujimoto 1987)) and in numerical (Fig. 5c) experiments.

The amplitude of oscillations excited by a disturbance with a period of contra-phasic oscillations is less than in the case when the period of the disturbance is equal to the period of co-phasic oscillations. This takes place both in laboratory (Fig. 2 (Nakano and Fujimoto 1987)) and in numerical experiments (Fig. 5).

In Nakano and Fujimoto (1987) it is indicated that when the waves in a laboratory installation were excited by a disturbance with a period corresponding to contra-phasic oscillations, then after the cessation of the disturbance, co-phasic oscillations occurred in both bays, which over time, through beatings, transformed into contra-phasic ones (Fig. 2b (Nakano and Fujimoto 1987)).

In our numerical experiment with a disturbance having a period of 2.81 s, co-phasic oscillations also initially appeared in bays A and B, then beats appeared (Fig. 5a), and the phase of mutual oscillations at points 1 and 2 began to shift (Fig. 6). But "pure" contra-phasic oscillations could not be obtained in a numerical experiment, and over time the phases of the oscillations became closer again.

As shown by the analysis of the calculation results, in cases of free oscillations that take place after the stopping of disturbances with periods equal to 3.16 and 3.62 s, the occurrence of a phase difference between the oscillations at points 1 and 2 was not noted; they were of co-phasic nature.

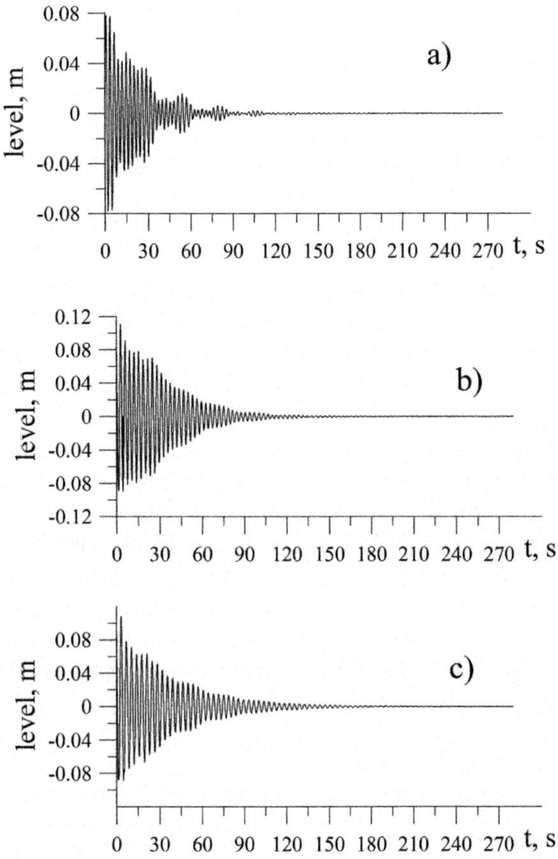

Fig. 5 Free level fluctuations in Bay A, point 1, caused by disturbances with periods equal to: **a** 2.81 s; **b** 3.16 s; **c** 3.62 s

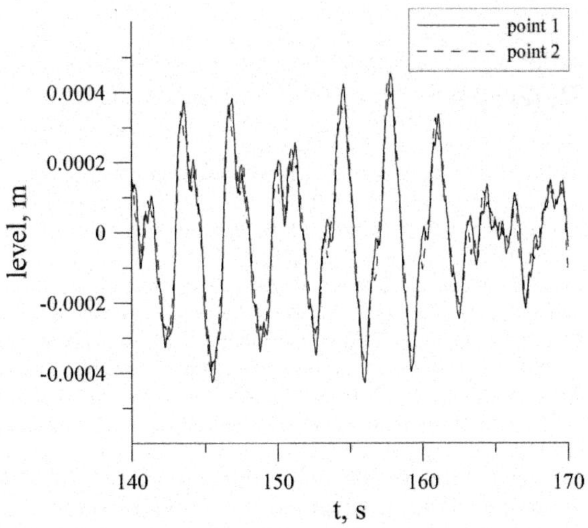

Fig. 6 Free level fluctuations in bay A, point 1 and bay B, point 2, caused by disturbance with a period equal to 2.81 s

5 Conclusion

Numerical experiments were carried out using the *FUNWAVE* model, repeating the laboratory experiments of Nakano and Fujimoto (1987), devoted to the study of the interaction of two connected bays. Analysis of the simulation results showed the following.

Depending on the period of disturbance, forced oscillations of various types occur in the bays. A type of oscillation has been identified when the amplitude of forced oscillations first quickly increases and then decreases, reducing in twice, reaching a stationary regime, at which it practically does not change. At the considered resonant periods, the oscillation structures are close to each other.

It was possible to qualitatively and numerically repeat the free fluctuations of the level in the apex parts of the bays that took place during laboratory experiments in Nakano and Fujimoto (1987): with and without beats. But in all our numerical experiments, oscillations in connected bays were co-phasic. Even in the case of beats, free oscillations did not turn into contra-phasic over time. Only a slight phase shift was observed, which disappeared over time and the oscillations became co-phasic.

The used model has shown its efficiency for studying long-wave oscillations in bays. It is advisable to continue numerical experiments using a computational domain that has parameters close to the parameters of real bays, and also to consider cases where the sizes and depths of connected bays differ, and to study the influence of the distance between the entrances to the bays on the parameters of the oscillations in them.

Acknowledgements The work was carried out within of the state task of Federal Research Center Marine Hydrophysical Institute of RAS (MHI RAS), topic No. FNNN-2024-0016. The simulation was carried out on the computing cluster of the MHI RAS.

References

Chen Q (2006) Fully nonlinear Boussinesq-type equations for waves and currents over porous beds. J Eng Mech 132:220–230. https://doi.org/10.1061/(ASCE)0733-9399(2006)132:2(220)

Gotlieb S, Shu C-W, Tadmore E (2001) Strong stability-preserving high-order time discretization methods. SIAM Rev 43(1):89–112

https://fengyanshi.github.io/build/html/basics.html

Kennedy AB, Chen Q, Kirby JT, Robert A, Dalrymple RA (2000) Boussinesq modeling of wave transformation, breaking and runup. I: one dimension. J Waterway Port Coastal Ocean Eng 126(1):726–726. https://doi.org/10.1061/(ASCE)0733-950X(2000)126:1(39)

Kennedy AB, Kirby JT, Chen Q, Dalrymple RA (2001) Boussinesq-type equations with improved nonlinear performance. Wave Motion 33:225–243. https://doi.org/10.1016/S0165-2125(00)00071-8

Kovalev DP, Kovalev PD, Borisov AS, Zarochintsev VS, Kirillov KV (2022) Features of seishe initiation in the water area near Poronaisk (Sakhalin Island). Geosistemy Perehodnykh Zon (Geosyst Trans Zones) 6(2):114–123 (in Russian, abstr. in Engl.). https://doi.org/10.30730/gtrz.2022.6.2.114-123; https://www.elibrary.ru/wmvfjq

Liu PL-F, Monserrat S, Marcos M, Rabinovich AB (2003) Coupling between two inlets: observation and modeling. J Geophys Res Oceans 108(C3):3069. https://doi.org/10.1029/2002JC001478

Manilyuk YuV, Lazorenko DI, Fomin VV (2019) Resonance oscillations in the system of adjacent bays. Phys Oceanogr 26(5):374–386. https://doi.org/10.22449/1573-160X-2019-5-374-386

Manilyuk YuV, Lazorenko DI, Fomin VV (2020) Investigation of seiche oscillations in the adjacent bays by the example of the Sevastopol and the quarantine bays. Phys Oceanogr [e-journal] 27(3):242–256. https://doi.org/10.22449/1573-160X-2020-3-242-256

Manilyuk YuV, Lazorenko DI, Fomin VV (2022) Study of the resonant response of an extended narrow bay to long wave perturbations coming from of the open sea, on the example the Sevastopol bay. Protsessy v Geosredakh (Process Geomedia) 34(4):1888–1895 (in Russian)

Nakano M (1932). Die seiches in gekoppeltes system formenden buchten. Geophys Mag 5:163–170

Nakano M, Fujimoto N (1987) Seiches in bays forming a coupled system. J Oceanogr Soc Jpn 43:124–134. https://doi.org/10.1007/BF02111888

Rabinovich AB (2009) Seiches and harbor oscillations. In: Kim YC (ed) Handbook of coastal and ocean engineering, chap 9. World Sci. Publ., Singapore, pp 193–236

Rabinovich MI, Trubetskov DI (1984) Introduction to the theory of oscillations and waves. Nauka, Moscow, 432 p (in Russian)

Shi F, Kirby JT, Harris JC, Geiman JD, Grilli ST (2012) A high-order adaptive time-stepping TVD solver for Boussinesq modeling of breaking waves and coastal inundation. Ocean Model 43–44:36–51. https://doi.org/10.1016/j.ocemod.2011.12.004

Tonelli M, Petti M (2009) Hybrid finite volume-finite difference scheme for 2DH improved Boussinesq equations. Coastal Eng 56:609–620. https://doi.org/10.1016/j.coastaleng.2009.01.001

Toro EF (2009) Riemann solvers and numerical methods for fluid dynamics: a practical introduction, 3rd edn. Springer, New York. https://doi.org/10.1007/b79761

Wave Climate of Koktebel Bay of the Black Sea

L. V. Kharitonova, V. V. Fomin, and D. V. Alekseev

Abstract Wave climate parameters determine the exploitation conditions of coastal infrastructure facilities and maximum loads on its. Investigation of statistical characteristics and spatial structure of wind waves in Koktebel Bay (the Black Sea, Crimea) was carried out in the paper using statistical analysis and mathematical modelling methods. Wave regime characteristic in the Koktebel Bay central part was obtained using retrospective calculation data of wind-wave parameters over the 1979–2020 period. Retrospective calculation was performed by V. V. Fomin in previous works on the basis of global atmospheric reanalysis ERA-Interim and ERA5 and numerical spectral SWAN model. It was found in the paper, that waves with significant wave heights less than 0.5 m and with mean periods range 2.5–3.0 s account for 55% of all cases. The repeatability of waves higher than 1.0 and 1.5 m is 14 and 5% correspondingly. Storm waves coming from the east and east-southeast are most probable in the study area. A long-term annual mean storm duration is 16.4 days while the total annual storm duration varies from 5.6 to 34.3 days. The longest storms occur from November to March. The heights, periods and wavelengths were determined in the system of calculated storms occurring once every *n* years. So, for the storm possible once every 25 years significant wave heights is 4.6 m, mean wave height is 2.6 m, mean period is 11.0 s, mean wave length is 108 m, wave height of 1% probability in the storm system is 5.4 m. For this storm, numerical simulation of wind wave fields was performed with high spatial resolution. It was found that for east, southeast, and south waves the configuration and details of the significant wave height fields are almost identical, starting at depths of 3 m and less due to refraction. Near the bottom at depths of 2–3 m, the amplitude of the velocity of orbital wave motions is 0.8–1.0 m/s.

Keywords Wind waves · Wave climate · Numerical modeling · SWAN model · The Black Sea · Coastline of Crimea · Koktebel Bay

---

L. V. Kharitonova (✉) · V. V. Fomin · D. V. Alekseev
Marine Hydrophysical Institute of the Russian Academy of Sciences, Sevastopol, Russia
e-mail: l.kharitonova@mhi-ras.ru

T. Chaplina (ed.), *Processes in GeoMedia—Volume VIII*, Springer Geology,
https://doi.org/10.1007/978-981-97-6627-7_28

1 Introduction

The study area is Koktebel Bay located on the north-eastern coast of the Black Sea. It is bounded by Cape Planerny in the west and Cape Khameleon in the east (Fig. 1a). The bay coastline is about 7 km long with about 3 km of anthropogenic coasts. Koktebel Bay is shallow, with the 5 m isobath at a distance of 200–300 m from the coastline and with depths of about 10–15 m at the outer boundary (Fig. 1b). On the coast, there is the urban-type resort settlement of Koktebel. Reconstruction of its seafront and beach nourishment were last carried out in the late 1980s; neither repair work nor additional nourishment were done thereafter. Currently, the coastal protection structures of the study area are mainly in disrepair (Goryachkin and Kharitonova 2017; Goryachkin and Dolotov 2019; Goryachkin and Markov 2023). Science-based recommendations for reconstruction of the coastal protection structures (determination of their location and characteristics) consider parameters of wind waves in the coastal water area. Therefore, it is important to describe the wave climate with high spatial resolution for the coastal water area.

Regional investigations of the wave regime in the study area are scarce. For a part of the Black Sea along the Crimean coast, data on the wind-wave regime are available in reference materials (Efimov and Komarovskaya 2009; Lopatoukhin et al. 2006), in which the entire sea is divided into large areas. The paper (Goryachkin and Repetin 2009) analyzes data from hydrometeorological stations. The most advanced method to study hydrodynamic processes with high spatial resolution is numerical modeling. Recently, a number of works on reanalysis of wind waves in the coastal zone of the Black Sea have been published, e.g. Arkhipkin et al. (2014), Divinsky and Kosyan (2018), Divinsky et al. (2019, 2020), Gippius and Myslenkov (2020), Kharitonova et al. (2021a, b), Fomin and Goryachkin (2022). However, in these studies, the nearest analyzed point is located in the central part of Feodosiya Bay (approximately 30 km from the study area).

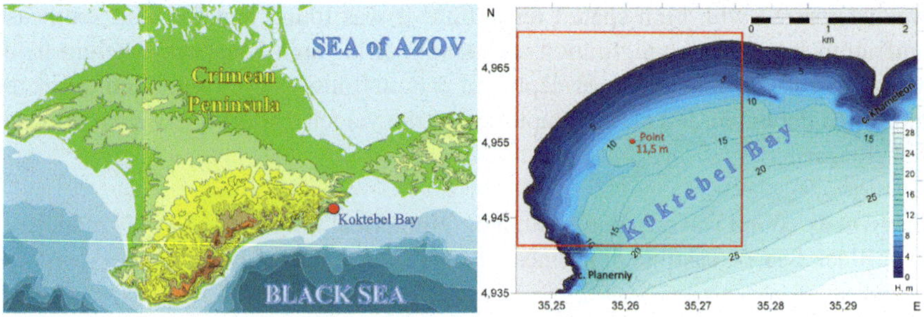

a b

Fig. 1 Location of the area under study (**a**) and bathymetry scheme of Koktebel Bay (**b**). Point 11.5 m is the place where the data of the SWAN–ERA array were analyzed. Red rectangle is the fifth computational grid for the SWAN model

The study area is poorly covered with in situ data. The nearest observation station is at the marine hydrometeorological station Feodosiya approximately 15 km to the north-east. Reference documents allow spectral-discrete or spectral-parametric models for calculation of wave elements in open and enclosed water areas. Therefore, numerical modeling data are used to analyze the regime-climatic features of wind waves.

The paper aims to study the statistical characteristics and spatial structure of wind waves in the area of Koktebel Bay using methods of statistical analysis and numerical modeling.

2 Materials and Methods

On the first stage, we used a data array of retrospective calculations of wind-wave parameters in the Black Sea over the 1979–2020 period with a time discreteness of 1 h to estimate statistical characteristics of waves in the area under study. The data array (hereinafter referred to as the SWAN–ERA array) was obtained using the numerical spectral SWAN (Simulating Waves Nearshore) model (Booij et al. 1999; SWAN User Manual 2018) on an unstructured computational grid with condensation in the coastal zone of the Black Sea (Divinsky et al. 2020). The atmospheric forcing of the model was the data of global atmospheric reanalysis ERA-Interim and ERA5 (https://www.ecmwf.int/en/forecasts). The model was validated on the basis of data of wave measurements on the offshore platform in Karkinitsky Bay of the Black Sea (Polonsky et al. 2011).

From the SWAN–ERA array, the computational grid node in the center of Koktebel Bay with a depth of ~ 11.5 m was chosen. For this point long-term series of wind and wave parameters were formed, including wind speed and direction on 10 m height, the significant wave height h_s, the mean wave period $\overline{\tau}$ the mean wave direction θ, and the peak wave period τ_p. On the basis of these series, operational and extreme characteristics of the waves were calculated, as well as storm durations and intensity. Operational wave characteristics determine the background operating conditions of coastal infrastructure facilities in Koktebel Bay whereas extreme ones determine the regime of maximum loads on these facilities.

As a criterion for storm identification, the (Amarouche and Akpınar 2021) condition was used

$$h_s \geq \overline{h}_s + 2\sigma, \tag{1}$$

where h_s is the significant wave height at a fixed time point measured in meters; $\overline{h}_s =$ 0.61 m is the mean annual value for time series of h_s; $\sigma = 0.48$ m is root-mean-square deviation of the h_s from $\overline{h}_s$. Thus, the minimal threshold value of the significant wave height is $h_{st} = 1.57$ m for storm. Therefore, the storm is defined as long time period during which h_s exceeds the minimal threshold value h_{st}.

A storm power index (*SPI*) was calculated according to the formula from (Amarouche and Akpınar 2021)

$$SPI = h_d^2 T_d, \tag{2}$$

where h_d (m) is a mean value of h_s for the storm period, T_d is a storm duration (h). When calculating h_d and T_d the values of h_s that satisfy the criterion (1), were used.

To estimate the extreme values of mean wave heights $\overline{h} = 0.63h_s$ and mean wave periods $\overline{\tau}$ possible once every *n* years, the extreme Gumbel distribution (Lopatoukhin et al. 2000), obtained on the basis of a sequence of annual maxima of $\overline{h}$ and $\overline{\tau}$ from SWAN–ERA series was applied. Then, using the distribution functions of wave elements for the sea of finite depth (Krylov 1966), which includes $\overline{h}$ and $\overline{\tau}$ maxima, the heights and periods of waves of different probability in the system of storms possible once every *n* years were determined.

On the second stage of the work, the space distribution of wind wave parameters of 4% regime probability in the Koktebel Bay was obtained using of the SWAN model (Booij et al. 1999; SWAN User Manual 2018). In order to correctly specify the characteristics of waves coming from the open sea, a 5-step method of nested grids was implemented: in step 1, wave fields for the whole Black Sea basin were modeled; in steps 2 and 3, wave fields were calculated for the areas including the Crimean Peninsula and the South Coast of Crimea, respectively; in step 4, wave fields for the area including Koktebel Bay and the adjacent water area were calculated; in step 5, wave fields for the western Koktebel Bay were calculated. In step 2 and all subsequent steps, the wind wave parameters at the liquid boundaries of the computational domain were determined by interpolation of the modeling data obtained in the previous step. To build the bottom relief of the calculation areas, bathymetric maps of the Black Sea were digitized. The model bottom relief in the fifth calculation area with a resolution of about 6 × 6 m was formed from measurement data in Koktebel Bay (Fig. 1b).

For numerical modeling of wave fields, the estimated surface wind speeds for the Feodosia area were used. The wind duration time was 16 h, which corresponds to the mean duration of calculated storms of 4% regime probability in the study area (Lopatoukhin et al. 2006). The angular resolution of the SWAN model was 10°. On the frequency coordinate, varying from 0.02 to 2 Hz, a non-uniform grid with 31 nodes was used. The calculations were performed on the computational cluster of the Marine Hydrophysical Institute of the Russian Academy of Sciences.

3 Discussion of Results

The analysis of operational characteristics of waves in the central region of Koktebel Bay at the 11.5 m isobath showed that about 55% of all h_s values do not exceed 0.5 m; waves with heights $h_s \geq 1$ m occur in 14% of cases and the repeatability of waves with heights $h_s \geq 1.5$ m is about 5% (Fig. 2b). Thus, there is an obvious dependence: as the significant wave heights increases, their frequency of occurrence decreases.

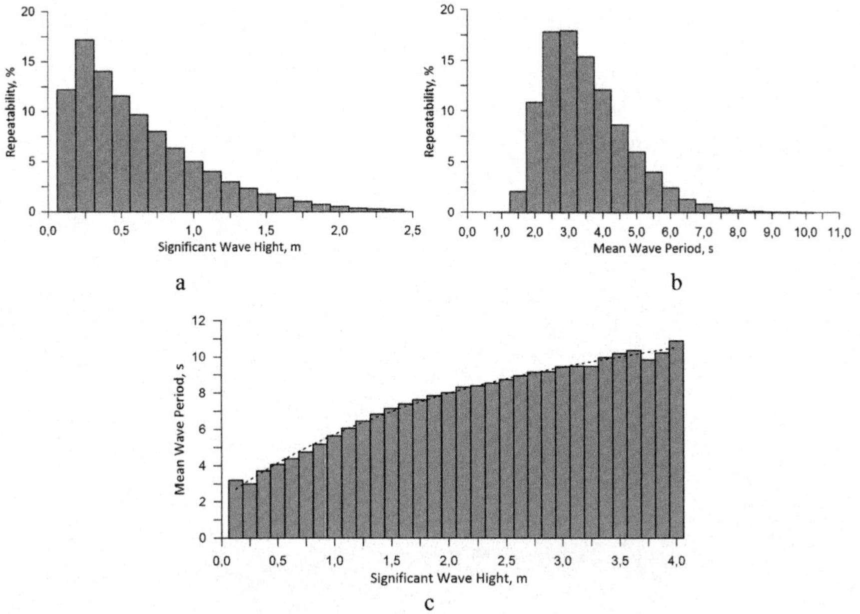

Fig. 2 Repeatability of the significant wave height h_s (**a**) and mean wave period $\overline{\tau}$ (**b**) relationship between the mean wave period and significant wave height (the dashed line is a polynomial of degree 3) (**c**) from the SWAN–ERA array in Koktebel Bay at a depth of 11.5 m

On the histogram of mean wave period repeatability (Fig. 2a), there is an interval of 2.5–3.5 s with increased repeatability (over 15%), with 51% of all cases falling into this interval.

Figure 2c shows the dependence of the mean wave period in the study area on the significant wave height. To build this dependence, the mean wave period was determined for each interval of h_s values by averaging all values $\overline{\tau}$ which fell within the given interval. As the calculations showed, this dependence is well approximated by a polynomial of degree 3

$$\overline{\tau} = a_0 + a_1 h_s - a_2 h_s^2 + a_3 h_s^3 \tag{3}$$

where $a_0 = 2.20$ s; $a_1 = 4.29$ s/m; $a_2 = 0.85$ s/m^2; $a_3 = 0.07$ s/m^3.

The highest repeatability (~ 39%) is for winds coming from the land side (sector 0°–90°). The maximum repeatability of strong winds ($W > 15$ m/s) corresponds to the north-eastern direction. From the sea side, the 90°–180° sector is the most storm-hazardous in the study area, and the wind repeatability from this sector is the lowest (~ 11%) (Fig. 3a).

At the bay entrance, waves coming from the east have the highest repeatability (18%) (Fig. 3b). Figure 3c shows the storm wave rose plotted against the values of significant wave heights h_s, which, according to (1), exceed the threshold value h_{st}

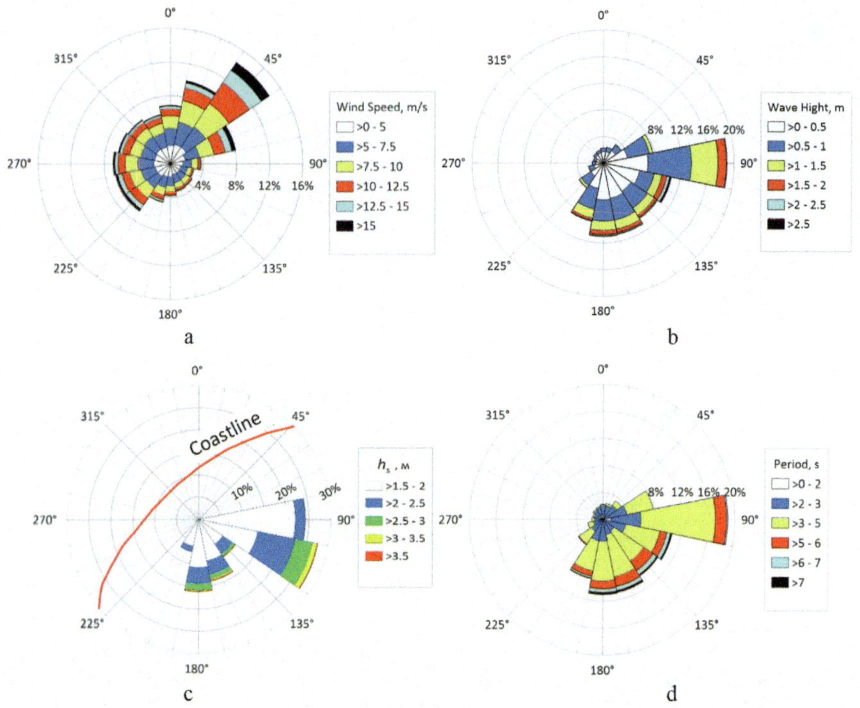

Fig. 3 Wind roses from ERA-Interim and ERA5 data (**a**), waves (**b**), storm waves (**c**) and mean wave periods (**d**) from the SWAN–ERA array in Koktebel Bay at a depth of 11.5 m. The red curve shows the coastline of the study area

= 1.57 m. It can be seen that over 50% of all storm waves come to the study area from the east (90°) and east-south-east (112.5°), with the highest storm waves ($h_s >$ 2.5 m) entering the bay from the east-south-east direction. The distribution of wave periods by direction and gradation (Fig. 3d) shows that waves with periods $\overline{\tau} \geq 7$ s approach the study area from the 112.5°–180° sector.

The analysis of the yearly distribution of the storm-event total duration (Fig. 4a) showed that the duration of storm events ($h_{st} \geq 1.57$ m) varies from 5.6 to 34.3 days, their mean value being 16.4 days. The monthly averaged storm duration varies from 0.6 to 9.8 days. The longest storm events (over 7 days) occur from November to March. The minimum duration of storms (under 1 day) is in May–August (Fig. 4b).

The storm power index *SPI* calculated from the SWAN–ERA array varies from 62 to 660 m^2 h, the mean storm duration T is 26 h and during the mature phase of the storm, the significant wave heights h_s varies from 2.3 to 3.6 m with a mean value of 2.6 m. Figure 4c shows the yearly distribution of the total storm power index *SPI* normalized to its mean annual value $SPI_m = 0.659 \times 10^6$ m^2 h. The histogram highlights eight years (1979, 1993, 1999, 2001, 2003, 2012 and 2014) with storm activity, which is half greater than average.

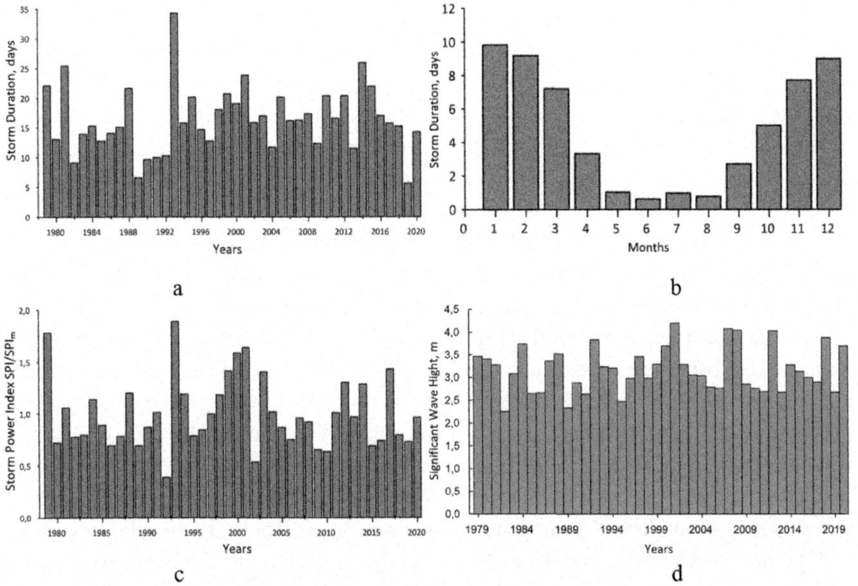

Fig. 4 Distribution by year of the total storm duration (**a**), by month of the average storm duration (**b**), by year of the total storm power index *SPI* normalized by its mean annual value SPI_m (**c**), annual maximum of significant wave heights (**d**) according to the SWAN–ERA array in Koktebel Bay at a depth of 11.5 m

Figure 4d and Table 1 present the results of calculations of extreme wave characteristics. It follows that the values of the annual maxima of significant waves vary from 2.25 to 4.20 m with a mean value of 3.17 m. For a storm of 4% regime probability (possible once every 25 years), the calculated value of the wave height of 1% probability is 5.4 m. This value, along with values $\overline{\lambda}$ and $\overline{\tau}$, is used in the design of coastal protection structures such as groynes and breakwaters.

Table 1 Significant wave heights h_s, mean wave height $\overline{h}$, mean wave period $\overline{\tau}$, mean wave length $\overline{\lambda}$, wave heights of 30, 13 and 1% probability in the system of storms possible once a year, in 5, 10, 25, 50 and 100 years (from the SWAN–ERA array for Koktebel Bay at a depth of 11.5 m)

T, years	Wave parameters						
	h_s, m	$\overline{h}$, m	$\overline{\tau}$, s	$\overline{\lambda}$, m	$h_{30\%}$, m	$h_{13\%}$, m	$h_{1\%}$, m
1	3.3	1.8	8.4	79	2.3	2.8	4.0
5	3.9	2.2	9.6	93	2.7	3.4	4.7
10	4.2	2.4	10.2	100	2.9	3.6	5.0
25	4.6	2.6	11.0	108	3.2	3.9	5.4
50	4.8	2.8	11.5	115	3.4	4.2	5.7
100	5.1	3.0	12.1	121	3.6	4.4	5.9

The wind wave fields in Koktebel Bay are determined by wind speed, direction and time of action, as well as by the features of the bottom relief, coastal configuration and values of effective fetches. The coastal area of Koktebel is characterized by a relatively straight coastline and parallel isobaths (Fig. 1b). Thus, the wave fields in the alongshore direction are relatively homogeneous here. All wave fields analyzed below were obtained for the calculated wind speeds of 4% regime probability (for speeds possible once every 25 years) and correspond to a time point of 16 h.

Figure 5 shows the spatial distribution of significant wave heights h_s and mean wave direction in the study area for the four most storm-hazardous wind directions: east, south, south-east, south-west. As can be seen, the strongest wave in Koktebel Bay is formed in easterly winds (Fig. 5d), which is caused by its highest speed. At wind waves from the east, south-east and south, a characteristic feature can be noted: the configuration and details of the fields of significant wave heights h_s, starting from the isoline of 3 m and less, are almost the same. Thus, at depths of 5 m and less, wave heights are determined mainly by the depth distribution.

Due to the relative shallowness of the bay, refraction has a significant effect on wave propagation. As waves approach the coast, wave fronts are parallel to isobaths and the mean wave direction turns normal to the isobaths (Fig. 5). Near the coastline of the study area, the waves mostly have a direction of 135°. The calculations of the mean wave direction for different wind directions showed that in all cases, except for the north-easterly and north-westerly winds, wind waves enter the bay from seawards.

Closer towards the coast, the wavelengths decrease monotonically for all calculated cases. Figure 6a shows an example of field of mean wavelengths $\overline{\lambda}$ for an easterly wind of 4% regime probability. At depths of 10 m and deeper, the $\overline{\lambda}$ values range 70–100 m. In the close vicinity of the water edge, the $\overline{\lambda}$ values decrease to 20–30 m. The maximum wavelengths correspond to the easterly and southerly winds.

An important factor determining the wave energy losses due to bottom friction is the amplitude of the orbital velocity of wave motion near the bottom V_b (m/s). As calculations for different wind directions (Fig. 6b) showed, values of 0.8–1.0 m/s are characteristic for V_b near the coast, at which all bottom soil particles up to 15 mm in size become suspended.

4 Conclusions

The paper studies parameters of the regional wave climate in the Koktebel Bay water area (the Black Sea, Crimea). Storm waves coming from the east and east-southeast are most probable in the study area, which is due to the predominance of easterly winds, the orientation of isobaths, and the influence of wave refraction.

For the point located at a depth of 11.5 m in the central part of the bay, a statistical analysis of retrospective wind wave calculations for 1979–2020 years was performed from the SWAN–ERA array. It was found that waves with significant wave heights $h_s \leq 0.5$ m and mean periods $\overline{\tau} = 2.5$–3.0 s account for 55% of all cases; the repeatability of waves at $h_s \geq 1.0$ m is 14%, and the frequency of waves at $h_s \geq 1.5$ m reaches

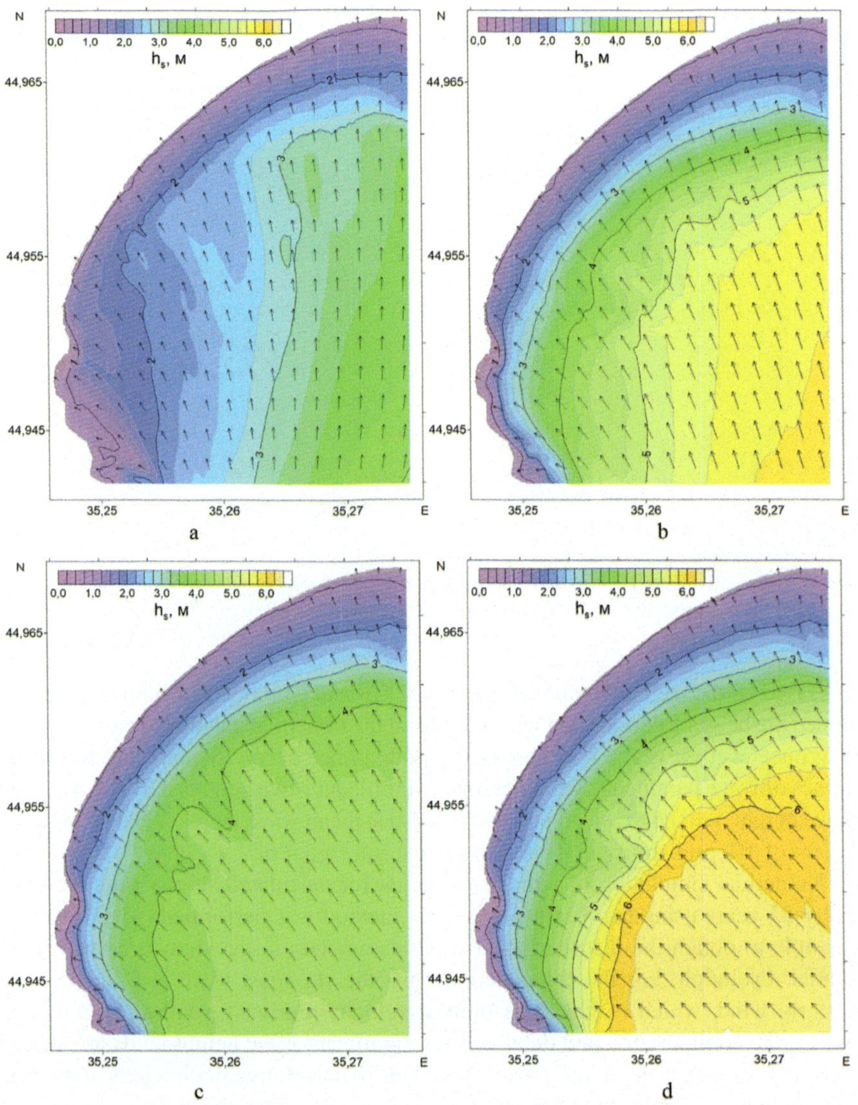

Fig. 5 Significant wave heights (h_s) and mean wave direction in south-westerly (**a**), southerly (**b**), south-easterly (**c**) and easterly (**d**) winds of 4% regime probability in Koktebel Bay

5%. Estimates of storm duration showed that the total annual storm duration varies from 5.6 to 34.3 days, with a long-term annual mean of 16.4 days. During the year, the longest storms (over 7 days per month) occur from November to March. The minimum duration of storms (under 24 h per month) is in May–August.

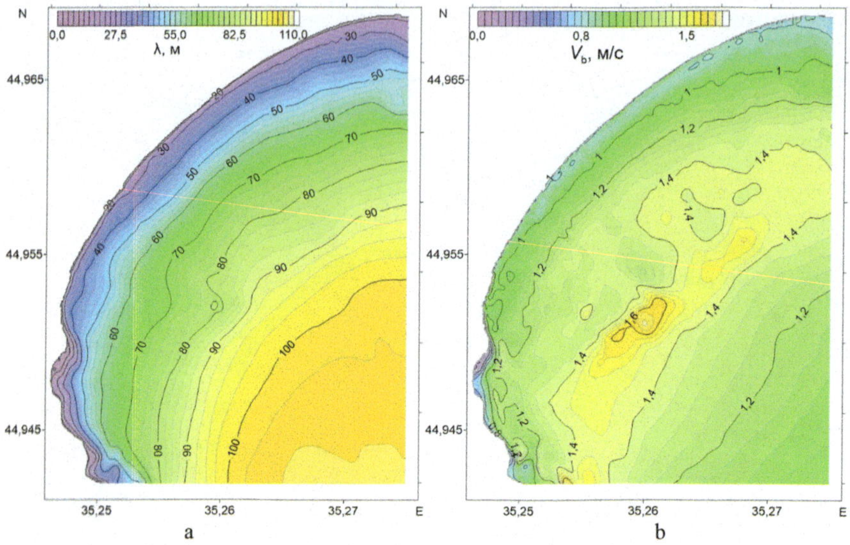

Fig. 6 Mean wavelength ($\overline{\lambda}$) (**a**) and amplitude of orbital velocity of wave motion near the bottom V_b (**b**) in the easterly wind of 4% regime probability in Koktebel Bay

Based on the method of annual maxima series and Gumbel distribution, we determined the heights, periods and wavelengths in the system of calculated storms occurring once every n years. For the storm possible once every 25 years, the following values of wave parameters were obtained: significant wave heights $h_s = 4.6$ m, mean wave height $\overline{h} = 2.6$ m, mean period $\overline{\tau} = 11.0$ s, mean wave length $\overline{\lambda} = 108$ m, wave height of 1% probability in the storm system $h_{1\%} = 5.4$ m.

Numerical simulation of wind wave fields for calculated storms possible once every 25 years using the 5-step nested grid method was performed, which made it possible to obtain the distribution of wave parameters in the study area with high spatial resolution. For the calculated storms, the strongest wave is formed in the easterly wind, which is due to its highest velocity. For east, southeast, and south waves, the configuration and details of the significant wave height fields are almost identical, starting at depths of 3 m and less. Due to the relative shallowness of the bay, refraction has a significant effect on wave propagation. Near the bottom at depths of 2–3 m, the amplitude of the velocity of orbital wave motions is 0.8–1.0 m/s.

Acknowledgements This research was funded under state assignment no. FNNN-2024-0016 of the Ministry of Science and Higher Education of the Russian Federation.

References

Amarouche K, Akpınar A (2021) Increasing trend on storm wave intensity in the Western Mediterranean. Climate 9(1):11. https://doi.org/10.3390/cli9010011

Arkhipkin VS, Gippius FN, Koltermann KP, Surkova GV (2014) Wind waves in the black sea: results of a hindcast study. Nat Hazard 14(11):2883–2897. https://doi.org/10.5194/nhess-14-2883-2014

Booij N, Ris RC, Holthuijsen LH (1999) A third-generation wave model for coastal regions. Model description and validation. J Geophys Res 104(4):7649–7666. https://doi.org/10.1029/98JC02622

Divinsky BV, Kosyan RD (2018) Wave climate of the coastal zone of the Crimean Peninsula. Phys Oceanogr [e-journal] 25(2):93–101. https://doi.org/10.22449/1573-160X-2018-2-93-101

Divinsky B, Fomin V, Kosyan R, Lazorenko D (2019) Maximum waves in the Black Sea. In: Proceedings of the fourteenth international MEDCOAST congress on coastal and marine sciences, engineering, management and conservation MEDCOAST 2019, vol 2, pp 799–810

Divinsky BV, Fomin VV, Kosyan RD, Ratner YD (2020) Extreme wind waves in the Black Sea. Oceanologia [e-journal] 62(1):23–30. https://doi.org/10.1016/j.oceano.2019.06.003

Efimov VV, Komarovskaya OI (2009) Atlas of extreme wind waves of the Black Sea. MGI NASU, Sevastopol, p 59 (in Russian)

Fomin, VV, Goryachkin YuN (2022) Accounting for the local wave and morphodynamic processes in coastal hydraulic engineering. Phys Oceanogr [e-journal] 29(3):271–290. https://doi.org/10.22449/1573-160X-2022-3-271-290

Gippius FN, Myslenkov SA (2020) Black Sea wind wave climate with a focus on coastal regions. Ocean Eng 218:108199. https://doi.org/10.1016/j.oceaneng.2020.108199

Goryachkin YuN, Repetin LN (2009) Storm wind and wave regime near the Black Sea coast of Crimea. Ecol Saf Coast Shelf Zones Compr Use Shelf Resour 19:56–69 (in Russian)

Goryachkin YuN, Kharitonova LV (2017) Changes of the Crimean coast in the last century. In: Proceedings of the thirteenth international MEDCOAST congress on coastal and marine sciences, engineering, management and conservation MEDCOAST 17, vol 2, pp 861–870

Goryachkin YuN, Dolotov VV (2019) Sea coasts of Crimea. Colorit, Sevastopol, p 256 (in Russian)

Goryachkin YuN, Markov AA (2023) Analysis of the effectiveness of reconstruction of coast protection structures in Crimea. Hydrotechnics 3(72):2–9 (in Russian)

Kharitonova LV, Alekseev DV, Fomin VV (2021a) Comprehensive study of wave and lithodynamic processes in the coastal area of the village of Morskoye (Eastern Crimea). Ecol Saf Coast Shelf Zones Sea 3:5–22 (in Russian). https://doi.org/10.22449/2413-5577-2021-3-5-22

Kharitonova LV, Alekseev DV, Fomin VV (2021b) Mathematical simulation of the Ay-Todor bay wave regime. InterCarto. InterGIS. GI support of sustainable development of territories. Proc Int Conf [e-journal] 27(3):16–31 (in Russian). https://doi.org/10.35595/2414-9179-2021-3-27-16-31

Krylov YuM (1966) Spectral methods for investigation and calculation of wind waves. Gidrometeoizdat, Leningrad, p 258 (in Russian)

Lopatoukhin LJ, Rozhkov VA, Ryabinin VE, Swail VR, Boukhanovsky AV, Degtyarev AB (2000) Estimation of extreme wind wave heights. World Meteorological Organisation. JCOMM technical report WMO/TD, no. 1041. https://doi.org/10.25607/OBP-1495

Polonsky AB, Fomin VV, Garmashov AV (2011) Characteristics of wind waves of the Black Sea. Rep National Academy of Sciences of Ukraine, 8, pp 108–112 (in Russian)

Lopatoukhin LI, Bukhanovsky AV, Ivanov SV, Chernyshova ES (eds) (2006) Reference data on the wind and wave regime of the Baltic, North, Black, Azov and Mediterranean Seas. Russian Maritime Register of Shipping: St. Petersburg, Russia, p 452 (in Russian)

SWAN User Manual (2018) Cycle III version 41.20. Delft University of Technology, Netherlands, p 121

To the Question of Geomechanics of the Island of Atlantis (on the Data of Plato's Dialogues "Timaeus" and "Critias")

S. Yu. Kasyanov and M. S. Kasyanova

Abstract The possibility of formation and evolution of the Atlantis Island described in Plato's dialogues and the ring structure of the city on it due to the exit to the day surface of the intramantle vortex is established. It is shown that the formation of the island of Atlantis, the Rishat structure in Mauritania and the Santorin volcano were determined by the emergence of the Yellowstone-Mediterranean Sea intramantle vortex (belonging to the system of three intensive intramantle vortices) to the day surface, which occurred during the evolution of the vortex. It is shown that the independent data presented by Plato about Atlantis are mutually confirmed, which speaks in favor of the fact that Plato describes events related to a real geophysical process. The information provided by Plato allowed to estimate the internal structure of material layers and the structure of circulation in the vortex. The commonality of the origin of Atlantis, the Santorin volcano and the Rishat structure in Mauritania allowed us to identify the stages of evolution of the intramantle vortex and to construct two independent estimates of the location of possibly preserved traces of Atlantis in the Atlantic Ocean west of Gibraltar. These estimates are in good agreement with each other and with the location of the Azores.

Keywords Intramantle vortex · Plato's dialogues · Atlantis · Santorin volcano · Rishat structure · Circulation cells

1 Introduction

Plato's dialogues contain useful geophysical information, which can be interpreted within the framework of the idea about the formation of the island of Atlantis and the ring structure of the city on this island due to the emergence of an intense intramantle

S. Yu. Kasyanov (✉)
N. N. Zubov's State Oceanographic Institute, Roshydromet, Moscow, Russia
e-mail: skas53@yandex.ru

M. S. Kasyanova
F. M. Dostoevsky Russian Christian Humanitarian Academy, Saint Petersburg, Russia

T. Chaplina (ed.), *Processes in GeoMedia—Volume VIII*, Springer Geology,
https://doi.org/10.1007/978-981-97-6627-7_29

vortex belonging to the system of three large-scale energy-carrying vortices in the Earth's mantle. In the present position, this intramantle vortex runs from Yellowstone to the eastern Mediterranean Sea.

With this interpretation, the independent information presented in Plato's texts, which allows us to estimate the angle of the vortex axis with the vertical in two different ways independently, appears to be internally consistent. Moreover, the values of the angle obtained close to each other practically coincide with the value obtained from cartographic data. Thus, our study confirms the fidelity of the geophysical information contained in Plato's dialogues.

The hypothesis that the island of Atlantis and the ring structure of the Atlantis city were formed by the output of an intense hot intramantle vortex, explains the evolution of the Atlantis structure described by Plato and its change in time, including: the formation of water and earth rings, then the appearance of a layer of stone on the boundaries of earth rings, then the appearance of layers of metals (orichalcum, tin and copper) on the walls of the rings. It was found out that the unknown metal, which Plato calls orichalcum, is aluminium with some impurities. Thanks to Plato's information there is an idea about radial structure of material layers and circulation structure in the intramantle vortex under Atlantis.

The hypothesis of the formation of Atlantis Island and the ring structure of the Atlantis city by the exit of an intense hot intramantle vortex allows us to link the formation and sinking of Atlantis with the explosion of the Santorin volcano in 1538 BC and with the formation of the Rishat ring structure in Mauritania as having a common origin.

The exit of the vortex to the day surface under the eastern part of the Mediterranean Sea caused the eruption of the Santorin volcano in 1538 BC, which led to the death of the Greek army, which had previously defeated the troops of Atlantis.

The ring structure of Rishat in Mauritania is related to the ring structure of Atlantis by the mechanism of formation. It was formed by the propagation into the body of Eurasia of vortex motion from the remaining section of the same intense intramantle vortex in the mantle after the vortex left the island of Atlantis at the turning of North America.

The position of the Santorin volcano at the turning of the North American mainland provides an estimate of the location of the city of Atlantis. A second independent estimate is given by the position of the Rishat structure in Mauritania at the turning of Eurasia together with Africa and North America as a whole. These estimates are close to each other and indicate the location of the city of Atlantis to the west of Gibraltar, which corresponds to Plato's description.

Estimates of the size of the island, derived from Plato's information, show that the Azores appear to be directly on the edge of the island, where Plato described high mountains.

2 Information About the Evolution of the Island of Atlantis According to Plato's Dialogues "Timaeus" and "Critias"

Plato begins his description of the evolution of the island of Atlantis from the moment when this island already existed and was located to the west of Hercules' Pillars.

Plato gives information on the geometry of the island: "from the sea to the middle of the island stretched a plain, if we believe the legend, more beautiful than all other plains and very fertile, and again in the middle of this plain, about fifty stadia from the sea, stood a mountain, low on all sides" (Losev et al. 1994, p. 507).

"The whole of this region lay very high and falls steeply to the sea, but the whole plain, surrounding the city and itself surrounded by mountains that stretched to the sea, was a flat plain, three thousand stadia in length, and in the direction from the sea to the middle - two thousand. The whole of this part of the island faced the south wind, and was enclosed by mountains on the north. These mountains are praised by tradition for the fact that they surpassed all the present ones in number, size and beauty: there was a large number of populous villages, there were rivers, lakes and meadows, providing food for all kinds of tame and wild animals, as well as huge forests, characterised by a variety of species, abundantly supplying wood for every kind of work." (Losev et al. 1994, p. 512).

It should be noted that the plain was later perimetered by a rectangular canal with a perimeter of ten thousand stadia. It follows that the size of the island in width, i.e. in the direction from the sea to the middle of the island, was more than four thousand stadia. The length of the island will be discussed a little later.

The text describes the evolution of the surface of the island. According to Plato, Poseidon strengthened the pre-existing hill in the middle of the plain, "by circumferentially separating it from the island and enclosing it alternately with water and earthen rings (there were two earthen rings and three water rings) of increasing diameter, drawn like a circular from the middle of the island and at an equal distance from each other. This barrier was insurmountable for people, for ships and navigation did not exist at that time. And the islet in the middle Poseidon without labour, as it befits a god brought into a comfortable form, drained from the earth two springs - one warm and the other cold - and made the earth give a variety and sufficient for life sustenance" (Losev et al. 1994, p. 508).

This circular structure was later surrounded by a wall 50 stadia from the outer of the three water rings, which reached the sea shore. A gap was made in the wall, through which a channel 50 stadia long was stretched from the sea to the outer of the three water rings surrounding the hill.

Plato reports important information from the geophysical point of view: "The largest in circumference water ring, to which the sea was directly connected, was three stadia wide, and the earth ring that followed it was equal to it in width; of the next two rings, the water ring was two stadia wide and the earth ring was again equal to the water ring; finally, the water ring that encircled the island in the middle was one stadia wide. The island on which the palace stood was five stadia in diameter; this island, as well as the earthen rings and the bridge, which was a pletre wide, the kings encircled it with circular stone walls and put towers and gates everywhere on the bridges at the passages to the sea. They quarried white, black and red coloured stone in the bowels of the middle island and in the bowels of the outer and inner earthen rings, and in the quarries, where on both sides (διπλούς) there were hollows covered with the same stone from above, they made berths for ships. ... The walls around the outer earthen ring, they all around the circumference of the copper, applying metal in the molten form, the wall of the inner shaft covered with tin casting, and the wall of the acropolis - orichalcum, emitting a fiery glitter" (Losev et al. 1994, p. 510).

According to Plato, orichalcum is a nugget metal, which in the times of Atlantis was mined in different places of the island: "Many things were imported to them from the subordinate countries, but most of the things necessary for life were given by the island itself, first of all, any types of fossil hard and fusible metals, and among them what is now known only by name, but then existed in reality: the nugget orichalcum, extracted from the bowels of the earth in different places of the island and by its value was then inferior only to gold" (Losev et al. 1994, p. 509). Note that in the last description of orichalcum as a metal Plato does not mention its fiery glitter. It was orichalcum that was characterised by its fiery shine, which made up the covering of the wall of the inner island (see above).

Thus, after the formation of a circular structure in the middle of the plain, the centre of this structure, located on a low mountain, was already at a distance of 63.5 stadia from the seashore. Consequently, the dimensions of the island of Atlantis have slightly increased during the formation of the ring structure.

According to Plato, the kings of Atlantis first of all "bridged the water rings surrounding the ancient metropolis, building a way from the capital and back to it. They built the palace from the beginning where the dwelling place of the god and their ancestors had stood, and then, taking it as their inheritance, one by one they embellished it more and more, each time endeavouring to surpass the predecessor, until at last they created a structure of astonishing size and beauty. From the sea they drew a canal three pletres wide and a hundred feet deep, and fifty stadia long, down to the outermost of the water-ring: thus they created an access from the sea into this ring, as if into a harbour, and prepared a sufficient passage even for the largest vessels. As for the earthen rings separating the water rings, near the bridges they dug channels of such width that from one water ring to another could pass one triera; above they planked the slabs, under which the navigation was to be made: the height of the earthen rings above the sea surface was sufficient for this" (Losev et al. 1994, p. 510). Thus, the height of the earth rings above the sea surface was greater than the height of the triera hull, which could reach 3.4 m (Ignatenko 2016).

In the centre of the island—inside the temple of Poseidon, there was an orichalcum stele (Losev et al. 1994, p. 513). The floor, pillars and inner surface of the walls of this temple were also covered with orichalcum. The outer surface of the walls of the temple was covered with silver.

Since the stele allowed to lead to its top the resisting wild sacrificial bulls (caught in the sacred grove of Poseidon adjacent to the temple without the use of iron with the help of only sticks and harness), it means that it should have had a slightly sloping side. The other side had to be steep, because it was marked with the texts of the laws of Atlantis and sacred writings and the blood of sacrificial animals flowed down it. It is clear that the metal of the stela was soft enough, as it allowed inscriptions on the surface.

The plain in the middle of the island had one side facing the sea. This side faced the south wind. The other sides of the plain were surrounded by high mountains that stretched all the way to the sea shore. The shores of the plain that faced the sea were located high above sea level and steeply steepened to the sea.

As can be seen from Plato's text, the rhythm of the island's life obeyed the 11-year cycle, which coincides with the cycle of solar activity (judgements and sacrifices took place alternately at intervals of five and six years: "In this temple they gathered on the fifth or sixth year, alternately measuring even or odd number, to confer about common concerns, to judge whether any of them had committed any violation, and to judge" (Losev et al. 1994, p. 513).

3 Relation of the Eruption of the Volcano Santorin to the Destruction of the Greek Army and the Sinking of Atlantis Described by Plato

Earlier we have shown (Kasyanov and Samsonov 2021a) that as a result of invasion in mantle of the temporary satellite captured by the Earth the African plate came in movement concerning the North Pole, as a result of which latitude and climatic conditions of Egypt constantly changed. In these conditions to use a solar year in Egypt in practice was useless and time could be counted only by lunar months. When movements of the African plate concerning the North Pole essentially slowed down, in Egypt again there was an opportunity to use a solar year. In Plato's dialogue "Timaeus" the Egyptian priest gives to Solon dating of the highest blossoming of the city (state) of Atlantis as 9000 years before the moment of their conversation (Losev et al. 1994, p. 428). If we compare this dating with the confirmed archaeological and historical data on the origin of the most ancient highly developed civilisations (Cretan-Mycenaean, Ionian, Doric), then in the specified period no traces of them were found. Evidence of the world's oldest city Jericho-1 dates back to the 8th–9th millennium BC. But Jericho was not a city or state in the ancient (and modern) sense, it is a Neolithic period, when known civilisation markers did not yet exist.

However, the dating of Plato's dialogue approximately coincide with modern ones, if the above-mentioned fact of transition from lunar months to solar year is taken into account in the calculation. Thus, from 9000 years, mentioned by the priest of the temple in Sais, a part of "years" refers exactly to lunar months, and the remaining part of years—to a solar year. So that from the time of Solon's conversation with the priest (about 590 BC) to the eruption of Santorin in 1538 BC. (Pearson et al. 2022) 9000 "years" were counted, the last part of these "years" had to be counted according to the solar year, and the main part of the "years"—according to the lunar months.

At the same time the change of the way of counting years should have taken place in 829 BC.

Plato reports that the Greek army was swallowed up by the earth 9000 years ago. Apparently, this event was caused by the army camped on the site of a future volcano and was destroyed when the Santorin volcano exploded in 1538 BC.

4 Intramantle Vortex as a Determining Factor of the Atlantis Island Evolution

In the present chapter, we put forward the hypothesis that the formation and evolution of the Atlantis Island was determined by a natural geophysical process associated with the approach of a hot intramantle vortex to the day surface. According to our hypothesis, the evolution of the Atlantis island, the formation of the Rishat ring structure in Mauritania and the eruption of the Santorin volcano on the island of Fira were determined by different stages of the evolution of the Yellowstone-Mediterranean Sea vortex (3), which is part of the system of three intense energy-carrying vortices Himalaya + Tian Shan-Kolyma (1), Kolyma-Yellowstone (2) and Yellowstone-Mediterranean Sea (3) (see Fig. 1), described in Kasyanov and Samsonov (2021b).

Vortex (3) contacts the Earth's liquid core under Iceland (Kasyanov 2023). At contact with the liquid core, the molten mantle matter moving in the vortex is additionally heated, rises upwards, and spreads under the ocean floor. This results in the

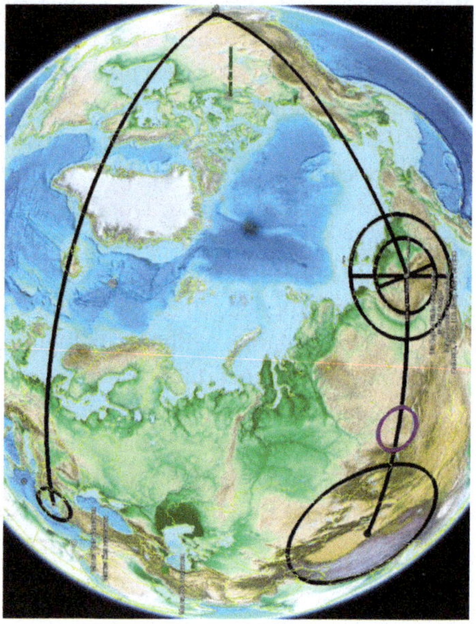

Fig. 1 A modern system of three energy-carrying vortices in the Earth's mantle vortices: (1) Himalayas + Tien Shan–Kolyma, (2) Kolyma–Yellowstone, (3) Yellowstone–Hudson Bay–the eastern part of the Mediterranean Sea (Kasyanov and Samsonov 2021b)

formation of the Icelandic hot spot with a characteristic relief of the lithosphere of the surrounding area. As the heated matter spreads out in the boundary layer beneath the ocean floor, Görtler vortices are formed, forming the relief of the lithosphere of the region surrounding Iceland (Kasyanov 2023).

In the radial structure of the vortex flow, similar to the previously studied intra-mantle vortices under the Mediterranean and the Sea of Japan (Kasyanov 2024a, b, c), azimuthal velocity discontinuities are present and zones with intense motion at the outer boundary of the vortex and in the vortex core in its central part are distinguished.

The vortex (3) in the mantle has a cylindrical shape. When the vortex exits to the day surface, the vortex axis forms an acute angle with the vertical. Therefore, the projection of the vortex exit to the Earth's sphere on the Earth's surface is shown as an ellipse with a large semi-major axis perpendicular to the vortex axis. High mountains on the periphery of the elliptical structure correspond to the zone with intensive motion at the outer boundary of the vortex, which agrees with Plato's description of the island of Atlantis.

From the natural-scientific point of view it seems that Plato attributes to Poseidon the natural process of creation of water rings separating earth rings. The process of lining the walls of earthen rings with stone, attributed by Plato to the kings of Atlantis, also appears to be natural and not man-made, since it was from the walls of earthen rings that stone was extracted.

The natural mechanism of appearance of the stone layer on the boundaries of the earth and water rings is that when an intensive hot intramantle vortex approached the day surface, the melt of the mantle substance came to the surface and formed a stone layer on the walls of the inner and outer earth rings during solidification.

In addition, Plato reports about the kings of Atlantis that "they covered the walls around the outer earthen ring with copper along the entire circumference, applying metal in molten form, they covered the wall of the inner shaft was covered with tin casting, and the wall of the acropolis itself - with orichalcum, emitting a fiery glitter" (Losev et al. 1994).

Plato attributes the appearance of layers of metal on the walls of the ring structure of Atlantis to man-made actions made by the will of the kings of Atlantis. However, the man-made origin of the metal layers on the walls of the Atlantis ring structure is not possible either in terms of the amount of metal needed to cover the walls, or in terms of the amount of energy (most likely obtained by burning wood) needed to smelt this metal, or in terms of the necessary labour input. Finally, how could one cover the surface of the wall around the central island with nugget orichalcum, and even so that it would retain its fiery glow? Therefore, the appearance of metal layers on the walls of the Atlantis ring structure could be caused only by some natural process possessing axial symmetry.

5 Preliminary Considerations on the Structure of Material Layers and Circulation Patterns in the Intramantle Vortex Beneath Atlantis

Based on the information provided by Plato, we can make the following considerations about the structure of the vortex under Atlantis.

The matter in the vortex outlet (3) moves cyclonically. The melt of the mantle substance contained in the hot intramantle vortex is electrically conductive and moves in the Earth's magnetic field. The projection of the Earth's magnetic field induction vector on the vortex axis, directed downward along the vortex axis, plays an important role in the process. Under the action of this component of the Earth's magnetic field, electric charges of metal oxides contained in the melt are separated and metal layers are separated. Metal layers are formed at certain radii of the vortex where two conditions are fulfilled. Firstly, the radial velocity of the vortex at this radius is zero, i.e. this surface is the boundary of two circulation cells in the plane passing through the vortex axis. Secondly, at this radius, the Lorentz force tending to shift the positively charged metal ions to the centre of the vortex is balanced by the centrifugal force.

In this way liquid metallic aluminium is extracted from the melt of the Al_2O_3 contained in the mantle substance, which is the main component of the unknown precious nugget orichalcum. The presence of small amounts of additives can diversify its properties—increasing density, imparting strength and the possibility of fiery brilliance. Liquid metallic aluminium forms the central circulation cell in the vortex. The upward movement of the substance on the vortex axis occurring in it forms a cone-shaped stele of orichalcum in the temple of Poseidon when the metal solidifies. Since the axis of the vortex comes out at an angle to the vertical, it explains the above-mentioned flatness of one side of the stele and steepness of the opposite side, which corresponds to the description of the stele in Plato (Losev et al. 1994).

Due to stratification of the substance in the field of centrifugal forces, the density of the substance in the circulation cells increased with increasing radius. The central cell, in which liquid pure aluminium with density 2698.9 kg/m^3 circulated, was followed by the cell of circulation of liquid $CaCO_3$ ("white stone" in Plato (Losev et al. 1994, p. 510)—it is, apparently, a kind of limestone). The density of calcium carbonate 2740 kg/m^3 (calcite) is slightly higher than the density of pure aluminium. On the outer boundary of the circulation cell of liquid $CaCO_3$, coinciding with the wall of the inner island, there was a denser layer of aluminium with additives increasing its density and causing a fiery glare, which could include copper. Its density is already slightly higher than that of $CaCO_3$. Next was a cell of molten Al_2O_3, which has a density of 3990 kg/m^3.

When the vortex approached the day surface, the melt of the mantle substance came to the surface of the earth, which formed a layer of stone on the walls of the inner and outer earthen rings during solidification.

Relatively thin layers of tin (7310 kg/m^3) and copper (8920 kg/m^3) on the outer stone walls of the inner and outer earthen rings, about which Plato speaks, could be formed only at the boundaries of the melt circulation cells of the substances composing the mantle. Since tin and copper are heavier than the surrounding oxides, from the melts of which they were extracted, they could be held near the upper surface of the melt only due to the upward directed tangential stresses arising from the movement of fluid flows in the surrounding layers of the vortex substance. Therefore, under the outer boundaries of the inner and outer earthen rings, sufficiently intensive upward movements should have been observed in both cells of the vortex substance circulation contacting the metal layer from opposite sides. This applies both to the tin layer at the outer boundary of the inner earthen ring and to the copper layer at the outer boundary of the outer ring.

6 Estimation of the Angle with the Vertical of the Axis of Exit of the Intramantle Vortex to the Day Surface at the Formation of the Annular Structure on Atlantis Island

If we accept the hypothesis that Atlantis was formed by the exit of the intramantle vortex, then the data cited by Plato allow us to estimate the angle formed by the axis of the intramantle vortex at its exit to the day surface in two independent ways.

Firstly, we know that at the point at which Plato begins his description of the island of Atlantis, there was a mountain on the island, the distance from which to the sea was about 50 stadia. Later, after the mountain was surrounded by water rings, the distance to the centre of the rings increased to 63.5 stadia. Let us assume that this mountain was formed as a result of the fact that a round flat tip of the vortex with a radius of 2.5 stages came to the day surface. In this case, the plane of the vortex that came to the surface and then solidified was located at an angle to the horizontal. The highest point of the mountain in this position would be observed at some point between the centre of the rings and the point of the circle furthest from the sea. The distance from the sea to this point was equal to 50 stadia.

Over time, the intramantle vortex approached the surface, resulting in the formation of a system of rings around the mountain, which became a small inner island. Differentially rotating axisymmetric material layers of ever larger radius were successively coming to the day surface. The molten matter coming to the surface solidified, forming a ring. Then the solidified substance of the formed ring, preserving its axisymmetric form, descended to a position close to horizontal. However, some height difference was still preserved. The distance from the sea to the upper point at lowering of rings was equal to 63.5 stages plus some part of the radius of the mountain, equal to 2.5 stages.

In this way the system of rings and a small inner island kept its round shape. Therefore the coast of the sea was moved away from the top of the mountain. The highest point of the mountain was observed a little further from the sea than the centre of the rings. If we take the distance from the highest point to the centre of the rings to be equal to one stage, we obtain an estimate of the angle θ of the vortex axis with the horizontal $\theta = \arcsin(50/(63.5 + 1.0)) = 50.82°$.

7 Estimation of the Angle with the Vertical of the Axis of Exit of the Intramantle Vortex to the Day Surface from the Geometrical Parameters of Atlantis Island

Based on the information given by Plato, let us construct estimates of the corresponding geometrical parameters of the Atlantis island and the intramantle vortex forming it.

Let us consider the section of the intramantle vortex from the point where the vortex touches the liquid core to the exit of the vortex axis to the day surface, schematically represented in Fig. 2. Let us denote the point of tangency of the vortex and the liquid core of the Earth as *C*, the point of the day surface located above it as *D*, and the point of exit of the vortex axis to the day surface corresponding to the centre of the ring structure of the Atlantis city as *A*.

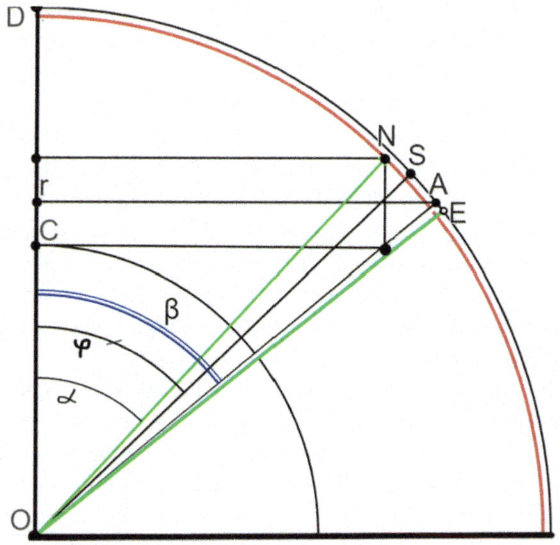

Fig. 2 Location of the intramantle vortex in the mantle and location of characteristic points of the vortex exit to the surface. C is the point of contact between the vortex and the Earth's liquid core, *D* is the point of the day surface above it, r is the radius of the vortex, R_e is the radius of the Earth, R_c is the radius of the Earth's liquid core, N is the location of the upper point of the vortex face lying at a depth h from the day surface (h = 100 km is taken in the calculation of estimates), on the day surface above this point is the border of Atlantis island on the opposite side of the island from the city, S is the projection of the vortex end centre on the day surface, A is the exit of the vortex axis on the day surface

Let us choose the ray from the Earth centre O to the point D as the axis of ordinates. Let there be β an angle between the rays OD and OA, equal to the angle of the vortex axis with the horizontal at the point of exit of the vortex axis to the day surface. Let $\delta = 180\pi(4000 - 63.5)s/\mathrm{Re} = 7.41°$—the angular distance between points A and N, known to us from Plato's data, where $s = 209.4$ m is the length of the pharaonic system stage.

Then, using the notations of Fig. 2, we obtain two relations:

$$r + Rc = \mathrm{Re}\cos\beta,$$
$$2 + Rc = (\mathrm{Re} - h)\cos\alpha = (\mathrm{Re} - h)\cos(\beta - \delta),$$

from which follows the equation for finding the angle β:

$$2\mathrm{Re}\cos\beta - Rc = (\mathrm{Re} - h)\cos(\beta - \delta).$$

From here are:

$$\beta = 50.81°, r = 530.60 \text{ km}, \alpha = \beta - \delta = 43.40°,$$
$$y = (2r + Rc)\mathrm{tg}\alpha = 430.83 \text{ km}, \varphi = \mathrm{a}\tan(y/(r + Rc)) = 46.94°.$$

Note that according to Fig. 2, the above described angle of the vortex axis with the horizontal should be equal to the angle β.

Comparing the obtained values of angles $\beta = 50.81°$ and $\theta = 50.82°$ we see that two independent estimates of the angle β obtained by us (based on the above information from Plato's dialogues) give practically coinciding results.

Thus, we can conclude that the independent information presented in Plato's dialogues are consistent with each other and, therefore, fairly truthfully reflect the geophysical process that actually took place.

The exit of the intramantle vortex to the day surface formed the rise of the island Atlantis, close in form to an ellipse with semi-axes of $e1 = 570.99$ km and $e2 = 419.85$ km. The distance from the centre of Atlantis to the centre of the ellipse was 404.45 km.

8 Traces Left by the Intramantle Vortex During Its Evolution in the Mantle and an Assessment of the Possible Location of Traces of the Atlantis City

In the present state, vortex 3, starting at Yellowstone, passes through the lithosphere of mainland North America (NA), then passes through the mantle beneath the North Atlantic. In doing so, it touches the Earth's liquid core. Vortex 3 then passes through the Eurasian (EA) lithosphere and ends by reaching the day surface under the eastern Mediterranean Sea (see Figs. 1, 3 and 5).

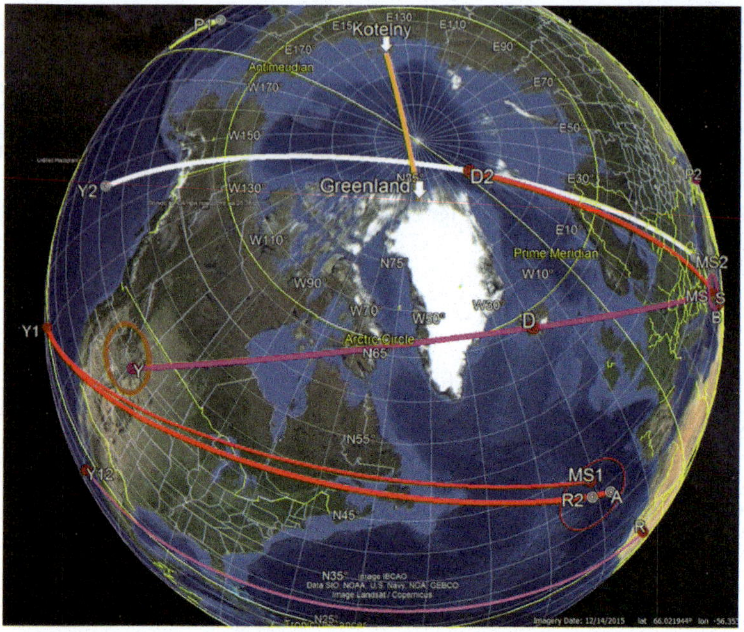

Fig. 3 Locations of the axis and outputs of vortex 3 at the present state (*Y-MS*), at 35.40° clockwise rotation around pole *P1* (*Y1-MS1*) and at 25.38° clockwise rotation around pole *P2* (*Y2-MS2*). *D*—Iceland—the point where the liquid core touches vortex 3 in the present state; *D2*—Iceland—the point where the liquid core touches vortex 3 when rotated 25.38° clockwise around pole *P2*; *MS*—vortex core exit under the Mediterranean Sea (with a radius equal to 2.5° of meridian arc); *MS1*—vortex core exit under the Mediterranean Sea when rotated 35.40° clockwise around pole *P1*; *MS2*—vortex core output under the Mediterranean Sea when rotated 25.38° clockwise around pole *P2*; *Y*—Yellowstone; *Y1*—Yellowstone when rotated 35.40° clockwise around pole *P1*; *Y2*—Yellowstone when rotated 25.38° clockwise around the *P2* pole; *Y12*—Yellowstone's position relative to the mantle at two CA rotations; *P1*—Eurasia-bound pole of rotation of the North American mainland (at 35.4° clockwise); *P2*—pole of rotation of Eurasia together with Africa and North America as a single whole; *S*—Santorin volcano; *B*—point on the extension of the *D2-S* beam, corresponding to the estimation of the position of the exit of the vortex axis 3 on the day surface; *A*—estimation of the position of Atlantis according to the position of Santorin volcano, i.e. point *B* at 35.40° clockwise rotation around the *P1* pole; *R*—position of the Rishat structure in Mauritania; *R2*—estimation of Atlantis position by the position of the Rishat structure at 25.38° rotation around the *P2* pole; *Y1-MS1*—position of vortex 3 at 34.5° clockwise rotation around the *P1* pole; *Y12-R*—position of extensions of the vortex section preserved in the mantle towards Yellowstone (at two SA rotations) and towards the Rishat structure

In the initial position, the North American continent in relation to Eurasia was rotated clockwise by 35.4° relative to the Eurasia-related pole of rotation *P1*, located south of Beringia (see Fig. 4).

The gyroscopic moment acts on the vortex, tending to align its axis of rotation with the Earth's axis of rotation. In doing so, the vortex rotating within the lithosphere of the continent affects the body of the continent and causes a change in the position of the continent on the Earth's sphere. As a result of their movements, the continents

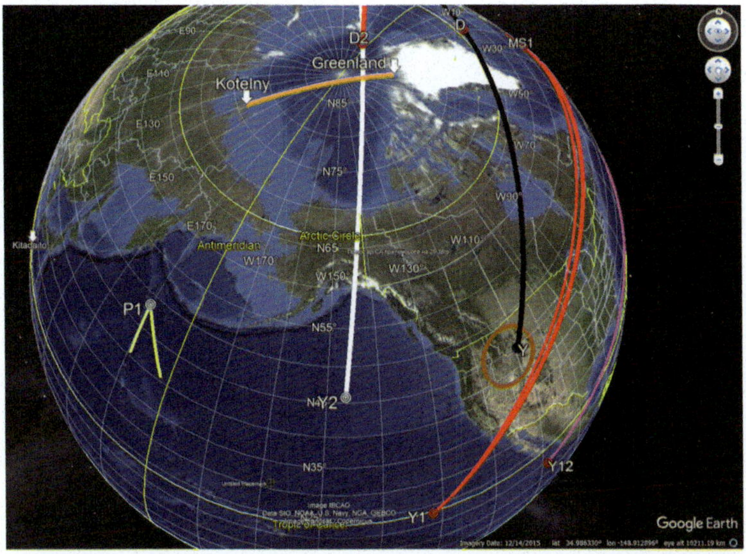

Fig. 4 Position of the North American rotation pole (at 35.4° clockwise) relative to Eurasia and the position of the vortex exit points 3. *P1* is the rotation pole of the North American mainland (at 35.4° clockwise) associated with Eurasia; *Kotelny* and *Greenland* are the points at which Eurasia and Greenland were in contact before Eurasia rotated with Africa and SA relative to the *P2* pole at 25.38° clockwise (see Fig. 5). In addition, the arrow-marked points on *Borneo* Island and *Kitadaito* Island were also in contact with each other before the indicated rotation. *D*—Iceland—the point of contact of the liquid core by vortex 3 in the present state; *D2*—Iceland—the point of contact of the liquid core by vortex 3 at a 25.38° clockwise rotation around the *P2* pole; *MS1*—the exit of the vortex core under the Mediterranean Sea at a 35.40° clockwise rotation around the *P1* pole; *Y1*—Yellowstone at a 35.40° clockwise around *P1* pole; *Y*—Yellowstone; *Y2*—Yellowstone at 25.38° clockwise rotation around *P2* pole; *Y12*—position of Yellowstone relative to the mantle at two CA rotations; *P1*—rotation pole of the North American continent (at 35.4° clockwise) associated with Eurasia

of North America and Eurasia, together with Africa, have now reached a quasi-equilibrium position.

In addition to the above-mentioned rotation relative to Eurasia in the initial position, the North American continent has been rotated as a unit with the Eurasian continent and Africa counterclockwise by 25.38° relative to the rotation pole *P2* located south of the Caspian Sea (see Fig. 5).

During the movements of the continent of North America and Eurasia together with Africa relative to the underlying mantle, Greenland Island remained stationary relative to the mantle. Before the rotation by 25.38° relative to the rotation pole *P2*, the cape on Greenland Island coincided with Precious Bay on Kotelny Island in the archipelago of the Novosibirsk Islands, and the coast of the bay on Borneo Island coincided with the coast of Kitadaito Island. The turn of Eurasia, first discovered by Academician V. A. Magnitsky in 1995 according to astronomical data of the very first edition of Ulugbek's zij of 1437, was discussed in Kasyanov (2023).

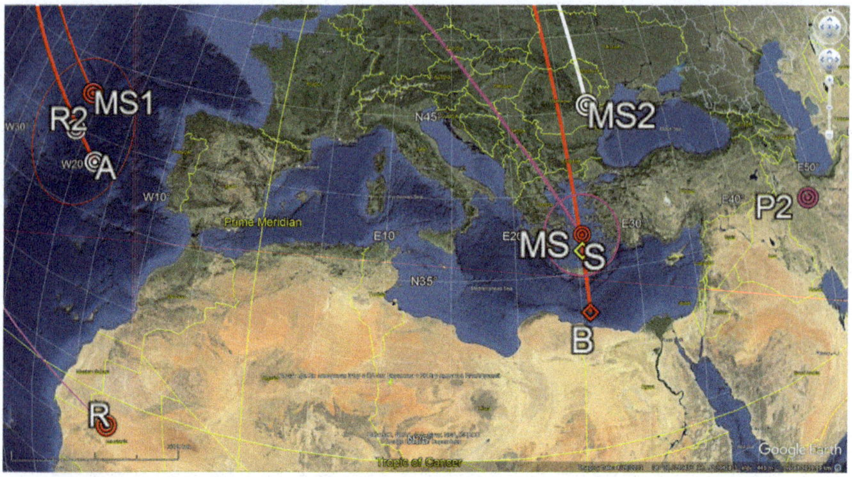

Fig. 5 Position of the pole of rotation of Eurasia together with Africa and North America *P2* as a single whole. Estimated position of Atlantis. Locations of the axis and outputs of vortex 3 in the present state (*Y-MS*), before the SA rotation (*Y1-MS1*) and before the Eurasian rotation (*Y2-MS2*). *MS*, vortex core output under the Mediterranean Sea (with a radius equal to 2.5° of meridian arc); *MS1*—vortex core output under the Mediterranean Sea at 35.40° clockwise rotation around the *P1* pole; *MS2*—vortex core output under the Mediterranean Sea at 25.38° clockwise around the pole *P2*; *P2*—pole of rotation of Eurasia together with Africa and North America as a single whole; *S*—Santorin volcano; *B*—point on the extension of the beam *D2-S*, corresponding to the estimation of the position of the exit of the vortex axis 3 on the day surface; *A*—estimation of the position of Atlantis according to the position of Santorin volcano, i.e. point *B* at rotation by 35.40° clockwise around the *P1* pole; *R*—position of the Rishat structure in Mauritania; *R2*—estimation of Atlantis position by the position of the Rishat structure when rotated by 25.38° around the *P2* pole

Thus, in order to understand the initial position of the vortex axis relative to the mantle, it is necessary to make two rotations of the vortex axis: first, a rotation around the *P1* pole of rotation of SA relative to Eurasia, and then a rotation of Eurasia together with Africa and SA as a whole relative to the mantle around the *P2* pole of rotation. At the same time, to understand the initial position of the vortex axis relative to Eurasia and Africa, it is necessary to make only one rotation of the axis—a clockwise rotation of 35.4° around the *P1* pole of rotation of SA relative to Eurasia.

It can be considered with certain accuracy that a part of vortex 3 is "frozen" in the body of North America and rotates in it as in a bearing. Under the action of the gyroscopic moment created by the vortex, North America, together with the associated part of vortex 3, rotated relative to Eurasia by 35.4° around the rotation pole *P1*.

In Fig. 3, the present-day position of the axis of vortex 3 is marked by the points of its exits as *Y-MS*. At a clockwise rotation of 35.4° relative to pole *P1*, vortex 3 takes the position *Y1-MS1* (relative to Eurasia).

However, the initial position of vortex 3 relative to Eurasia is slightly different from the *Y1-MS1* position for the following reason. After the rotation of SA (by 35.4° anticlockwise around pole *P1*) from the initial position, vortex 3 exerted the gyroscopic momentum on North America and Eurasia with Africa and rotated them as a whole by an angle of 25.38° around pole *P2* relative to the underlying mantle. At the same time, the vortex partially melted the surrounding substance of the mantle and lithosphere of the continents and slightly moved relative to it.

Let us consider the initial position of vortex 3 relative to Eurasia. We assume that in the initial position of vortex 3 and SA, the exit of vortex 3 to the day surface formed the territory of the entire island of Atlantis. The exit of the axis and core of vortex 3 formed the ring structure of the city on Atlantis Island described by Plato (in other words, the city of Atlantis). Later, when the rotation of SA around the pole *P1* began, vortex 3 left from under the island of Atlantis, as a result of which the surface of Atlantis Island began to sink. We assume that after the rotation of vortex 3 together with SA around the pole *P1*, vortex 3 again went to the free surface, which caused the formation and catastrophic eruption of the Santorin volcano. At the same time, the Santorin volcano was formed directly above the centre of the end face of vortex 3. The formation and eruption of the Santorin volcano caused the death of the Greek army, described by Plato, which by tragic accident was located on the site of the future volcano.

Thus, the position of the Santorin volcano allows us to obtain some rough estimate of the location of the Atlantis city relative to Eurasia by rotating the position of the Santorin volcano clockwise around the *P1* pole by 35.4°. We assume that the ring structure of the Atlantis city is formed by the vortex core exit with the vortex axis at its centre. Therefore, we need to find an estimate of the position of the vortex 3 axis exit to the surface corresponding to the position of the Santorin volcano as a projection on the ground surface of the centre of the cylindrical vortex end face at the exit of vortex 3 to the day surface. To do this, we will need to approximate the angular distance between the point at which vortex 3 touched the liquid core and the Santorin volcano at the time of Santorin volcano formation.

Note that the Santorin volcano was formed before the clockwise rotation of Eurasia with North America and Africa as a whole around the *P2* pole by 25.38°. In the present state, vortex 3 is in contact with the liquid core of the Earth at the point *D* lying under Iceland. Therefore before the specified rotation the point of contact between vortex 3 and the liquid core was located at point *D2*, which is obtained by rotating point *D* around pole *P2* by 25.38° clockwise.

For estimation, we approximate the angular distance between the point of tangency of vortex 3 of the liquid core and the Santorin volcano at the time of formation of this volcano as equal to the angular distance between points *S* and *D2*, which turns out to be equal to 47°.

This estimate corresponds with good accuracy to the estimate of the angle φ in Fig. 2 of 46.94° obtained above from Plato's data. This confirms the validity of our assumption that the island of Atlantis is formed by the exit of the intramantle vortex 3, and the ring structure of Atlantis is formed by the exit of the vortex core and vortex axis to the day surface.

Due to a sufficiently accurate correspondence of the angular distance between points *S* and *D2* obtained from cartographic data and calculated from Plato's data, we can obtain an estimate of the position of the ring structure of the Atlantis city relative to Eurasia. Let the point *B*, located on the extension of the ray *D2-S* in Figs. 3 and 5, corresponds to the estimation of the position of the exit of the vortex axis 3 to the day surface. Then we can take as our unknown angular distance from the point *D2* to the point *B* the value of the angle (see Fig. 2) obtained above according to Plato's data. Thus the position of point *B* is determined by us. Turning the point *B* relative to the rotation pole *P1* clockwise by 35.4°, we obtain the point *A*—an estimate of the position of the ring structure of the Atlantis city relative to Eurasia.

In the initial position, vortex 3 passed in the body of the North American lithospheric plate, then continued in the mantle thickness, where it came in contact with the Earth's liquid core, then exited the mantle and ended by exiting to the day surface near the boundary of the Eurasian and African lithospheric plates. At clockwise rotation of Eurasia, Africa and North America as a whole around the rotation pole *P2* by 25.38°, the sections of the vortex associated with the North American, Eurasian and African lithospheric plates moved together with them, while the section passing in the thickness of the underlying lower mantle remained in place. After the plates rotated, it began to contact the new areas of the lithospheric plates and led to the formation of vortex motions in their thickness.

Thus, the vortex section preserved in the lower mantle formed extensions in both directions, schematically shown in Fig. 3 by the line (Y12-R). The exit of the formed vortex extension to the day surface in the African plate formed the Rishat ring structure in Mauritania, shown in Fig. 6, similar in formation mechanism to the ring structure of Atlantis. This provides an independent second estimate of the position of the Atlantis city ring structure by rotating the centre of the Rishat ring structure around the rotation pole *P2* clockwise by 25.38°. In Fig. 6, the position of the *Rishat* structure in Mauritania is labelled as R, and the location estimate of Atlantis city obtained from the position of the Rishat structure is labelled as *R2*. The point *R2* is located near the vortex axis. The discrepancy with the estimate of the location of the Atlantis city based on the position of the Santorin volcano is about 350 km and can be partially explained by the change in the angular position of the vortex continuation axis under the action of the gyroscopic moment during the formation of the vortex continuation in the African plate body.

9 Conclusions

1. Our study confirms the correctness of the geophysical information contained in Plato's dialogues, since the estimates of the angle formed by the axis of the intramantle vortex with the vertical obtained from two independent data sets contained in Plato's text coincide with each other with good accuracy. In addition, the values of these two angle estimates approximate the value of the corresponding angle obtained from the cartographic data.

Fig. 6 Structure of Rishat, or "Eye of the Sahara" in Mauritania (https://bestmaps.ru/map/yandex/hybrid/12/21.1127/-11.3708)

2. It is established that the unknown metal, which Plato calls orichalcum, is aluminium with some impurities.
3. The hypothesis that the island of Atlantis and the ring structure of Atlantis city were formed by the output of an intense hot intramantle vortex, explains the evolution of the Atlantis structure described by Plato and its change in time, including: the formation of water and earth rings, then the appearance of a layer of stone on the boundaries of earth rings, then the appearance of layers of metals (orichalcum, tin and copper) on the walls of the rings.
4. The information given by Plato gives an idea about the radial structure of material layers and the circulation structure in the intramantle vortex under Atlantis.
5. The island Atlantis appeared as a result of the exit of the intensive hot intramantle vortex to the day surface in its initial position connected with the position of the North American continent before its rotation by 35.4° around the *P1* pole relative to Eurasia and Africa. In its present position, this intramantle vortex passes from Yellowstone to the eastern Mediterranean Sea.
6. Later, North America together with the section of the vortex passing in its body rotated by 35.4° around the *P1* pole relative to Eurasia and Africa under the action of the gyroscopic moment created by the vortex. Coming out of the body of the North American continent, the vortex formed its continuation in the mantle and further in the Eurasian lithosphere. The vortex reached the day surface under the eastern part of the Mediterranean Sea and caused the eruption of the Santorin volcano in 1538 BC, which led to the death of the Greek army.
7. The ring structure of Rishat in Mauritania is related to the ring structure of Atlantis by the mechanism of formation. It is formed by the propagation into the body of Eurasia of the vortex motion from the remaining in the mantle section of

the same intramantle vortex after the vortex left from under the Atlantis island during the North American rotation.

8. The position of the Santorin volcano at its 35.4° rotation around the *P1* pole relative to Eurasia and Africa gives an estimate of the location of the city of Atlantis. The second independent estimation gives that position of structure Rishat in Mauritania concerning Eurasia which is received at rotation of structure Rishat on 25.38° clockwise around a pole P2. These estimates are close to each other and indicate the location of the city of Atlantis to the west of Gibraltar, which corresponds to Plato's description.
9. In addition, based on the information given by Plato, we obtained an estimate of the size of the island as a figure close to an ellipse with semi-major axes of $e1 = 570.99$ km and $e2 = 419.85$ km. The distance from the centre of Atlantis to the centre of the ellipse is 404.45 km. Using these parameters, the Azores appear directly on the border of the island, where according to Plato's description there were high mountains.

References

https://bestmaps.ru/map/yandex/hybrid/12/21.1127/-11.3708

Ignatenko AA (2016) Triera—a warship of the ancient Greeks. In: Ignatenko AA (ed) Interaction of science and society: problems and prospects: collection of articles of the international scientific and practical conference: in 3 parts, Volgograd, 05 Nov 2016, vol Part 3. Limited Liability Company "Aeterna", Volgograd, pp 16–18

Kasyanov SY (2023) About the possible relationship of the Iceland hot spot and the surrounding ocean bottom relief with the system of intense intra-mantle vortexes. In: Karev VI (ed) Physical and mathematical modeling of earth and environment processes—2022, PMMEEP 2022. Springer proceedings in earth and environmental sciences. Springer, Berlin, pp 613–623. https://doi.org/10.1007/978-3-031-25962-3_61

Kasyanov SY (2024a) On the possibility of using lithospheric topography to estimate the velocity at the horizontal upper boundary of an intense vortex in the mantle beneath the sea of Japan region. In: Karev V. (ed) Proceedings of the 9th international conference on physical and mathematical modelling of earth and environmental processes—2024. Springer proceedings in earth and environmental sciences. Springer, Berlin. https://doi.org/10.1007/978-3-031-54589-4_56

Kasyanov SY (2024b) On the possible influence of an intense vortex in the mantle beneath the sea of Japan region on the formation of lithospheric relief and sea of Japan water circulation. In: Karev V (ed) Proceedings of the 9th international conference on physical and mathematical modelling of earth and environmental processes—2024. Springer proceedings in earth and environmental sciences. Springer, Berlin. https://doi.org/10.1007/978-3-031-54589-4_60

Kasyanov SY (2024c) On the possibility of using the lithosphere relief to estimate the velocity at the horizontal upper boundary of an intensive vortex in the mantle under the eastern part of the Mediterranean sea. In: Karev V (ed) Proceedings of the 9th international conference on physical and mathematical modelling of earth and environmental processes—2024. Springer proceedings in earth and environmental sciences. Springer, Berlin. https://doi.org/10.1007/978-3-031-54589-4_52

Kasyanov SY, Samsonov VA (2021a) Egyptian pyramids as indicators of angular displacement of the African plate. In: Chaplina T (ed) Processes in GeoMedia, vol III, pp 203–216. Springer geology. Springer, Berlin. https://doi.org/10.1007/978-3-030-69040-3_20

Kasyanov SY, Samsonov VA (2021b) About the modern system of three energy-carrying intensive vortices in the earth's mantle. In: Chaplina T (ed) Processes in GeoMedia, vol II. Springer geology. Springer, Berlin, pp 273–291. https://doi.org/10.1007/978-3-030-53521-6_31

Losev AF, Asmus VF, Takho-Godi AA (1994) Plato collected works in 4 vols. T. 3/Trans. from ancient Greek. Auto. entry art. and art. in note A. F. Losev; Note A. A. Taho-Godi. M.: Mysl, 654 p (philosophical heritage)

Pearson C, Sigl M, Burke A, Davies S, Kurbatov A, Severi M, Cole-Dai J, Innes H, Albert PG, Helmick M (2022) Geochemical ice-core constraints on the timing and climatic impact of Aniakchak II (1628 BCE) and Thera (Minoan) volcanic eruptions. PNAS Nexus 1(2):pgac048. https://doi.org/10.1093/pnasnexus/pgac048

Organic Matter and Rate of Modern Carbon Accumulation in Marsh Soils of Southern Sakhalin Island

V. V. Afanas'ev, V. V. Demin, and Yu. A. Zavgorodnyaya

Abstract It has been established that in the coastal zone of Aniva Bay (Sakhalin Island, 46.74° N; 142.73° E) the following soil type change occurs with increasing location above sea level: marsh algal-detrital → marsh detrital → marsh detrital-peat → marsh peat → peat under forest. The results describing for the first time carbon accumulation in marsh soils of the temperate zone under plant communities dominated by *Triglochin* and *Carex* genera were obtained. Modern sediment accumulation in Aniva Bay marsh soils proceeds at a high rate, averaging 2.1 cm year^{-1}, accompanied by intensive carbon accumulation, averaging 720 g C m^{-2} year^{-1}. The found values of carbon accumulation rates are several times higher than the global average rate calculated for marshes soils—244.7 g C m^{-2} year^{-1}, but do not exceed the known maximum values—1713 g C m^{-2} year^{-1}.

Keywords Marshes · Marsh soils · Carbon accumulation rate · SAR · CAR · *Triglochin* · *Carex*

1 Introduction

Marshes are tidal wetland ecosystems and dominate sheltered shores and estuarine margins in a wide range of climatic conditions from subarctic to tropical, but are most common in temperate latitudes (Ouyang and Lee 2014). Critical to their existence is the rate of global sea level rise. By 2100, mean sea level could increase by various estimates from 60 to 110 cm (Best et al. 2018) or 20–180 cm (Schile et al. 2014): 0.7–1.2 cm year^{-1} or 0.2–2.0 cm year^{-1}, respectively. Salt marsh ecosystems are able to resist sea-level rise through a so-called 'biogeomorphological positive feedback

V. V. Afanas'ev (✉)
Sakhalin State University, Yuzhno-Sakhalinsk, Russia
e-mail: vvasand@mail.ru

Institute of Marine Geology and Geophysics FEB RAS, Yuzhno-Sakhalinsk, Russia

V. V. Demin · Yu. A. Zavgorodnyaya
Lomonosov Moscow State University, Moscow, Russia

T. Chaplina (ed.), *Processes in GeoMedia—Volume VIII*, Springer Geology,
https://doi.org/10.1007/978-981-97-6627-7_30

loop'. Marsh vegetation captures sediment from the sea, which in turn allows plants to grow better, trap sediment even more actively, and accumulate more organic matter in living and dead biomass (Bouma et al. 2016).

To calculate the present-day sediment accretion rate (SAR, cm year^{-1}), either data on the depth of the layer of maximum ^{137}Cs concentration (associated with the maximum of radioactive fallout in 1963 (DeLaune et al. 1978) or the position in the profile of a marker layer with a known date are used. Sediment accumulation rates in the marsh areas range from 0.07 to 4.2 cm year^{-1}; sedimentation rates exceeding the critical marks of 0.2, 0.7, 1.2, and 2.0 cm year^{-1} were found in 83, 27, 12, and 6% of the studied areas, respectively (Ouyang and Lee 2014).

Simultaneously with precipitation accumulation, carbon burial in marsh soils occurs. The global average carbon accumulation rate (CAR, gC m^{-2} year^{-1}) in marsh soils is 244.7 gC m^{-2} year^{-1} according to recent data, and for the area lying between 40S and 69.7 AeS ($\sim$ 42,000 km^2). The carbon burial rate is estimated to be $\sim$ 10.2 TgC year^{-1} (Ouyang and Lee 2014). Earlier estimates give a mean carbon storage rate of 218.7 gC m^{-2} year^{-1} for an area lying between 22.4S and 55.5S ($\sim$ 22,000 km^2) and a carbon burial rate of $\sim$ 4.8 TgC year^{-1}, with values varying greatly between areas: the standard deviation is $\pm$ 235 gC m^{-2} year^{-1}, with a range of variation of 18–1713 gC m^{-2} year^{-1} (Chmura et al. 2003). The rate of carbon accumulation in marsh soils can be very high, even slightly higher than carbon accumulation in mangrove soils, and 50 times higher than in forested continental ecosystems (Chmura et al. 2003; Macreadie et al. 2013).

At present, data on the areas of coasts occupied by marshes, sediment accumulation rates and carbon accumulation remain estimates, and large areas of coasts are practically unstudied. The coastline of the Russian Federation is a striking example. The coastline of Russia is $\sim$ 9.5% of the world's coastline, slightly inferior to the length of the US coastline—11.5% (World Resources Institute (WRI)), but while about 50% of the data on carbon accumulation rates refer to the US marshes, the current literature lacks data on carbon accumulation rates in soils of the Russian Federation marshes and, consequently, on the total annual carbon sink in coastal landscapes.

The purpose of this study was to investigate the distribution and morphology of organic matter in soil profiles of saline marches of Southern Sakhalin and to estimate the current rate of organic carbon accumulation in them.

2 Objects and Methods

Surveys were conducted in June 2023 at a carboniferous test site located in the coastal zone of Salmon Bay in Aniva Bay (Sea of Okhotsk), Sakhalin Island (46.74 N; 142.73 E). Five soil profiles were uncovered 500 m north-east of the mouth of the Susuya River along a transect running from the shoreline towards a forested area under vegetation communities formed under temperate climate conditions at different stages of development of tidal zone ecosystems (Table 1).

Table 1 Description of soil profile locations

Point	Height, m (EGM08)	Distance from the water edge, m	Tidal frequency	Ecosystem type	Plant community
1	0.768	2	Daily	Low march	Sedge-sedge meadow
2	0.759	70	Daily	Low march	Grass-sitnik-sedge meadow
3	1.091	230	14–15 a month	Middle march	Sedge-reed meadow
4	1.371	300	Sisygian	High march	Reed grass meadow
5	3.056	550	–	Mixed forest	Birch-pine moss-fern-broadleaf forest

Soil profiles were opened and soil samples were collected using a steel percussion sampler with a rectangular cross-section of 20 cm^2. The soil monolith was extracted to the whole depth of sampling, after morphological description the monolith was divided into separate samples corresponding to soil horizons. Sampling was carried out in twofold repetition. Soil volumetric weight was determined by the cutting ring method (Ø5 cm).

The samples were dried to air-dry condition, crushed in a porcelain mortar with a rubber-tipped pestle and passed through a system of sieves, separating the following fractions: > 3 mm for determining the content of large plant residues (ground litter, roots, wood, charcoal) and mineral inclusions (stones, pebbles, shells); 1–3 mm for determining the content of small inclusions of different genesis; < 1 mm for chemical analysis of soil. The content of fractions was determined by weight method.

The morphological composition of all fractions was studied at magnification ×10–×20. For the < 1 mm fraction, the content of biophilic elements CNS was determined on the Vario ELIII elemental analyser (Elementar), by potentiometric method, the content of anions Cl^-, SO_4^{2-}, by ion chromatography on the ICS-2000 chromatograph (Dionex). All analytical determinations were carried out in duplicate.

The carbon accumulation rate was calculated using the formula (Ouyang and Lee 2014):

CAR [gC m^{-2} $year^{-1}$] = SAR [cm $year^{-1}$] × ρc [gC cm^{-3}] × 10,000, where ρ_c is the bulk density of carbon [gC cm^{-3}].

3 Results

The driving force behind the evolution of soils of accumulative seashores in the Arctic and temperate zones is the gradual change of subaqual conditions to subaerial and aerial conditions. As the altitude rises above sea level, plant communities and soil organoprofile types change. The process of burial of morass of terrestrial plants by marine sediments containing organic matter in the zone of action of tides and storm surges is gradually replaced by the process of peat accumulation, the synlithogenic type of soil formation is replaced by the postlithogenic one, and organogenic allochthonous soils are replaced by organogenic autochthonous soils for which their precursors become soil-forming rock (Gubin and Lupachev 2022).

The lower layers of all studied soils (Fig. 1) were formed during the accumulation of marine sediments in the intertidal zone and are Watt deposits. Compared to the overlying horizons, they have a lighter (sandy loam) granulometric composition, contain a large number of organic inclusions represented by wood remains, as well as inclusions of pebbles and shells (Table 2). Higher up the profile, the granulometric composition becomes heavier and the proportion of organic inclusions represented by the remains of modern vegetation and buried tidal sediments increases. The upper part of the profiles is the zone of formation of organogenic horizons represented by turf, peat and peat layers or organomineral mass formed by overlapping of the soil surface by the tidal wave. Mineral inclusions are absent here, but layers with a noticeable content of coals are found (Fig. 1).

The general scheme of soil formation process in the studied area is as follows (Fig. 1): marsh algal-detrital soil (T.1) → marsh detrital soil (T.2) → marsh detrital-peat soil (T.3) → marsh peat soil (T.4) → peat soil (T.5).

The first stage of development of coastal ecocenoses is phytocenoses of low marshes (Table 1), dominated by Sitnik and sedge (genera *Triglochin* and *Carex.*) Soils under these phytocenoses undergo the stage of synlithogenic soil formation under conditions of regular inflow of marine sediments forming horizons of different granulometric composition (Fig. 1). For the soil in T.1, the organoprofile in the upper thickness is formed by allochthonous organic matter, mortmass of plant remains, mainly roots, and in the uppermost horizon—also by algal mortmass. S_{orgh} content in the profile does not exceed 4%, C/N ratio is about 10 (Fig. 2). In the upper part of the soil profile in T.2, formation of low thickness weakly corrugated horizons with high content of mineral mass is already noticeable (Figs. 1 and 2), which is reflected in the increase of carbon concentration up to 5–10% (C/N = 11).

Dense reed vegetation (genus *Phragmítes*) in T.3 marks the zone of middle marshes (Table 1). They are characterised by elements of the initial stages of postlithogenic soil formation, the input of allochthonous mineral and organic material from the sea is strongly limited. The transition from low to middle marshes is associated with a decrease in the accumulation of soluble salts (Cl^- and SO_4^{2-}) in the soil profile (Table 2) and with an increase in carbon stocks during the formation of peat mass on the soil surface (Figs. 1 and 2). There is a sharp increase in the carbon content up to 30% with the C/N > 15 value typical for peats.

Meadow phytocenoses with a smaller proportion of reed vegetation and a wide distribution of bloodroot (genus *Sanguisórba*) develop in the zone of high marshes (T.4), which is characterised by the lowest frequency of tidal events (Table 1). In soils under these phytocenoses there is an intensive accumulation of organic matter in the upper part of the profile in the form of peat layers (Fig. 1). The process of synlithogenic soil formation is practically completed, probably also due to the fact

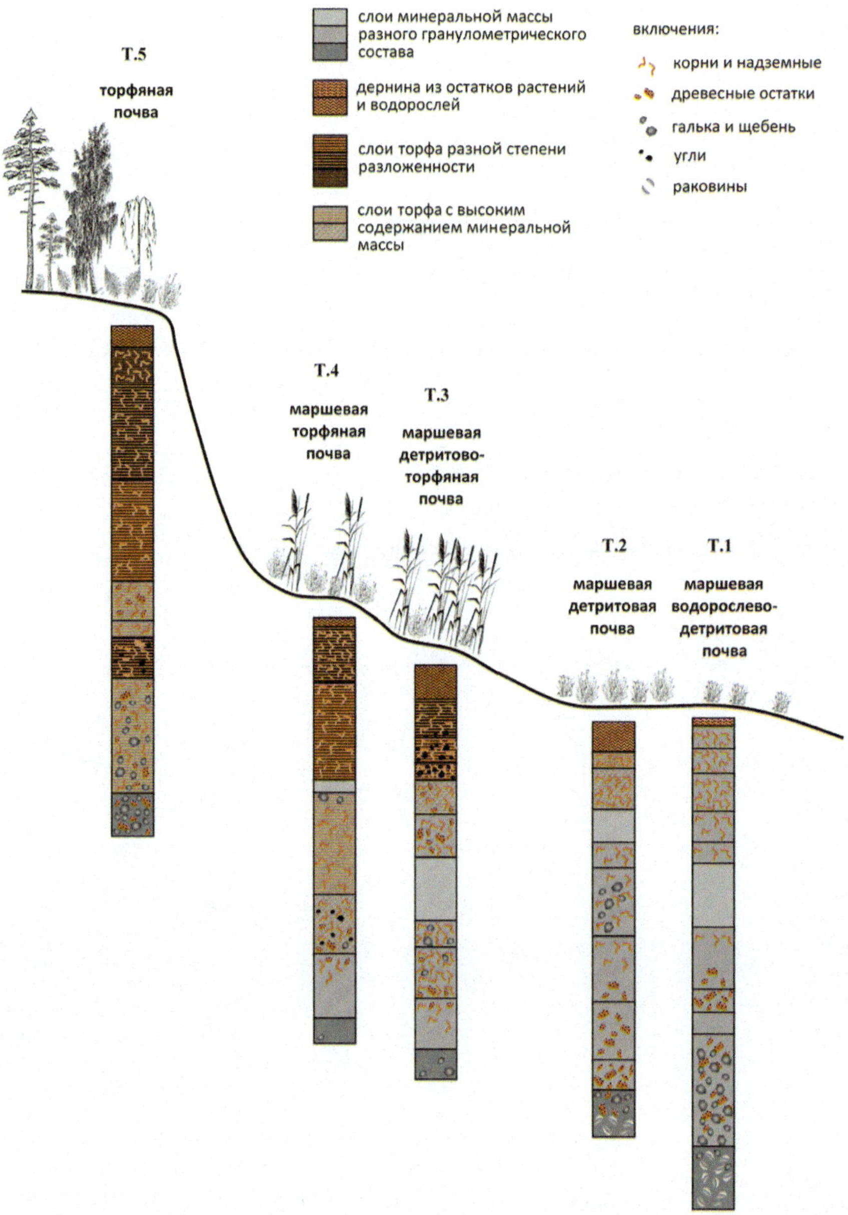

Fig. 1 Soil profiles along the transect

Table 2 Soil properties

Point	1	2	3	4	5
Organic inclusions content in the upper part of the profile, g[a]	15.5	6.5	36.5	11.1	30.6
Content of mineral inclusions in the lower part of the profile, g	56.2	29.9	24.0	9.8	88.4
Average Cl—content in the upper part of the profile, mg/g	7.77	4.60	3.79	0.02	0.23
Average Cl—content in the lower part of the profile, mg/g	3.58	1.35	0.84	0.02	0.09
Average S content in the upper part of the profile, mg/g	12.0	17.8	14.3	8.5	8.0
Average S content in the lower part of the profile, mg/g	7.8	6.7	7.2	11.3	19.7
SSO_4^{2-}/S_{total}, at the top of the profile, %	36	23	23	0.1	1.0
SSO_4^{2-}/S_{total}, at the bottom of the profile, %	24	11	43	1.6	32

[a]Upper part of the profile—soil layer above the tsunamigenic horizon

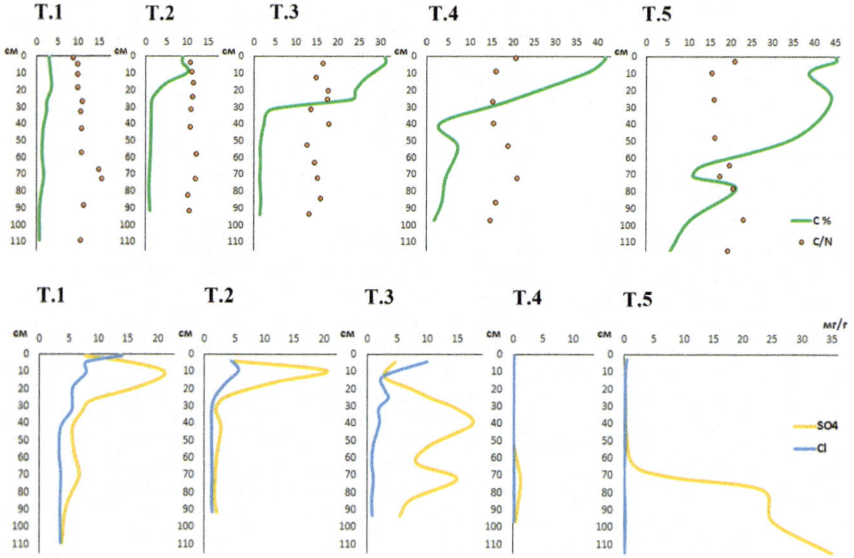

Fig. 2 Changes in profile distribution of organic carbon and soluble salts in soil profiles

that the dense vegetation cover of low and middle marshes is an effective filter for marine suspension. In the upper part of the soil profile there is a well-defined accumulation of reed peat with different degrees of decomposition and mineral content ($S_{or}g$ = 24–42%; C/N = 15–20) (Fig. 2).

Soil formation in T.5 takes place in conditions when the vegetation of salt marsh meadows and bogs is completely replaced by the community of mixed pine-small-leaved forest with developed herbaceous and moss cover (Table 1). The soil here is also formed on sediments of marine genesis—this is indicated by the high content of total and sulfate sulfur in the lower part of the profile (Table 2), but a significant part of the soil column is the result of postlithogenic soil formation with accumulation of upper sphagnum peat, typical for soils of forest phytocenoses ($S_{org} = 35–45\%$; C/N = 19) (Fig. 2).

In general, in soils from T.1 to T.5 there is an increase in organic matter content first in the uppermost part of the profile and then in deeper layers. At the same time, the C/N ratio increases, which indicates a gradual change in the type of accumulated organic matter—from marine to terrigenous with a low degree of decomposition of plant remains. The C/N ratio increases observed in different parts of the profiles coincide with the high content of woody or charcoal inclusions in soil horizons (Figs. 1 and 2).

The present-day sediment accumulation rates and carbon accumulation rates were calculated for the studied soils. A well-defined marking layer at a depth of 20–60 cm in all profiles below the marshes, representing tsunami deposits caused by the earthquakes of 15.11.2006 (with magnitude Mw = 8.3) and 13.01.2007 (with magnitude Mw = 8.1) in the region of the Central Kurils (Pinegina et al. 2008; Pinegina and Razhigayeva 2013; Kulikov et al. 2013), was used to calculate SAR. The tsunamigenic horizon was distinguished in soil profiles by its light grey colour, higher density, heavier granulometric composition (heavy loam) and lack of organic inclusions. This horizon differs from the tidal sediments of the coastal zone of Salmon Bay as it is composed of marine sediments brought up from greater depths. In calculating the average rate of carbon accumulation in the entire layer above the tsunamigenic sediments, weighted average values of volumetric carbon density in individual layers were used (Table 3).

The CAR values obtained for Aniva Bay marshes soils are 2–3 times higher than most CAR values obtained for other coastal areas (Chmura et al. 2003; Ouyang and Lee 2014), which may be related to local climatic and hydrological peculiarities of the study site.

In the Carboniferous polygon area (about 15.5 km^2), marshes ecosystems are located between elevations of 0.5 and 1.5 m, occupying an area of 3 km^2 (20% of the total area). The average rate of carbon accumulation in marsh soils is 720 gC m^{-2} $year^{-1}$ or for the entire marsh area 2.2×10^9 gC $year^{-1}$ (8.8×10^3 CO^2 $year^{-1}$ or 8.8×10^3 carbon units per year). Soils of the postlithogenic stage of soil formation (peat soil, T.5) also accumulate carbon at a very high rate

Table 3 Carbon accumulation rate in soils of the South Sakhalin coastal soils

Point	1	2	3	4	5
SAR, cm $year^{-1}$	1.9	1.3	2.8	2.4	3.8
CAR, gC m^{-2} $year^{-1}$	576	449	929	919	2143

(Table 3), much higher than the known average estimates of peat accumulation rates—20–30 gC m^{-2} year^{-1} (Roulet 2000).

It should be emphasised that the CAR values obtained refer to the layer that has formed in the last 17 years after the passage of the tsunami and are estimates of 'net' carbon accumulation during this period. Organic residues contained in the underlying layers may now be subject to mineralisation, especially when subaqual conditions change to subaerial conditions. The possibility of these processes is indirectly indicated by the high proportion of oxidised sulphur in these horizons (Table 2). CO_2 emission from deep horizons of marsh soils can partially offset the present-day process of carbon sequestration in these soils.

4 Conclusion

In the coastal zone of Aniva Bay (territory of the Carboniferous polygon) during the change of subaqual conditions to subaerial and aerial conditions with increasing location above sea level the following change of soil type occurs: marsh algal-detrital → marsh detrital → marsh detrital → marsh detrital-peaty → marsh peaty → peaty under forest.

Modern sediment accumulation within the Carboniferous polygon has proceeded at a high rate, averaging 2.1 cm year^{-1}, and seashore evolution in upper Aniva Bay has resulted in an increase in the area of salt marshes, accompanied by intensive carbon accumulation in marshes soils, averaging 720 gC m^{-2} year^{-1} (2.2 × 10^9 gC year^{-1} for the entire marshes area).

The found values of C accumulation rates are several times higher than the average C accumulation rate in marsh soils—244.7 gC m^{-2} year^{-1}, but do not exceed the known maximum values—1713 gC m^{-2} year^{-1}. Previously obtained estimates of C accumulation in marsh soils refer to soils developed under plant communities of the genera *Distichlis*, *Spartina*, *Phragmites*, *Juncus*, and *Halmione* (Chmura et al. 2003; Ouyang and Lee 2014). The presented results describe for the first time carbon accumulation in marsh soils of the temperate zone under plant communities dominated by the genera *Triglochin* and *Carex*.

Acknowledgements The work was carried out within the framework of the State Task FEFF-2024-0004 'Creation of scientific bases for management of carbon uptake and storage processes in biomorpholithosystems of coastal-marine wetlands'.

References

Best ÜSN, Van Der Wegen M, Dijkstra J, Willemsen PWJM, Borsje BW, Roelvink Dano JA (2018) Do salt marshes survive sea level rise? Modelling wave action, morphodynamics and vegetation dynamics. Environ Model Softw 109:152–166. https://doi.org/10.1016/j.envsoft.2018.08.004

TJ Bouma J Belzen Van T Balke J Dalen van P Klaassen AM Hartog DP Callaghan Z Hu MJF Stive S Temmerman PMJ Herman 2016 Short-term mudflat dynamics drive long-term cyclic salt marsh dynamics Limnol Oceanogr 61 2016 2261 2275 https://doi.org/10.1002/lno.10374

GL Chmura SC Anisfeld DR Cahoon JC Lynch 2003 Global carbon sequestration in tidal, saline wetland soils Global Biogeochem Cycles 17 4 1111 https://doi.org/10.1029/2002GB001917

DeLaune RD, Patrick WHJr, Buresh RJ (1978) Sedimentation rates determined by ^{137}Cs dating in a rapidly accreting salt marsh. Nature 275:532–533

Gubin SV, Lupachev AV (2022) Approaches to classification of soils of accumulative seashores of the eastern sector of the Russian Arctic. Soil Sci 1:25–32. https://doi.org/10.31857/S0032180X22010051

Kulikov EA, Ivashchenko AI, Lobkovsky LI, Rabinovich AB, Yakovenko OI (2013) Kuril earthquakes and tsunamis on 15 Nov 2006 and 13 Jan 2007. Landslide tsunami modelling. In: World ocean, vol 1. Geology and ocean tectonics. Catastrophic phenomena in the ocean, pp 459–487

PI Macreadie AR Hughes DL Kimbro 2013 Loss of 'Blue Carbon' from coastal salt marshes following habitat disturbance PLoS ONE 8 7 e69244 https://doi.org/10.1371/journal.pone.0069244

Ouyang X, Lee SY (2014) Updated estimates of carbon accumulation rates in coastal marsh sediments Biogeosciences 11:5057–5071. https://doi.org/10.5194/bg-11-5057-2014

Pinegina TK, Razhigayeva NG (2013) Studies of palaeotsunami on the Far Eastern coast of Russia. In: World ocean, vol 1. Geology and ocean tectonics. Catastrophic phenomena in the ocean, vol 1. Nauchny Mir, Moscow, pp 488–498

Pinegina TK, Bourgeois D, Razhigayeva RG, Levin BV, Kaistrenko VM, Krachunovskaya EA, McInness B (2008) The tsunami of 15 November 2006 on the central Kuril Islands and the recurrence of similar events in the past (based on paleoseismological data). Geophysical monitoring and problems of seismic safety of the Russian Far East, vol 1. GS RAS, Petropavlovsk-Kamchatsky, pp 200–204

NT Roulet 2000 Peatlands, carbon storage, greenhouse gases, and the Kyoto Protocol: prospects, and significance for Canada Wetlands 20 605 615

Schile LM, Callaway JC, Morris JT, Stralberg D, Parker VT, Kelly M (2014) Modeling tidal marsh distribution with sea-level rise: evaluating the role of vegetation, sediment, and upland habitat in marsh resiliency. PLoS ONE 9(2). https://doi.org/10.1371/journal.pone.0088760

World Resources Institute. Coastal and marine ecosystems—marine jurisdictions: coastline length. http://earthtrends.wri.org/text/coastal-marine/variable-61.html

SPRINGER NATURE

GPSR Compliance

The European Union's (EU) General Product Safety Regulation (GPSR) is a set of rules that requires consumer products to be safe and our obligations to ensure this.

If you have any concerns about our products, you can contact us on ProductSafety@springernature.com

In case Publisher is established outside the EU, the EU authorized representative is:

Springer Nature Customer Service Center GmbH
Europaplatz 3
69115 Heidelberg, Germany

The manufacturer's authorised representative in the EU is Springer Nature Customer Service Centre GmbH, Europaplatz 3, 69115 Heidelberg, Germany. If you have any concerns regarding our products, please contact ProductSafety@springernature.com

Printed and bound by CPI Group (UK) Ltd, Croydon, CR0 4YY
01/10/2026
02225907-0011